Pesticide Formulation *and* Adjuvant Technology

Edited by
Chester L. Foy
David W. Pritchard

CRC Press
Boca Raton New York London Tokyo

Pesticide Formulation *and* Adjuvant Technology

Acquiring Editor:	Marsha M. Baker
Editorial Assistant:	Jean Jarboe
Project Editor:	Carrie L. Unger
Marketing Manager:	Susie Carlisle
Direct Marketing Manager:	Becky McEldowney
Cover design:	Dawn Boyd
PrePress:	Carlos Esser
Manufacturing:	Sheri Schwartz

Library of Congress Cataloging-in-Publication Data

Catalog record available from the Library of Congress.

Direct all inquiries to CRC Press, Inc., 2000 Corporate Blvd., N.W., Boca Raton, Florida 33431.

International Standard Book Number 0-8493-7678-5
Printed in the United States of America 1 2 3 4 5 6 7 8 9 0
Printed on acid-free paper

Preface

Formulations Forum '94, a comprehensive first-ever international symposium, specifically devoted to reviewing developments and issues in agricultural formulations, was held in Washington, D. C. on June 30 and July 1, 1994. Speakers represented global specialists from every phase of formulations chemistry including academia, industry, regulatory organizations, and the legal profession. More than 200 formulations and agricultural chemists and related experts, representing 32 countries and 170 companies, were in attendance.

This book is based on presentations at the symposium and covers the following important issues: regulation of inert ingredients and adjuvants in the U.S. and Europe; targeting and efficacy enhancement technologies (drift reduction, microencapsulation, leaching inhibition, penetration and translocation, and protection against inactivation); packaging technology and trends; water dispersible granules, emulsifiable concentrates, microemulations, emulsion/dispersion concentrates, and gels; and adjuvants, including pyrrolidone and silicone surfactants. Many of the topics are comprehensive reviews of the literature from all relevant fields. This book will be of interest to agricultural and pharmaceutical formulation chemists, agrichemical industry personnel, regulatory agencies and legal experts, agrichemical and pharmaceutical researchers, agricultural specialists, and others interested in the latest developments, new technologies, issues, and trends in pesticide formulation.

Chester L. Foy
David W. Pritchard

Editors

Chester L. Foy, Ph.D., received his B.S. degree from the University of Tennessee, his M.S. degree from the University of Missouri, and his Ph.D. from the University of California-Davis.

Dr. Foy is now Professor of Plant Physiology and Weed Science at Virginia Polytechnic Institute and State University in Blacksburg, Virginia. He joined the university as Associate Professor in 1966, was promoted to Professor in 1968, and served from 1974 to 1980 as Head of the department now known as Plant Pathology, Physiology and Weed Science. Formerly, he was Associate Botanist and Associate Professor of Agricultural Botany at the University of California-Davis.

Dr. Foy is a charter member of the Weed Science Society of America (WSSA) and the International Weed Science Society (IWSS). He has served as President for both societies. He is Vice-President and Secretary of the newly incorporated International Association of Formulation Chemists. His affiliations with other professional organizations, past and present, include the American Society of Agronomy, American Society for the Advancement of Science, American Institute of Biological Sciences, American Society of Plant Physiologists, Council of Agricultural Science and Technology, International Congress of Plant Protection, Plant Growth Regulatory Society of America, Southern Weed Science Society (SWSS), Torch International, and Virginia Academy of Sciences. Dr. Foy's recognitions include election to membership in several academic honorary societies and several "Who's Who" listings. Other awards for professional achievement include National Academy of Sciences Resident Research Associate Award, Gamma Sigma Delta Faculty Research Award, WSSA Outstanding Paper Award, WSSA Fellow, the first SWSS "Weed Scientist of the Year" Award, and "Outstanding Researcher" Award from the WSSA.

Dr. Foy chaired and hosted the Second International Symposium in Adjuvants for Agrichemicals held in Blacksburg, Virginia, was Program Chairman for Formulations Forum '94, and has organized and participated in a number of other prominent symposia for agrichemicals. In addition, he has served as Editor of *Reviews of Weed Science* and is currently serving a third 3-year term as Editor of *Weed Technology*.. He is a charter (and current) member of the editorial board of the international journal *Pesticide Biochemistry and Physiology* and has served many years as WSSA Associate Editor and Reviewer for *Weed Science*, as well as Reviewer for several other scientific journals.

Dr. Foy conducts and directs field, greenhouse, and laboratory studies in the following areas: crop production and protection; vegetation management in agronomic and fruit crops, and control of specific perennial weeds; routes and mechanisms of absorption, translocation, accumulation, and exudation of herbicides and growth regulators, surfactants, and other adjuvants; metabolism and fate of these substances; physiological, biochemical, and morphological changes induced by exogenous chemicals; modes of action and selectivity of herbicides and growth regulators; minimizing pesticide residues in the biosphere; allelopathy; and parasitic weeds.

David W. Pritchard, Ph.D., is the President of the Association of Formulation Chemists (AFC). The AFC was incorporated in 1994 in direct response to a survey of attendees at Formulations Forum '94. The survey indicated a need for an organization that focused on formulation technology regardless of industry. There are substantial similarities in the technology of making dispersions, emulsions or other delivery systems regardless of whether they are developed in the pharmaceutical, personal care, food, coatings, or agricultural industries. Respondents indicated that an organization promoting cross-industry fertilization of formulation developments could significantly promote technical synergies. The AFC will thus carry forward with the Formulations Forum '94 concept, only on a much more expanded basis.

Dr. Pritchard is currently the Worldwide Director of Marketing, Agricultural Products at International Specialty Products (ISP) in Wayne, New Jersey. He has been employed by ISP and its predecessor GAF since 1979, where he held a variety of positions including Research Scientist, Manager of Commercial Development, and Manager of New Business Development and Planning. Prior to joining GAF, Dr. Pritchard owned and operated a consulting service and analytical laboratory, Consulting Services of Idaho, which he established after completing his postdoctoral work at Ohio State University in 1974. He earned his B.S. and M.S. degrees at Utah State University and obtained a Ph.D. in virology from Cornell University in 1972. Dr. Pritchard holds four patents and is the author of numerous publications.

Contributors

David Z. Becher
Monsanto Corporation
St. Louis, Missouri

George B. Beestman
E.I. DuPont de Nemours, Inc.
Wilmington, Delaware

Farhad Dastgheib
Department of Plant Science
Lincoln University
Canterbury, New Zealand

Jeremy N. Drummond
International Specialty
Products/Europe
Guildford, England
United Kingdom

Roger J. Field
Department of Plant Science
Lincoln University
Canterbury, New Zealand

Robert D. Fox
Application Technical Research Unit
USDA/ARS/OARDC
Wooster, Ohio

Chester L. Foy
Virginia Polytechnic Institute and State University
Blacksburg, Virginia

Bruno Frei
Ciba Crop Protection
Ciba-Geigy Corporation
Basel, Switzerland

Steven I. Gleich
E.I. DuPont de Nemours, Inc.
Wilmington, Delaware

Franklin R. Hall
Laboratory of Pesticide Control Application Technology
Agricultural Research and Development Center
Ohio State University
Wooster, Ohio

Edward G. Hochberg
Hochberg and Co., Inc
Chester, New Jersey

Edwin L. Johnson
TS6 Inc.
Washington, D.C.

Tina E. Levine
U.S. Environmental Protection Agency
Washington, D.C.

Alvin J. Lorman
Mintz, et al., P.C.
Washington, D.C.

Kolazi S. Narayanan
Research and Development
International Specialty Products
Wayne, New Jersey

David W. Pritchard
International Specialty Products
Wayne, New Jersey

Peter Schmid
Ciba Crop Protection
Ciba-Geigy Corporation
Greensboro, North Carolina

Megh Singh
University of Florida
Citrus Research and Educational Center
Lake Alfred, Florida

Alan J. Stern
Monsanto Corporation
St. Louis, Missouri

Siyuan Tan
University of Florida
Citrus Research and Educational Center
Lake Alfred, Florida

Joseph R. Winkle
Dow Elanco
Indianapolis, Indiana

Table of Contents

Acknowledgments

International Specialty Products (ISP) is recognized for its early endorsement of Formulations Forum '94 and its funding of the formative stages of the forum. Warranting special recognition is David W. Pritchard of ISP. Without his dedication and key role as the principal organizer, Formulations '94 would not have been a reality. Thanks are also extended to J. Potter, J. Amdursky, M. Brown, J. Pierson, and L. Ruitt for organizational assistance as well as manuscript editing. Appreciation is extended to Chester L. Foy, who chaired the forum, and to two members of the support staff at Virginia Tech: Harold Witt for valuable editorial assistance, and Judy Fielder for typing.

Chapter 1

The Regulation of Inert Ingredients in the United States

Tina E. Levine

CONTENTS

I. BACKGROUND

The Federal Insecticide, Fungicide, and Rodenticide Act (FIFRA)[6,7,8,9] requires that all pesticide products sold or distributed in commerce be registered by the Environmental Protection Agency (EPA). Prior to the establishment of the EPA, the Department of Agriculture (USDA) registered pesticides under FIFRA. Although the EPA registers pesticide products, most of the data requirements and regulatory activities under FIFRA have traditionally focused on the active ingredient.

In addition to its mandate under FIFRA, EPA has authority to regulate pesticide products under the Federal Food, Drug, and Cosmetic Act (FFDCA).[8,9] Section 408 of FFDCA authorizes the EPA to establish tolerances or safe levels of pesticide residues in raw agricultural commodities; Section 409 similarly authorizes EPA to promulgate food additive regulations for pesticide residues in processed foods. Prior to the establishment of the EPA, the Food and Drug Administration (FDA) had the responsibility for establishing tolerances and food additive regulations for pesticide residues.

In 1961, the FDA published a notice in the Federal Register[1] stating that USDA had determined that each component of registered pesticide products, including the inert ingredients, were pesticide chemicals and thus subject to the requirement of tolerances or exemption under FFDCA. Several years later, in 1969, the FDA established a policy regarding data requirements and review procedures for clearance of pesticide inert ingredients used on food.[2]

1-8493-7678-5/96/$0.00+$.50

This guidance provided the basic framework for the regulation of inert ingredients in the United States until 1987.

In 1987, the EPA announced the Inert Strategy.[4] This strategy was designed to reduce the potential of adverse effects from chemicals used as inert ingredients contained in pesticide products and to make sure all inert ingredients were supported by valid data. EPA divided the extant inert ingredients into four toxicity categories. List 1 contains "Inerts of Toxicological Concern", chemicals that have been found to produce cancer, adverse reproductive effects, developmental toxicity, other chronic effects, ecological effects, or that have the potential for bioaccumulation (Table 1). In general, chemicals were placed on the list based upon a well-documented peer review, such as a bioassay from the National Toxicology Program or International Agency for Research on Cancer (IARC) review.

Table 1
List 1 - Inerts of Toxicological Concern

CAS No.	Chemical Name
62-53-3	Aniline
1332-21-4	Asbestos fiber
1332-21-9	1,4-Benzenediol
7440-43-9	Cadmium compounds
56-23-5	Carbon tetrachloride
67-66-3	Chloroform
106-46-7	p-Dichlorobenzene
103-23-2	Di-(2-ethylhexyl)adipate
78-87-5	1,2-Dichloropropane
117-87-8	Di-ethylhexylphthalate
66-12-2	Dimethylformamide
123-91-1	Dioxane
106-89-8	Epichlorohydrin
110-80-5	2-Ethoxyethanol
111-15-9	Ethanol ethoxyacetate
107-06-2	Ethylene dichloride
109-86-4	Ethylene glycol monomethyl ether
140-88-5	Ethyl acrylate
110-54-3	n-Hexane
302-01-2	Hydrazine
78-59-1	Isophorone
7439-92-1	Lead compounds
568-64-2	Malachite green
591-78-6	Methyl n-butyl ketone
74-87-3	Methyl chloride
75-09-2	Methylene chloride
25154-52-3	Nonylphenol
127-18-4	Perchloroethylene
108-95-2	Phenol

Table 1 (continued)
List 1 - Inerts of Toxicological Concern

CAS No.	Chemical Name
90-43-7	o-Phenylphenol
75-56-9	Propylene oxide
8003-34-5	Pyrethrins
81-88-9	Rhodamine B
10588-01-9	Sodium dichromate
26471-62-5	Toluene diisocyanate
79-00-5	1,1,2-Trichloroethane
56-35-9	Tributyl tin oxide
79-01-6	Trichloroethylene
1330-78-5	Tri-orthocresylphosphate (TOCP)
78-30-8	Tri-orthocresylphosphate (TOCP)

List 2 covers "Inerts With a High Priority for Testing" that are generally closely related by structure or chemical class to compounds on List 1 (Table 2). Many of the chemicals on this list had been targeted for testing under the Toxic Substances Control Act (TSCA).[7,8,9] List 3 covers "Inerts of Unknown Toxicity". The chemicals in this group are those that do not fit into any of the other three groups. List 4 comprises the "Minimal Risk Inerts". These chemicals are generally regarded as safe. In 1989, List 4 was subdivided into 4A and 4B.[6,7,8,9] List 4A covers those substances judged to be of minimal risk based on their inherent nature, such as food substances like corn cobs and cookie crumbs. List 4B contains those chemicals which the EPA has judged to be of minimal risk based on a review of the data submitted and knowledge of their use in pesticides.

Table 2
List 2 - Potentially Toxic Inerts/High Priority for Testing

CAS No.	Chemical Name	CAS No.	Chemical Name
85-68-7	Butyl benzyl phthalate	120-32-1	2-Benzyl-4-chlorophenol
84-74-2	Dibutyl phthalate	75-00-3	Chloroethane
84-66-2	Diethyl phthalate	88-04-0	p-Chloro-m-xylenol
131-11-3	Dimethyl phthalate	97-23-4	Dichlorophene
117-84-0	Dioctyl phthalate	100-41-4	Ethyl benzene
95-49-8	2-Chlorotoluene	149-30-4	Mercaptobenzothiazole
1319-77-3	Cresols	74-83-9	Methyl bromide
95-48-7	o-Cresol	75-45-4	Chlorodifluoromethane
106-44-5	p-Cresol	75-43-4	Dichloromonofluoromethane
108-39-4	m-Cresol	75-45-6	Chlorodifluoromethane
108-94-1	Cyclohexanone	75-37-6	1,1,-Difluroethane
95-50-1	o-Dichlorobenzene	75-68-3	1-Chloro-1,1-difluoromethane
112-34-5	Diethylene glycol monobutyl ether	25168-06-3	Isopropyl phenols
			Petroleum hydrocarbons

Table 2 (continued)

List 2 - Potentially Toxic Inerts/High Priority for Testing

CAS No.	Chemical Name	CAS No.	Chemical Name
111-90-0	Diethylene glycol	1330-20-7	Xylene
	monoethyl ether	100-02-7	p-Nitrophenol
111-77-3	Diethylene glycol	106-88-7	Butylene oxide
	monomethyl ether	79-24-3	Nitroethane
34590-94-8	Dipropylene glycol	75-05-8	Acetonitrile
	monomethyl ether	71-55-6	1,1,1-Trichloroethane
111-76-2	2-Butoxy-1-ethanol	102-71-6	Triethanolamine
5131-66-8	1-Butoxy-2-propanol	111-42-2	Diethanolamine
124-16-3	1-Butoxyethoxy	97-88-7	Butyl methacrylate
	-2-propanol	80-62-6	Methyl methacrylate
107-98-2	1-Methoxy-2-propanol		Xylene-range aromatic solvents
29387-86-8	Propylene glyco l	95-82-9	2,5-Dichloroaniline
	monobutyl ether	95-76-1	3,4-Dichloroaniline
25498-49-1	Tripropylene glycol	626-43-7	3,5-Dichloroaniline
	monomethyl ether	554-00-7	2,4-Dichloroaniline
141-79-7	Mesityl oxide	608-27-5	2,3-Dichloroaniline
108-10-1	Methyl isobutyl ketone	608-31-1	2,6-Dichloroaniline
96-29-7	Methyl ethyl ketoxime	101-84-8	Diphenyl ether
108-90-7	Monochlorobenzene	76-13-1	Trichlorotrifluoroethane
75-52-5	Nitromethane	75-69-4	Trichlorotrifluoromethane
108-88-3	Toluene	75-71-8	Dichlorodifluoromethane
2938-43-1	Tolyl triazole	76-14-2	Dichlorotetrafluoroethane
95-14-7	1,2,3-Benzotriazole		

The EPA has never published Lists 3 and 4 in the Federal Register. A version of List 4 has been available through the Freedom of Information Office along with copies of Lists 1 and 2 and a complete list of all pesticide inert ingredients by name and Chemical Abstracts Service (CAS) Registry Number. In the near future, we will be publishing our current List 4A in conjunction with a proposed rule to exempt certain active ingredients from the registration process.

The Inert Strategy also required that all new inert ingredients have a minimum base set of data which characterizes the material. The data requirements are dependent upon the use pattern, with food use inerts requiring the most extensive testing. The data required for all new food use and non-food use inert ingredients are shown in Table 3.

Table 3
Data Required to Evaluate Risks Posed by Inert Ingredients in Pesticide Products

Data	Guideline No.
Residue Chemistry	
• Description of the pesticide or type pesticide formulation(s) in which the inert will be used and the maximum percent by weight it can occupy in any formulation#	
• Description regarding the range of use patterns and range of concentrations of the inert material	171-3
Product Chemistry	
• Description of the chemical or chemical mixture including structural formula	61-1
• Chemical Abstracts Service (CAS) Registry Number and file	61-1
• Any technical bulletins available on the inert: purpose of inert in pesticide formulation (i.e., solvent, emulsifier, etc.)	61-1
• Discussion of possible toxic contaminants such as nitrosamines, polynuclear aromatics, or dioxins	61-3
• Batch analyses	62-1
• Density/specific gravity	63-7
• Solubility	63-8
• Vapor Pressure	63-9
• Dissociation constant	63-10
• Octanol/water partition coefficient	63-11
• pH	63-12
Toxicology	
• 90-day feeding study: rodent and dog	82-1
• Subchronic dermal toxicity	82-2; 82-3
• Teratology study: rodent	83-3
• Gene mutation test	84-2
• Structural chromosomal aberration test	84-2
• Other genotoxic effects	84-4
Ecotoxicology	
• Acute 96-Hr fish EC_{50} (preferably in rainbow trout or bluegill)	72-1
• 48-Hr LC_{50} or EC_{50} in *Daphnia*	72-2
• Avian oral LD_{50} (preferably in mallard or bobwhite)	71-1
• 8-Day avian dietary LC_{50} (preferably in mallard or bobwhite)	71-2

Table 3 (continued)
Data Required to Evaluate Risks Posed by Inert Ingredients in Pesticide Products

Environmental Fate	
• Hydrolysis	161-1
• Aerobic soil metabolism	162-1
• Photodegradation in water	161-2
• Photodegradation in soil	161-3
• K_{oc} or K_d	163-1
• Anaerobic aquatic metabolism	163-1

Some or all of these data requirements may be waived depending upon the toxicity of the chemical and the use pattern of the pesticide product containing the inert ingredient. For example, ecotoxicity and environmental fate testing are required for only those inert ingredients used in pesticide formulations intended for outdoor use. Conversely, additional data may be required following EPA review of the base set data. In recent years we have established finite tolerances for several herbicide safeners; residue field trial data and additional chronic/oncogenicity data were required in order to establish these tolerances. Since 1987, several chemicals were deleted from all lists because either they were no longer contained in pesticide products or were found to be impurities or active ingredients rather than intentionally added inert ingredients (i.e., benzene; formaldehyde and paraformaldehyde; pentachlorophenol and sodium pentachlorophenate; thiourea; carbon disulfide; 1,1-dimethylhydrazine (UDMH); mercury oleate). Lists 1 and 2 were last updated in 1989 via a Federal Register notice.[5]

II. THE INERT STRATEGY

The strategy outlined in the Federal Register notice of April 22, 1987[3] was based on the use of the inerts lists as a prioritization tool for regulatory and data gathering activities. The EPA was attempting to increase the significance of toxicity as a selection criterion for pesticide registrants when determining which inert ingredient to use in the formulation of their products. The first goal was to eliminate the use of List 1 inert ingredients if less toxic substitutes were available. The EPA noted that it was encouraging the substitution of less toxic inert ingredients in products containing List 1 inerts. All products containing a

List 1 inert were required to bear a statement indicating that the product contained a toxic inert ingredient and the identity of the inert ingredient.

Since the chemicals on List 1 had been determined to be biologically active, the EPA decided to treat them as it would treat pesticide active ingredients. Therefore, all products retaining List 1 inert ingredients were issued a comprehensive data call-in notice (DCI) requiring the submission of the same data necessary to support the registration of an active ingredient for the product's use pattern(s).

DCIs were issued in 1988 and 1989 for 30 of the remaining 33 chemicals on List 1. Following the issuance of these DCIs, the majority of the chemicals on List 1 were eliminated by means of reformulation, cancellation, or suspension. Recently, DCIs were issued on the remaining List 1 inert ingredients. Most registrants have elected to reformulate their products rather than generate the data being required. Several consortia have been developed to generate a subset of data for particular products which would allow some decisions to be made about the risks posed by the inert ingredients in those products. For example, data are about to be developed to determine whether there is dietary exposure to Rhodamine B from its use as a dye in seed treatment pesticide products. Although technically a product can still be registered containing a List 1 inert, the EPA discourages the use of these inerts by notifying the registrant that they will receive a DCI requiring them to either reformulate or submit data to support use of that inert prior to registration of the product. It is EPA policy that products containing List 1 inert ingredients cannot be reregistered.

The EPA has also begun to evaluate extant data for List 2 chemicals to determine whether they should be elevated to List 1 or judged to be minimal risk (List 4B). Some of these data have been collected under the TSCA. The EPA will issue DCIs for chemicals on List 2 for which there are insufficient data to evaluate the toxicity of these chemicals. In addition, many of the chemicals on List 3 have been reviewed to determine whether there now exists sufficient information on these chemicals, perhaps at other agencies such as the FDA, to enable classification of these inerts to one of the other lists. As a result of this review, approximately 500 chemicals currently on List 3 will shortly be proposed to be moved to List 4B. The EPA may in the future issue DCIs for the base set of data currently required for new inert ingredients for some of the unclassifiable compounds on List 3.

III. NEW INERT INGREDIENTS

As noted above, the announcement of the Inert Strategy also included a set of data required to obtain approval of a new inert ingredient. In order to develop a consistent policy regarding inert ingredients and to aid in the review of these ingredients, the EPA formed the Inerts Review Group. It originally consisted of members of the Hazard Evaluation Division (HED), and so it was nicknamed the HIRG, the HED Inerts Review Group. The group is now composed of scientific reviewers from the various technical branches and also members of the Registration Division. The HIRG is the focal point of our preliminary process for approving new inert ingredients. The group's main focus is to develop consistent and comprehensive risk assessments for new inert ingredients in order to help ensure that minimal risk inert ingredients are available to pesticide formulators.

To obtain approval of a new inert, the applicant should submit a request to the EPA. For approval of a non-food use inert this request would be in the form of a letter; for an exemption from the requirement of a tolerance under 40 CFR 180.1001(c), (d), and/or (e), the request would be considered a tolerance petition. There are no fees associated with inert ingredient petitions. It is important to specify the scope of the approval requested and the expected use(s) of the inert ingredient. In addition, the applicant should submit the base set of data required or available product chemistry, as well as toxicological, ecotoxicological, and environmental fate data on the chemical with a request and justification for waiver of other data requirements. This information is generally given a preliminary review by the HIRG. The HIRG decides whether: (1) additional data are required; (2) the available data should be put into formal review; (3) the chemical is innocuous and we can proceed with a tolerance exemption proposal; (4) in the case of a non-food use inert, they can assign a pesticide chemical code without any additional review. If the data are formally reviewed, the HIRG may meet once the reviews are completed to determine whether an exemption from the requirement of a tolerance may be proposed. If this determination is made, a proposed tolerance exemption is published in the Federal Register; 30 days are allotted for public comment and the tolerance exemption is then finalized.

IV. TRENDS AND ISSUES IN THE NEXT DECADE

Based on the patterns seen in recent inert submissions, it is clear that the development of novel formulation types is a trend for the next decade. This trend is driven by a number of factors: the desire to maximize the benefits derived from investment in the reregistration of existing active ingredients, the need to develop safer technologies to deal with a variety of regulatory concerns such as the Clean Air Act's prohibition against volatile organic solvents and ozone depleting agents, groundwater and storage/disposal issues, as well as the safer pesticide initiative.

One trend we have noted is the increased development of safeners and greater regulation of these safeners. Safeners are herbicide antidotes that protect desirable crops while allowing the herbicide to act on the intended weed targets. They allow the use of active ingredients where they previously could not be used because of crop damage. While safeners are currently classified as inert ingredients by the Office of Pesticide Programs, they are frequently similar in structure to the active ingredient. Most of them act by changing the plant metabolism such that the active ingredient is detoxicated rapidly by the crop. A number of safeners were granted tolerance exemptions many years ago, before the inert strategy was developed. Since the inert strategy took effect, the minimum base set of data required for inert ingredients is also required for safeners. But, because of their similarity to active ingredients, the EPA decided to establish finite tolerances for these inert ingredients. Therefore, additional chronic/oncogenicity studies as well as the necessary residue chemistry studies have been required of all safeners approved in the past several years. The other long-term toxicity studies generally required for the establishment of a tolerance have not been required. We anticipate that the EPA will continue to establish finite tolerances for safeners and will continue to require chronic/oncogenicity and residue studies for this purpose. We also anticipate that in cases where an inert ingredient is found to possess some degree of toxicity or exposure it must be restricted in order to assure that its use in a pesticide formulation does not result in unreasonable adverse effects on the environment. The EPA will establish a tolerance for residues of the inert ingredient on raw agricultural commodities.

Another trend we expect to see continued is the increased use of polymeric materials as inert ingredients. Over half of the inert ingredient tolerance exemptions developed last year were for polymers meeting the

OPPT Polymer Exemption Rule.[3] Several years ago the HIRG determined that polymers meeting the criteria outlined in the rule could be exempted from tolerance without requiring the submission of the base set data because of the low expected toxicity of polymers meeting the rule. Therefore, processing of polymers meeting the rule is considerably faster than other types of inert exemption requests. These polymers are generally very large molecules without reactive groups. Many of the polymers have uses that address the regulatory concerns outlined above. Examples include use as microencapsulating agents, a reduced exposure technology, and use in water soluble packaging which reduces worker exposure as well as solving container disposal problems.

Provisions of the Clean Air Act[9] have also impacted the inerts program. Efforts to decrease the use of volatile organic solvents in order to lessen smog have occasionally run up against FIFRA regulations. In addition, the phase-out of CFCs and other ozone-depleting agents has spurred the search for alternatives that are acceptable for use in pesticide products; agents that are listed by the Air program as acceptable alternatives may not have been reviewed from the FIFRA perspective. We have endeavored to ease these transitions to the extent possible, but we anticipate that the need to develop and obtain approval of new compounds for use as propellants in pesticide products will continue into the next decade. Having familiarity with our review process and allowing sufficient time for review are our best recommendations for the future.

As the formulators of pesticide products deal with the increased emphasis on the reduced use of toxic active ingredients, the decreased use of volatile organic solvents, ozone-depleting agents, and groundwater contaminants, as well as the concurrent encouragement of technologies that reduce exposure to the mixer/loader/applicator and alleviate disposal issues, it is clear the there is much to encourage development of inert ingredient technology. On the other hand, the EPA has increased its scrutiny of inert ingredients and, as a result, innovative inert technologies may face considerable data development costs. There has recently been increased concern for the need for the data generation costs of these inert ingredients to be compensable, as such costs are for active ingredients. Traditionally, there has not been data compensation for inert ingredients because requests for approval and exemptions from tolerance come to the EPA before the inert ingredient is incorporated into a pesticide product. Others on this program will discuss this in greater detail. There are many issues to be dealt with in the development of a data compensation program for inert ingredients; probably the largest is the guarantees of confidentiality currently required by the law. We have just begun to tackle these issues. But we want to leave you with the knowledge that the development of reduced risk formulations and reduced risk inert ingredients is clearly consistent with the goals of our program and we hope to see great activity in this area over the next decade.

REFERENCES

1. **Food and Drug Administration**, Certain inert ingredients in pesticide formulations, *Federal Register*, 10640, 1961.
2. **Food and Drug Administration**, Tolerances and exemptions from tolerances for pesticide chemicals in or on raw agricultural commodities, *Federal Register*, 6041, 1969.
3. **United States Environmental Protection Agency**, Premanufacture notification exemptions; exemptions for polymers, *Federal Register*, 46066, 1984.
4. **United States Environmental Protection Agency**, Inert ingredients in pesticide products; policy statement, *Federal Register*, 13305, 1987.
5. **United States Environmental Protection Agency**, Inert ingredients in pesticide products; policy statement; revision and modification of lists, *Federal Register*, 48314, 1989.
6. 7 U. S. C. § 136
7. 15 U. S. C. § 2601
8. 21 U. S. C. § 7401
9. 42 U. S. C. § 301

Chapter 2

European Directives on Inert Ingredients and Adjuvants

Edwin L. Johnson

CONTENTS

ABSTRACT

The regulation of inert ingredients and adjuvants, just as for active substances, is undergoing change in the European Community (EC). The EC is moving toward a harmonized, community-wide approach toward adjuvants and inert ingredients. While a Directive (91/414/EEC) and implementing regulations have been issued for active substances, the approach to adjuvants and inert ingredients was still being debated in spring 1994, although issuance of a revised Directive 91/414/EEC covering these substances was originally expected by the end of 1993. Until issuance of the Directive, these substances will continue to be regulated by individual Member States.

I. INTRODUCTION

The European Directive on regulation of inert materials and adjuvants was due to be formally proposed by the end of 1993. The Directive was

1-8493-7678-5/96/$0.00+$.50

proposed in a working document to take the form of amendments to the agrochemical registration Directive (91/414/EEC of 15 July 1991),[1] extending its scope to the regulation of inert ingredients and adjuvants. The first Commission working document was prepared and circulated for Member State and trade association discussion and comment in 1992. Subsequently, on 30 January 1993 a new working draft (Revision 3)[2] was issued that dealt with some of the problems identified in the first working draft. However, the draft Directive has not been issued as a final proposal.

According to information received informally from Commission staff, the timing is currently uncertain, but is likely more than a year off.[4] It can also be expected that there may be substantial changes in the requirements from the working drafts and the final proposal, although it is not entirely clear what these will entail.

Given the current state of development of this Directive, this chapter will discuss the various draft documents and their provisions and industry reaction. It will also provide an overview of the procedures for development of Directives, their implementation, and the roles of various bodies in the European Union, including Member States, in this process. Finally, it will briefly address current applicable regulation.

The term "inert" is the commonly used U.S. term for ingredients in pesticide formulations other than the active ingredient(s). In Europe, the term used for the same purpose is "coformulant" or "formulant". Therefore, "inert" is used interchangeably with the term "coformulant" or "formulant". Similarly, the terms "active ingredient" and "active substance" are used interchangeably.

II. THE EUROPEAN UNION REGULATORY PROCEDURES

This section will provide a brief overview of the European Union procedures for developing community-wide regulatory documents, mechanisms for implementation, and the roles of various organizations in the process.

The European Union may issue any of four types of legislative or regulatory pronouncements. These are:

- Directives
- Regulations
- Decisions
- Recommendations and Opinions

Directives are binding on the Member States, but must be implemented through national laws approved by national parliaments on a schedule dictated by the Directive. These national laws must be consistent with the Directive but may also contain elements that differ from the laws promulgated by other

Member States to implement the same Directive. This bears some similarity to U.S. federal environmental laws which often allow states flexibility in implementation but the state may be no less stringent than the federal requirement.

Regulations are directly applicable to all Member States upon issuance and take precedence over national law.

Decisions are binding for Member States, companies, or individuals to which they are directed. Those imposing financial obligations are enforceable in national courts.

Recommendations and opinions have no binding force, being advisory in nature.

The Commission of the European Union is the administrative bureaucracy for the European Union, being the administrator of the treaties binding the Union together and all subsidiary legislation; it also conducts the day-to-day running of the European Union. The Commission works through 23 Directorates-General, equivalent to U.S. Departments, with a civil service staff totalling some 14,000 persons. For pesticides, the two most important Directorates-General are VI (Agriculture) and XI (Environment). The Commission is the sole originator of proposals, directives, regulations, and other forms of regulation.

The Council, comprised of Ministers from the Member States, considers and adopts, with or without amendments, or rejects, proposals.

The European Parliament separately considers proposals, submits amendments, and can block certain measures. The formal opinion of the Parliament is required on any proposals before they may be adopted by the Council.

The Member States have the sole responsibility for implementing Directives through national legislation. As noted earlier, regulations have direct effect.

Finally, the European Court, sitting in Luxembourg, provides rulings on the interpretation and application of Community laws. The court may request modification of national laws and decisions incompatible with European Union Treaties and Directives and may fine Member States that refuse to implement European Union law. Unlike U.S. courts, a specific matter need not be ripe for decision before being submitted to the court, rather requests for interpretation of specific provisions may be requested.

III. EUROPEAN UNION REGULATION OF INERT INGREDIENTS AND ADJUVANTS – THE FIRST PROPOSAL

The current Directive, 91/414/EEC, Concerning the Placing of Plant Protection Products on the Market, indirectly deals with inert ingredients, but not with adjuvants. Annex III, Requirements for the Dossier to be Submitted for the Authorization of a Plant Protection Product, requires the submission of test data on the formulation which contains inert substances, and notes that: "In individual cases it may be necessary to require information as provided for in Annex II, Part A (Requirements for the Dossier to be Submitted for the Inclusion of an Active Substance in Annex I, Part A Chemical Substances) for formulants (e.g., solvents and surfactants)."

In August 1992, the Commission circulated for comment to Member States and interested parties a first draft working document entitled "Suggested Amendments with Regard to the Extension of the Scope of Directive 91/414/EEC Concerning the Placing of Plant Protection Products on the Market".[1] This amendment proposed to extend the Directive to include adjuvants and inert substances (coformulants intended for incorporation either in a plant protection product or in an adjuvant). It was anticipated to lead to a formal proposal in late 1993 but the schedule has slipped and the amendment is currently on hold. This section will summarize the provisions proposed in the first draft amendments to Directive 91/414/EEC.

Definitions: Definitions for several materials were added to Article 2: [Numbering below conforms to paragraph numbers in the working document.]

1.a Adjuvants – Coformulations and preparations containing one or more coformulants put up in the form in which they are supplied to the user and placed on the market with the objective as shown by the label to be mixed by the user with one or more plant protection products for the purpose of changing its or their properties or effects.

4. Active substances – Definition extended to also include in the meaning of the Directive synergists and safeners as active substances.

4.a Synergist – A substance, which, whilst formally inactive or weakly active, can give enhanced activity to an active substance or the active substance in a preparation.

4.b Safener – A substance added to a preparation to eliminate or reduce phytotoxic effects of the preparation to certain species.

4.c Coformulant – Any substance other than an active substance, which is used or intended to be used in a plant protection product or in an adjuvant, such as solvents or surfactants.

Article 3 (General Provisions), Article 4 (Granting Review and Withdrawal of Authorization of Plant Protection Products and Adjuvants) Article 9 (Application for Authorization), Article 10 (Mutual Recognition of Authorizations), Article 12 (Exchange of Information), Article 13 (Data Requirements, Data Protection, and Confidentiality) are modified to place adjuvants into the same category as plant protection products and subject to the same data requirements.

However, Article 14, dealing with confidentiality of industrial and commercial secrets, departed from this parallel treatment of plant protection products and adjuvants by declaring that: "Confidentiality shall not apply to: ...the names and content of the coformulant or coformulants in the adjuvant..." Thus the formula of a plant protection product may be granted confidential treatment, but that of an adjuvant may not.

Article 15 (Packaging and Labelling of Plant Protection Products and Adjuvants) also treats the two differently, making plant protection products subject to the requirements of Directive 78/631/EEC, Article 5(1) whereas Article 6 of Directive 88/379/EEC would apply to adjuvants.

Article 16 is amended by adding a new section, Article 16a, specifically addressing adjuvants and requiring *inter alia* that: "All packaging must show clearly and indelibly the following: ... the chemical name of the coformulant or coformulants present in an adjuvant in accordance with the provisions as referred to in Article 7(1)(c) and (3) of Directive 88/379/EEC."

Article 7 (Information on Potentially Harmful Effects) is amended by addition of a major new section, Specific Measures Concerning Coformulants, which essentially establishes a list of prohibited inerts and a mechanism for prohibiting or restricting adjuvants. Specifically, Article 7a provides:

> 1. When information is submitted by a Member State to the Commission in the framework of Article 7 or Article 11, showing that the coformulant will not fulfill one or both of the conditions mentioned in Article 5(1) and (b) [essentially absence of harmful effects on human or animal health or the environment] it may be decided according to the procedure provided for in Article 18 [Administrative Provisions], that the coformulant will be included in a new Annex VII [Coformulants prohibited or restricted in use], with the effect that its use will be prohibited or restricted to the conditions specified in this Annex VII.
> 2. Such decision may also be taken according to the same procedure where information is submitted by a Member State to the Commission in the framework of Article 7 or Article 11, showing that for an adjuvant it may be expected that mixtures prepared by the user and containing the adjuvant will not fulfill one or both of the conditions mentioned in Article 5(1) (a) and (b).
> 3. The Member States shall ensure that the use of the coformulant is prohibited or restricted, as appropriate, within the period determined in the Commission decision.
>
> Before a decision as referred to in paragraph (1) is taken, the Commission may invite the interested manufacturers or importers to present within a specified delay any information or test results required for the decision to be taken.

Article 23 (Implementation of the Amending Directive) requires Member States to bring into force laws, regulations, and administrative provisions necessary to comply within one year and that uniform principles in relation to adjuvants be adopted in the same time frame.

IV. INDUSTRY REACTION TO THE FIRST DRAFT AMENDMENT

The pesticide industry represented by the European Crop Protection Association (ECPA) responded to the first draft amendment to Directive 91/414/EEC on several significant points:[3]

1. Article 2.4.2 a and b – Synergist and Safener – Under the proposed revised definitions, synergists and safeners would be considered active substances. As such they would be subject to full Annex II data requirements. It has been suggested that data requirements for synergists and safeners should be flexible and determined on a case-by-case basis, keeping in mind the usually inactive or weakly active nature of such substances.
2. Authorization of Adjuvant – The new requirements would subject an adjuvant to all the same data requirements as a plant protection product, including appropriate methods for determination of the nature and quantity of its coformulants. "This across-the-board requirement for Annex III data for authorization of an adjuvant, which might be a corn or soybean oil, a complex polymer or a distillation fraction, is clearly excessive, as is the need to develop analytical methods for coformulants."[3] The cost and effort of developing such data, the comment continues, would not be commensurate with the information gained, and in many cases, it would be impossible to perform the work. Clearly, the data required for approval of an adjuvant must be flexible and conditioned on the nature and properties of the adjuvant.
3. Lack of confidentiality of the composition of an adjuvant – As noted above, new Article 14 does not permit the names and content of the coformulants in the adjuvant to be treated as confidential, and new Article 16a requires listing of the names and quantities of coformulants on the labelling. While the names and quantities of coformulants in plant protection products may be considered confidential, those in adjuvants may not under the draft amendments. This appears illogical and industry recommends that the composition of an adjuvant be given confidentiality when requested by the manufacturer and agreed to be warranted by the Member State or the Commission.
4. Article 7.a.1 – Prohibited list of coformulants – Industry did not object in principle to the concept of creating such a list, but suggested that further information is needed on which data trigger the process leading to negative listing. It further suggested that consultation with industry should be mandatory rather the discretionary before taking a Commission decision.

Adjuvant manufacturers were not invited to comment on the working document, according to Smith,[5] since such documents are circulated only to "interested bodies" and not to individual companies. The European Adjuvant Association (EAA) was formed in July 1993 to get around this problem;

another group of adjuvant producers not belonging to other associations formed the Independent Adjuvants Association (IAA).

The EAA is in favor of uniform registration requirements for adjuvants. They consider efficacy testing important to support the credibility of adjuvant claims and prevent manufacturers from making unrealistic claims. EAA also recommends testing to assure safety for users, the environment, and the public.

EAA has prepared a White Paper which details its specific comments on the working draft and suggestions for an approach to regulation of these products. The overall approach would mirror that which is currently used in France.

The IAA takes a somewhat different stance from the EAA. They agree that toxicology and environmental fate data are essential for adjuvants, but consider other aspects of the Commission working paper to unfairly discriminate against these products by imposing excessive testing requirements.

V. THE COMMISSION REACTION TO COMMENTS – REVISION 3 OF THE WORKING DOCUMENT

Revision 3 of the working document was issued on 30 January, 1993.[2] Significant changes were made which address industry comments as well as several minor changes.

First, the term "coformulant" was changed to "formulant" and the definition of adjuvant changed as follows [numbering of paragraphs below conform to that of the working document, Revision 3]:

1.a Adjuvants – Formulants and preparations containing two or more formulants, put up in the form in which they are supplied to the user and placed on the market with the objective as shown by the label to be added by the user to a plant protection product or a mixture of plant protection products at a diluted or ready to use stage, for the purpose of changing its or their properties or effects.

The definition of active substances to include synergists and safeners was unchanged from Revision 1, but slight, nonsubstantive modifications were made to the definitions of some specific terms:

4.a Synergist – Definition unchanged from Revision 1.

4.b Safener – A substance which, when added to a preparation, eliminates or reduces phytotoxic effects of the preparation to certain species.

4.c Formulant – Any substance other than active substance incorporated in the preparation.

In addition, Revision 3 substantially affects the issue of confidentiality of an adjuvant formulation in Article 14 which now reads:

"Confidentiality shall not apply to:

- the name of the plant protection product or adjuvant.
- the name and content of the active substance in the plant protection product or the name and content of the formulant or formulants in the adjuvant *mentioned on the label*" (emphasis added).

Article 16 (Packaging and Labelling of Plant Protection Products) was also amended to clarify that: "Article 6 of Directive 88/379/EEC shall apply to all adjuvants covered by Directive 88/379/EEC". That is, Article 6 shall apply to those adjuvants whose toxicological properties make them subject to the dangerous preparations provisions.

Aside from the changes noted above, Revision 3 is essentially unchanged from Revision 1.

The amendments contained in Revision 3 of the working document appear to largely deal with industry concerns over confidentiality of adjuvant ingredients. However, the concerns expressed regarding the data requested for approval of an adjuvant, safener, or synergist do not appear to be addressed by the proposed changes.

VI. OTHER DIRECTIVES APPLICABLE TO INERTS AND ADJUVANTS

There are two other Directives applicable to inert substances and adjuvants. The first is Directive 88/379 relating to the classification, packaging, and labeling of dangerous preparations. Adjuvants whose composition is more than one substance fall under this preparations Directive according to Article 16 of the working paper. The person responsible for placing the adjuvant on the market is obliged to review the known physico-chemical properties and toxicological data and to compare them to criteria set out in the preparations Directive. If the adjuvant is deemed "dangerous" then it must also be packaged and labeled in accordance with the preparations Directive using appropriate danger symbols and risk phrases.

According to the working documents, coformulants of both plant protection products and of adjuvants are subject to the requirements of Directive 67/548 as amended by the Seventh Amendment. This Directive relates to the classification, packaging, and labeling of substances as well as the notification of new substances. Analogous to preparations, the properties and toxicological data must be reviewed to determine the classification, if any, and subsequent labeling and packaging requirements. Of course, any substance which is not found on the EINECS inventory must go through the notification procedure given in the Seventh Amendment.

VII. CONCLUSION

Inert substances are currently covered in an indirect manner in Annex III of Directive 91/414/EEC as part of the plant protection product to be tested, and Annex III of that Directive notes that additional data commensurate with that for an active substance under Annex II may be required on occasion. Amendments proposed by the Commission in its working draft of amendments to 91/414/EEC would add the provision for creation of a prohibited or restricted list of inert substances as Annex VII, but have not specified the criteria to trigger an inert material for inclusion in this list. Synergists and safeners would be considered the same as active substances in terms of data requirements.

Adjuvants are currently regulated by existing Member State laws and regulations and not under a European Union-wide Directive. Existing Member State laws vary widely from those which require toxicity data only, such as Spain and Portugal; to moderate requirements, Ireland;, to very stringent requirements for toxicology, efficacy, and residue test data, Germany (Table 1). The U.K. considers adjuvants to be adequately covered by national law, the EC Dangerous Substantive Directive (Seventh Amendment) and the Dangerous Preparations Directive; it considers toxicological and residue information to be important, but not efficacy data.

The first working document would treat adjuvants for most purposes the same as plant protection products under 91/414/EEC, subjecting them to the data requirements of Annex III. In other respects, however, the draft amendments would have required disclosure of the name and quantity of coformulants both by denying confidentiality from public information requests and requiring their inclusion on labelling. In addition, it would have applied the same procedures to prohibiting or restricting adjuvants as apply to coformulants.

Various industry groups commented on the first draft amendments, making points that seem reasonable and logical.

The third version of the working document appears to deal with issues of confidentiality, but not with concerns over excessive data requirements. However, the timing and content of the final directive is uncertain. The project is temporarily on hold, perhaps because of the significant workload on the Commission staff to implement the reregistration activity under 91/414/EEC. One notion expressed by Commission staff is that the future direction may be to provide a positive list for synergists and safeners and a negative list for other inert substances.[4] Hopefully, it would also incorporate industry suggestions regarding flexibility in testing requirements and taking maximum advantage of regulation under other EC Directives for these compounds under the Seventh Amendment. Under this scenario, adjuvants would be treated in a separate proposal, with the possibility of having a code of conduct approach rather than data requirements.

Perhaps next year the provisions of a final European Union Directive on inert substances and adjuvants will be available.

TABLE 1.
Current EC Country Requirements for Adjuvants

COUNTRY	TOXICITY	EFFICACY	RESIDUES
Germany	Y	Y	Y
France	Y	Y	N
U.K.	Y	N	Y
Ireland	Y	N	N
Spain	Y	N	Some
Portugal	Y	N	N
Greece	Y	Y	N
Netherlands	Y	Y	Y
Denmark	Y	N	N

Source: Secretary, European Adjuvant Association.

REFERENCES

1. **Commission of the European Communities**, Suggested amendments with regard to the extension of the scope of Directive 91/414/EEC concerning the placing of plant protection products on the market, Working Document 2772/VI/92 -EN rev. 1, 26 Aug., 1992.
2, **Commission of the European Communities**, Suggested amendments with regard to the extension of the scope of Directive 91/414/EEC concerning the placing of plant protection products on the market, Working Document 2772/VI/92-EN rev. 3, 30 Jan., 1993.
3. **European Crop Protection Association**, Interim ECPA response to commission Working Document 2772/VI/92-EN (26.08.1992) dated 23 Nov., 1992.
4. Personal communications with Commission Staff and consultants.
5. **Smith, Ashley**. Adjuvants in Crop Protection, PJB Publications Ltd., Richmond, Surrey, U.K., 1993.

Chapter 3

Encouraging Improved Pesticide Formulation: Data Protection for Inert Ingredients

Alvin J. Lorman

CONTENTS

Abstract

Sophisticated new inert pesticide ingredients can play an important role in improving the performance of pesticide formulations and in mediating their environmental effects. While the Environmental Protection Agency has recently recognized that the Federal Insecticide, Fungicide, and Rodenticide Act affords to inert ingredient manufacturers the same exclusive use and data compensation rights available to manufacturers of active ingredients, taking advantage of those rights is cumbersome. To provide the necessary incentive to develop useful new inert ingredients, a new procedure must be developed which recognizes the realities of the market for inert ingredients.

I. Introduction

The role of pesticides in American agriculture has received close scrutiny in recent years. In June 1994, the U.S. Environmental Protection Agency (EPAs) Office of Pesticide Programs conducted a major conference on, among other things, reducing the risk associated with pesticide use. This conference was a follow-up to an earlier conference sponsored jointly by EPA, the Food and Drug Administration, and the Department of Agriculture on the same subject. That same week, a House Agriculture Subcommittee chaired by Rep. Charles Stenholm opened hearings on the Clinton Administration's proposed revisions to the nation's pesticide and food safety laws. What the EPA program and the Clinton Administration legislative proposal had in common was their failure to recognize the role that improved inert ingredients can play in helping farmers, consumers, and the environment. Advances in chemistry and manufacturing processes enable manufacturers to produce dramatically improved inert ingredients for pesticide formulations. This presents an opportunity not only for inert ingredient manufacturers, but also for formulators working on new delivery systems. Some examples of the variety of roles new inert ingredients can play in pesticide

1-8493-7678-5/96/$0.00+$.50

formulations, and how those roles can be used to maximize safety, should be noted:

- By reducing the quantity of active ingredient needed in a formulated product, or by helping that active ingredient remain at the site of intended pesticidal action, there is an associated reduction in the need for reapplication and a reduction in the amount which enters the environment.
- While dry formulations of pesticide products are often the preferred formulation, dry products can be less efficacious and thus require more active ingredient to be applied in order to control the target pest. To achieve the goal of creating more efficacious pesticides with reduced concentrations of active ingredients, specialized inert ingredients can be used to improve delivery and thus reduce the quantity of the active ingredient needed.
- Aerial application of pesticides can result in volatilization of 60 to 70% of the active ingredient used, thus substantially increasing the amount that must be used to achieve the desired result. Certain polymers, which can be used as encapsulants or protective colloids, can materially reduce volatilization of active ingredients, thereby reducing the amount required.
- Drift during application to non-target pests and crops is a significant problem. Spray patterns, the size of droplets, and their potential to drift can be modified by the addition of certain polymeric inert ingredients.
- Leaching of pesticides into groundwater and contamination of surface water are serious problems posed by many pesticides. A 50% reduction in the leachability of several pesticides can be achieved by the addition of specialized polymers to commercially available pesticide formulations. With preemergence herbicides, restricting leaching keeps more herbicide in the zone in which seeds germinate, lessening the total amount of active ingredient needed.
- Microencapsulation can be used to reduce human toxicity and to extend pesticide efficacy, thus reducing the number of required applications and the total amount of pesticide applied.

Despite the great progress that has been made, the principal impediment to the full realization of these advances has come from the interplay of the necessity of obtaining a tolerance or exemption from a tolerance under Section 408 of the Federal Food, Drug, and Cosmetic Act (FD&C Act) and the data compensation provisions of the Federal Insecticide, Fungicide and Rodenticide Act (FIFRA).

While FIFRA grants pesticide registrants a 10-year period of "exclusive use" of data submitted in support of pesticides containing active ingredients initially registered after September 30, 1978, data submitted in support of an exemption from a tolerance under Section 408 is not similarly protected. For active pesticide ingredients, this distinction is not important. For inert ingredient manufacturers, however, a very important difference results: a company which conducts the tests EPA now requires to obtain an exemption from a tolerance for a new inert pesticide ingredient typically faces immediate competition from companies that have not incurred the research and regulatory costs -- so-called "free riders." This failure to protect the data submitted under Section 408 is especially significant because of the way the market for inert ingredients operates.

Because a company which chooses to devote the time and resources necessary to develop a new inert ingredient is not typically a pesticide manufacturer, it is not in a position to use that new ingredient in a formulation of its own. Thus, the new inert ingredient can reach the market only when a basic producer agrees to include it as part of a registered product. Understandably, basic producers are unwilling to risk the pending registration of a new active ingredient by including a new inert ingredient for which no exemption from a tolerance has been granted. After all, the basic producer probably has a greater interest in the new active ingredient or combination of active ingredients in the formulation than it does in the new inert. So the basic producer typically expects the inert ingredient manufacturer to obtain the Section 408 exemption *before* it will consider using the new ingredient in its product. When the inert ingredient manufacturer does so, it helps create a generic exemption available to all, including its free-rider competitors.

II. Data Protection

In order to establish the principle that data supporting new inert ingredients are entitled to some measure of protection, International Specialty Products (ISP), in collaboration with a pesticide manufacturer, arranged to have the data in support of a new inert ingredient submitted for the first time to EPA as part of the manufacturer's pesticide registration application. That application made an explicit claim that the data in support of the inert ingredient were entitled to the same protection as the data in support of the active ingredients.

In response to that claim, EPA has recently confirmed in writing -- for the first time that we are aware of -- that the data in support of the safety of new inert pesticide ingredients which are initially submitted as a part of a pesticide registration are indeed entitled to the same exclusive use and data protection afforded to active ingredient studies. While that is an eminently correct legal conclusion, the practicalities of the marketplace still demand that a better system be developed to encourage the production of advanced inert ingredients.

In order to see why EPA's decision is both correct and, unfortunately, of limited usefulness, it will help to examine the data protection provisions in FIFRA Section 3(c)(1)(D).

FIFRA's requirement that a pesticide registration applicant submit safety and efficacy data in support of the product endows that data with considerable value. The value inherent in the data is what has created the thorny problem of resolving its status. Today, FIFRA contains a complex scheme of exclusive use and data compensation provisions designed to assure that a company which develops the data necessary to secure a Section 3 pesticide registration is given the chance to obtain a return on its investment, for a fixed period, free from competition from "me too" imitators. This scheme is the result of long experience and the careful balancing of competing interests by Congress.

When first enacted in 1947,[7] FIFRA specifically prohibited disclosure of any information about the formulas of products, but it was silent with respect to the disclosure of any health and safety data submitted with an application for registration.* In 1972, Congress substantially amended FIFRA to add, in part, a new section governing public disclosure of data submitted in support of an application for registration. The legislative history of the 1972 amendments indicates that Congress intended to streamline pesticide registration procedures, increase competition, and avoid unnecessary duplication of data-generation costs.[6]

To accomplish these goals, many members of Congress believed that recognizing a limited proprietary interest in data submitted to support pesticide registrations would provide an added incentive beyond statutory patent protection for research and development of new pesticides.[9]

The 1972 amendments permitted a data submitter to designate any portions of the submitted material as "trade secrets or commercial or financial information" and prohibited EPA from disclosing such information.** In addition, the amendments instituted a mandatory licensing scheme: EPA could consider data submitted by one applicant for registration in support of another application involving a similar chemical if the subsequent applicant offered to compensate the original applicant. If the parties could not agree, EPA was to determine the amount of compensation. However, trade secrets or commercial or financial information were specifically exempt from disclosure.

Congress's failure to define what constituted "trade secrets or commercial or financial information" frustrated much of what it intended to accomplish with the 1972 amendment, and resulted in considerable

* FIFRA was originally administered by the U.S. Department of Agriculture (USDA). In 1970, USDA's FIFRA responsibilities were transferred to the then-newly created Environmental Protection Agency. For convenience, all references will be to EPA or its Administrator.

** In the event that EPA disagreed with the submitter's designation, the submitter could institute a declaratory judgment action in a U.S. district court to resolve the controversy.

litigation.* After a series of lawsuits, EPA was prevented from disclosing much of the data on which it based its decisions to register pesticides and from considering the data submitted by one applicant in reviewing the application of a later applicant.**

In response to the "logjam of litigation that resulted from controversies over data compensation and trade secret protection,"[10] Congress in 1978 again amended the data compensation and trade secret provisions of FIFRA.[8] The 1978 amendments replaced the 1972 compulsory license approach with a 10-year period of "exclusive use" for data submitted in support of pesticides containing active ingredients initially registered after September 30, 1978.[11] During that time the data may not be cited to support another registration without the original submitter's permission. In addition to 10 years' exclusive use, the original data submitter is entitled to an offer of compensation from an applicant wishing to use its data during the following 5 years.*** After the expiration of any period of exclusive use or compensation, EPA may consider such data in support of another application without compensation or permission from the original data submitter.

EPA now publicly agrees that, for purposes of FIFRA's exclusive use provisions, no principled distinction can be drawn between a Section 3 registration application which contains data in support of both an active ingredient and an inert ingredient, and one which contains data only on the active ingredient.

The language Congress chose to express the 10-year period of exclusivity encompasses the data supplied in connection with inert ingredients as well as active ingredients. That protection is afforded to

* *See, e.g., Chevron v. Costle*, 641 F.2d 104 (3d Cir.), *cert. denied*, 452 U.S. 961 (1981). EPA maintained that the exemption from disclosure only applied to a narrow range of information, including formulas and manufacturing processes. Pesticide firms claimed that the trade secret exemption from disclosure applied to any data, including health, safety, and environmental data.

** In *Ruckelshaus v. Monsanto Co.*, 467 U.S. 986 (1984), the Supreme Court ruled that the trade secret exemption of the 1972 amendments provided Monsanto and similarly situated companies an "explicit governmental guarantee" sufficient to warrant a reasonable investment-backed expectation that data marked "trade secret" and submitted to support a product registration would not be used or disseminated without the permission of the data submitter.

*** 7 U.S.C. § 136a(c)(1)(D)(iii). FIFRA Section 3(c)(1)(F)(ii) provides that, for data submitted after December 31, 1969 by an applicant for registration, EPA may, without permission of the original data submitter, consider the data in support of an application by any other person for a period of 15 years following the date after which the data were originally submitted, only if the applicant has made an offer to compensate the original data submitter. The effect of this provision on data submitted after 1978 is to provide, after the expiration of the 10-year exclusive use period, an additional 5-year period of data compensation protection.

"*pesticides* containing active ingredients initially registered . . . after September 30, 1978"; the language does not apply solely to active ingredients contained in pesticides initially registered after September 30, 1978. Similarly, the bar to use of the data refers to data "submitted to support the application for the original registration," and not to some subset of that data, such as data concerning only the active ingredient. Since Congress specifically defined "active ingredient," "inert ingredient," and "pesticide," it can be concluded that it knew how to distinguish between them when it chose to do so. And, for purposes of exclusive use, it chose to protect data in applications for "pesticide" registrations, thus affording protection to a broader universe than active ingredients alone.

The role of inert ingredients in pesticide products has begun to attract increasing attention from EPA and industry alike. While EPA has begun to focus attention on "hazardous" inert ingredients and is now requiring a battery of tests for all new inert ingredients, a less obvious aspect of pesticide safety is the role which inert ingredients can play in producing finished pesticide products which are safer for users, consumers, and the environment.

The sophisticated technology required to achieve these desirable goals, coupled with the ever-higher costs of health and safety testing, means that the policy adopted by Congress applies with equal force to inert, as well as active, ingredients. In amending the data compensation and exclusive use provisions of FIFRA in 1978, Congress understood "the need to assure the continued research and development of new pesticides by recognizing the limited proprietary interest of those who have incurred the expense of developing health and safety data . . . ."[5] A statutory interpretation which creates a disincentive to produce safer and healthier products and the very data which EPA believes are important in order for it to protect the public makes little sense.

In 1987, EPA articulated a new policy on inert ingredients.[1] EPA conceded that prior to this issuance of its new policy, "[m]ost of the data requirements and regulatory activities under FIFRA were focused on the active ingredient."[2] The new policy, however, "requir[es] the development of data necessary to determine the conditions of safe use of products containing inert ingredients."[3] In addition to creating a risk-ordered scheme of inert ingredient categories, EPA announced that it would require specific data for any *new* inert ingredient:

> "Any inert ingredient proposed for use in a pesticide product is considered to be a new inert ingredient if it is not currently identified as present in some approved pesticide formulation or has never been in a previously approved product. The minimal data generally required to evaluate the risks posed by the presence of a new inert ingredient in

a pesticide product is a subset of the kinds of data typically required for active ingredients under 40 CFR Part 158."[4]

The adoption of this policy and its impact on new inert ingredients, we believe, triggers the provisions of Section 3(c)(1)(D) even if they were not thought previously to apply. EPA now agrees, but having agreed that data supporting the registration of new inert ingredients are entitled to the same protection as new active ingredients, it is incumbent upon the agency to design administrative procedures that make that protection meaningful. EPA itself has conceded that the realities of the current data protection scheme leave something to be desired.

Unfortunately, EPA's procedures at 40 C.F.R. §§ 152.80-152.99, concerning the protection afforded data submitters, do not appear to cover inert ingredients. For example, an exclusive use study is described as one that, *inter alia*, "pertains to a new active ingredient . . . or new combination of active ingredients." And it is not clear how the submitter of data in support of a new inert ingredient gets listed on the data submitters list.

The process confirming the existence of this right is instructive because it shows that while the right may be there, obtaining it takes considerable effort. In order to confirm the existence of the right, ISP developed two new inert ingredients for which it did not, initially, file a petition for an exemption from a tolerance under Section 408. Instead, it sought the assistance of a pesticide manufacturer which would include the new inert ingredients in pesticide registrations it intended to file. At the same time, the registration applications would make an explicit claim for data protection for the inert ingredients, something that had never been done before, as far as was known. Negotiating this agreement took some time and was recorded in a fairly complicated contract. And, since ISP was not the registrant, it needed to obtain special permission simply to talk to EPA about its own new inert ingredients.

III. Conclusion

Lynn R. Goldman, M.D., Assistant Administrator for Prevention, Pesticides and Toxic Substances, EPA, has stated that "the Agency fully intends to create a mechanism to assure protection of data rights for those who submit data on inert ingredients in support of a pesticide application or registration." Dr. Goldman also noted that one of the problems associated with protecting inert ingredients is that their existence in some products is a trade secret. Industry welcomes and appreciates EPA's recognition of the commercial value represented by the data supporting new inert ingredients.

While no solution is likely to be simple, we believe that any solution should face up to the realities of the inert ingredient marketplace. The simplest solution to the problem is to attack it head on and extend data protection to data submitted in support of a Section 408 tolerance exemption. If that is done, the appropriate incentive will be provided without requiring an inert ingredient manufacturer to engage in the indirect and onerous process which was required for EPA to grant data protection for ISP's two new inert ingredients. Such an onerous process is only likely to be worthwhile for inerts which offer dramatic improvements over the prior art. Smaller, but still meaningful improvements, are not likely to rate that level of resource commitment. Furthermore, changing the status of Section 408 data would also make it easier to continue to treat inert ingredients as trade secrets in appropriate cases.

The massive rewrite of the nation's pesticide and food safety laws which the Clinton Administration transmitted to Congress in 1994 simply does not address this issue. H.R. 1627, which was favorably reported out of the Agriculture Subcommittee on Department Operations and Nutrition in the last Congress, explicitly extended to data in support of tolerances the same protections available to data submitted in support of registrations. That is a reasonable and useful way to proceed, and one that promises the greatest improvements for farmers, consumers, and the environment. In the meantime, we are faced with the interesting question of how to enforce data protection rights for inert ingredients under a statutory scheme that grants the rights and a regulatory scheme that does not mention them.

REFERENCES

1. 52 Fed. Reg. 13305-09 (April 22, 1987).
2. 52 Fed. Reg. 13305 (April 22, 1987).
3. 52 Fed. Reg. 13305 (April 22, 1987).
4. 52 Fed. Reg. 13308 (April 22, 1987).
5. H.R. Rep. No. 343, 95th Cong., 2d Sess., pt. II at 18 (1977).
6. S. Rep. No. 838, 92d Cong., 2d Sess. 72-73 (1972).
7. 61 Stat. 163 *et seq.*, Pub. L. No. 86-139, June 25, 1947.
8. 92 Stat. 819, Pub. L. 95-396 (1978).
9. *Thomas v. Union Carbide*, 105 S.Ct. 3325, 3328 (1985) (citing H.R. Rep. No. 633, 95th Cong., 1st Sess. 17-18 (1977), S. Rep. No. 334, 95th Cong., 1st Sess. 7, 34-40 (1977)).
10. *Thomas v. Union Carbide*, 105 S.Ct. 3325, 3328 (1985) (*quoting* S. Rep. No. 334, 95th Cong., 1st Sess. 3 (1977).
11. 7 U.S.C. § 136a(c)(1)(D)(i).

Chapter 4

Development Trends in Pesticide Formulation and Packaging

Bruno Frei and Peter Schmid

CONTENTS

1-8493-7678-5/96/$0.00+$.50

I. ABSTRACT

There is an increasing need for the crop protection industry to continue to develop new products and delivery systems which are optimized with regard to safety, environmental behavior, biological performance, and cost. Innovative formulations and packaging systems will have to make a major contribution for industry to reach these goals. In the last 10 years, much progress has been made in the design and development of new formulations and packaging options. In particular, by reduction of organic solvents in liquid formulations, the environmental impact and toxicity of products may be reduced. A reduction of dermal toxicity may be obtained by encapsulation of the active ingredient. Improved worker and environmental safety is also achieved by replacing powders with water dispersible granules which eliminates dusting during handling, or by packaging them in water soluble film, which offers a variety of advantages. Seed treatment provides an elegant way to minimize the amount of chemical needed by placing it directly at the spot where it is needed at the right time.

In terms of packaging strategy, waste reduction and improved handling safety are prime objectives. Refillable container programs aimed at larger growers and custom applicators will contribute to the reduction of one-way packaging and associated secondary packaging. For smaller growers and also for the use of highly active compounds, the introduction of solid formulations and gels, which can be packaged in water soluble film, completes the strategy to achieve a reduction of packaging waste.

In this chapter, examples of new formulation and packaging concepts will be presented and an overview of future trends will be given.

II. INTRODUCTION

Over the last decade the crop protection industry has been faced with many changes and today it has to meet not only economic, but also social and environmental challenges. With an overall stagnating market and increasing regulatory requirements concerning user safety and environmental compatibility of their products, innovative solutions to emerging customer needs become all-important if industry players are to reach their economic objectives. From a social point of view it is necessary to inform the public of the benefits and risks of products, and to aim for continuous improvement of product handling systems with regard to safety for the user. The main targets concerning the protection of the environment include saving resources with better products and manufacturing processes, introducing environmentally compatible products, and minimizing waste by optimizing production processes and packaging designs. To achieve these goals, combined efforts in research, development, production, and marketing will be necessary. As it will be shown by several examples, the development of new formulations and packaging systems can make an important contribution to the progress in the pursuit of this goal.

The core of each product is the chemical compound which is responsible for the intrinsic biological activity. For a safe and optimally targeted application and the development of the full inherent biological efficacy of the product, the active ingredient has to be formulated and packed properly. In the past a formulation chemist could take an active ingredient, put it in a solvent system with an emulsifier, and hand the formulation off to an engineer who would select the appropriately sized container from the shelf, add a label, and offer the finished product for sale. The process is much more complicated today, and as Bailey[1] states, "perhaps the most challenging aspect of delivery systems development is regulatory compliance". Therefore, the design of the formulation and its packaging are of increasing importance.

The formulation and packaging concept for a product depends on many factors. The physicochemical properties of an active ingredient, such as the physical state, the chemical stability, and the solubility in water and organic solvents determine technically and commercially feasible types of formulations for which the appropriate packaging has to be defined. Other important factors influencing the design of formulations and packagings include the toxicological properties of the active ingredient and, of course, user's needs. Farmers want economic, safe, and reliable solutions to problems and no or minimal waste disposal problems.

In today's European market most of the crop protection products are still sold as classical formulations such as emulsifiable concentrates (EC), soluble liquids (SL), suspension concentrates (SC), and wettable powders (WP), which add up to over 90% of the market.[3] This picture is similar to that in the U.S., where liquid formulations represented some 80 to 90% of the products applied to the major row crops over the past 10 years (including corn [*Zea mays* L.], soybeans [*Glycine max* (L.) Merr.], sorghum [*Sorghum bicolor* L.], cotton [*Gossypium hirsutum* L.], etc.). In the realm of solid formulations there has been a clear shift away from wettable powders to water dispersible granules (WGs or dry flowables). The market share of herbicide WG formulations has increased from around 1 to 3% in the early to mid-1980s to perhaps 15 to 20% today, again in the major row crop markets.[5]

Several active ingredients with low dose rates have recently been developed which are typically used at only a few g ha^{-1} instead of kg ha^{-1}. This certainly opens new opportunities for formulation and packaging development. Indeed, there has been a rapid increase in the farmer's interest in advanced types of formulations and delivery systems. Furthermore, the high development costs associated with bringing new active ingredients to the market may also push companies to investigate new formulation and packaging technologies in order to extend the life of existing products.

III. FORMULATION AND PACKAGING CONCEPTS

A. LIQUID FORMULATIONS

1. General

Liquid formulations are preferred by the farmer for preparing spray solutions for several reasons. They can be measured volumetrically, they are easy to handle, they spontaneously form stable emulsions or dispersions and, given appropriate container design, most formulations are easy to rinse out of the package. Liquid products are also easy to handle in today's bulk handling systems and generally don't cause application problems.

2. Emulsifiable Concentrates (EC)

Although they are the most widely applied liquid formulations, ECs also have disadvantages. Some of the organic solvents used in ECs may be harmful because of their toxicity and their flammability. ECs are also coming more and more under regulatory pressure due to the organic solvents, many of which contribute to volatile organic compounds (VOC) emissions. As a consequence, formulation chemists have to look for new solvent systems. These alternatives do not always have the same performance characteristics, such as active ingredient loading capacities, miscibility with water, and stability profiles.

3. Suspension Concentrates (SC)

These formulations have an advantage over ECs because they are generally water based and normally contain only small amounts of glycols or similar materials as antifreeze agents. However, the preparation of SCs is limited to solid active ingredients having low water solubility. The solid active ingredient is generally heavier than water, and thus tends to sediment during storage. Residues of the often viscous products are not easy to rinse out of the packaging.

It is also possible to prepare such formulations using an organic material such as an oil or even a liquid active ingredient as the liquid phase in which the solid is suspended.

4. Emulsions in Water (EW)

EWs are stabilized emulsions in water. For the formulation of hydrolytically stable liquid compounds, they are an attractive alternative to ECs. Because they are water based they are generally less hazardous for the user and the environment. Notably, they contain fewer VOCs, and the advantage of an EW over an EC increases as the concentration of the a.i. decreases, because more solvent is replaced by water.

There is also the possibility of combining an emulsion of a liquid active ingredient and a suspension of a solid active ingredient in one formulation. Such combinations represent special challenges, particularly in terms of long-term physical stability, and are known as "suspoemulsions".

5. Capsule Suspensions (CS)

A further step towards increased safety for the user may be obtained by encapsulation of the active ingredient. As has been proven for some insecticides, capsule suspensions (CS) show a remarkable reduction in oral and dermal mammalian toxicity as compared to the ECs. Encapsulation of the active ingredient may also offer an advantage for compounds which are volatile.

6. Gels (GL)

Gel formulations are innovative products which can be described as thickened ECs packed in water soluble bags, as reported by Dez et al.[2] The viscosity is increased with thickeners, the final gels viscosity being a compromise between the transport stability in the water soluble bag and the dispersibility in water. This concept offers the crop protection market a new form of a product/packaging combination. The first fungicide formulated as a gel is the propiconazole ((+/-)-1-[2-(2,4-Dichlorophenyl)-4-propyl-1,3-dioxolan-2-ylmethyl]-1H-1,2,4-triazole) GL 62.5, which was launched under the trade name PRACTIS®* by CIBA-GEIGY France in 1991. Because of the higher concentration of the GL 62.5 compared to the EC 500, there are less organic solvents. There are several new gel products in development, and this year the BANNER®* gel will be introduced in the U.S. market for turf and ornamentals, primarily for use on golf courses. Another example is the bromoxynil (3,5-dibromo-4-hydroxybenzonitrile) gel introduced by Rhône Poulenc under the trade name BUCTRIL®.

Gel products provide many benefits that are highly appreciated by farmers. The premeasured doses in water soluble bags offer advantages in easy handling and increased user safety, and the outer package is not contaminated with product and can be easily disposed of.

B. SOLID FORMULATIONS

1. General

Solid formulations have several advantages over liquid ones, especially in regard to their environmental impact. They are generally free of organic solvents, easy to pick up in case of spillage, and, in general, there is less packaging waste.

2. Wettable Powders (WP)

These are the most common solid formulations. To obtain a stable suspension upon dilution with water, WPs have to be ground to a very fine particle size. This makes them dusty and, therefore, less safe to use, particularly when measuring out. However, worker safety may be dramatically improved by packaging the WPs into water soluble bags, thereby allowing the farmer to use premeasured doses contained in a secondary package that is not contaminated by the product.

---

* Registered trademark of Ciba-Geigy Limited, Basel, Switzerland

3. Water Dispersible Granules (WG)

WGs are slowly becoming established in the crop protection industry. This formulation type combines the advantages of liquid and solid formulations. Thus, WGs are easy flowing products with constant bulk density and can therefore, in principle, be measured volumetrically. They are much less dusty, two to three times less voluminous, and leave less residues in empty packagings as compared to WPs. A drawback for products with high use rates may be the high price of WGs; nevertheless, the high processing costs may be balanced by reduced storage costs due to the higher bulk density. Thus, WGs are the formulation of choice for highly active solid products such as sulfonylureas. With some WG products, the disintegration time may also be a critical issue. These formulations often do not behave like standard EC or SC formulations, and will thus require training of pesticide users on the characteristics of this new generation of formulation types.

C. SEED TREATMENT FORMULATIONS

Tailor-made formulations for the treatment of seeds became established at Ciba during the last 10 years, and are becoming an increasingly important aspect of crop protection agents application practice. The underlying principle is to place the chemical as near as possible to where it is required to control seed or soil-borne pests and for uptake by the underground parts of the plants. Thanks to the ideal placement of the product directly on the seed, benefits include a more efficient use of product, less environmental contamination, and reduced exposure of non-target organisms. Therefore, from an environmental point of view, seed treatment demonstrates clear advantages over granules and soil sprays. Seed treatment advantages also include indoor application under controlled conditions by skilled operators, allowing the use of more sophisticated formulations and avoiding the variations caused by weather and different expertise of applicators.

D. PACKAGING STRATEGIES

1. Container Designs

A big problem for the farmer is the disposal of contaminated primary packaging waste, which is particularly pronounced for liquid products. It is important, therefore, to ensure that single-trip containers are rinsed immediately after emptying (triple rinsing is usually required to reach the 0.01 % level required by a Dutch covenant, and similar regulations will likely be introduced in the U.S.) and the rinsate be added back to the spray tank. This requires that containers are designed to be rinsed easily. Clean, rinsed containers are easier to be disposed of through municipal waste channels (where allowed) or to be collected for controlled recycling or energy recovery. Industry is in the process of agreeing on container performance standards and specific design criteria aimed at improving handling and rinsability. Handling convenience, safety, and new regulations are making packaging design a key factor in dealer product selection.[4]

2. Refillable Containers

Strategies aimed at reducing the number of single-trip containers include the development and eventual introduction to the market of refillable containers. Refillable containers such as Ciba's FarmPak®* and FieldPak®*, Monsanto's Shuttle® and FMC's U-Turn® have long been used in the USA and Canada and have contributed to a substantial reduction in the number of single-trip containers requiring disposal. The Ciba FarmPak®* was originally designed and launched as a mini-bulk (110-gallon or 420-liter) container for liquid herbicides in 1987. This closed filling system has continuously been improved by Ciba in the U.S. to meet the needs of their customers. Today, more than 60,000 Ciba mini-bulk containers of different sizes are in circulation in North America. An adapted version of the 60-gallon or 230-liter FieldPak has just recently be introduced in the banana market in Central American countries. A 15-gallon or 60-liter refillable container has been introduced in the US, primarily in the turf and ornamental market, under the name TurfPak.®*

For more than five years, Ciba has been considering the possibilities of developing small, i.e. 20- to 30-liter, refillable containers for fungicides in Europe. Emerging packaging waste legislation, i.e., EEC Directive on Packaging Waste, encourages the use of refillable packaging. During the last few years, a number of agrochemical companies had development programs with small volume refillable (SVR) containers using stainless steel kegs from 10- to 60-liters sizes. Efforts are being made by industry to standardize fittings attaching SVR containers to sprayer transfer systems

Ciba is also involved in the development of a 10-liter refillable, closed dispensing container called the LinkPak®*. The advantages of this system stem from the fact that no investments are required in transfer systems or for major spray equipment modifications. The only modification required is the fitting of a small adapter either to the top of the spray tank or to the lid of the induction bowl or hopper. The LinkPak®* is quick and easy to use; all that is required is to invert the container and engage the dispensing head into the adapter on the sprayer. Rotating the LinkPak®* opens the valve and the required amount of product can be dispensed. When the LinkPak®* is empty it is returned to the bulk site for re-filling. Following two years of trials in Canada and European countries, Ciba UK is ready to move into commercial production of its LinkPak®* and commercial launch could follow in 1995.

Clearly, the success of SVR containers will also depend on effective logistic systems. We believe there is an opportunity for refillable containers and that farmers will be quick to see the benefits of such systems when they become available.

3. Water Soluble Packaging

In particular for wettable powders, water soluble packaging is becoming increasingly important. Presently, Ciba has more than 60 products packed in water soluble bags, and the use of PVA film has increased substantially in the last years.

A crucial point for the packaging of gels in water soluble bags was the close and intense collaboration of formulation and packaging specialists. The challenge for them was to identify a polyvinyl alcohol film for the primary package which was compatible with the solvents in the gel formulation and still had a short dissolution time in the spray tank. Furthermore, a multicompartment secondary package had to be developed to provide mechanical protection for the pouch and could be sealed against moisture. Most commercially used water soluble films are based on polyvinyl alcohol type resins. These materials are, as a rule, not soluble in brine solutions. Thus, they cannot be used in liquid fertilizer solutions, a widespread practice in the U.S. Efforts are under way to develop film materials which will be soluble in fertilizer solutions as well as in plain water.

4. Packaging of Seed Treatment Formulations

Seed treatment customers also have a need for improved systems for handling chemical products. Although their needs are quite different from those of the large field crop farmers, they share the problem of disposal of empty single trip containers which most often are stained. In general, seed treaters require high volumes (500 to 5000 liters per season) and they need a closed system that allows the product to be pumped directly from the container to the seed treatment machine without dilution. A 500-liter tank for ready-to-use products has been developed and is ready for introduction into the market.

IV. FUTURE TRENDS

In the last few years, much progress has been made in the design and development of new formulation and packaging systems towards meeting requirements for safe and convenient use and reduction of packaging waste. During the next years, until the year 2000 and beyond, these developments will be continuously introduced into the market. New product forms - either new formulations or new formulation/packaging combinations - may extend the life of a product and/or make it more competitive in the market. Due to concerns about environmental contamination, there will be a trend to use water based EW and CS instead of solvent based EC formulations, and there will also be a general move away from liquid to solid formulations. In particular, compacted forms, such as WGs are increasing over the traditional WPs. Water soluble packaging provides improved user safety by eliminating dusting during handling of powders. Water soluble bags are also used for WGs and they are essential for packaging gels. Gels in water soluble bags offer an elegant solution for the handling of a viscous liquid, providing the farmer with a premeasured dose, and for leaving packaging waste which is not contaminated.

In terms of overall packaging strategy, the industry is committed to waste reduction and improved handling safety. A prime objective is to reduce the number of one-way containers which end up in the solid waste stream. Refillable container programs aimed at larger growers and custom applicators will contribute to the reduction of one-way packaging and associated secondary packaging. For smaller growers and also for the use of highly active compounds, the introduction of solid formulations and gels which can be packed in water soluble film, completes the strategy to achieve a reduction of packaging waste. There should be no illusions that developing innovative solutions and bringing them to market in a short time frame requires substantial resources and effort. It also requires legislative support to encourage the implementation of emerging packaging waste reduction alternatives.

In the US major recycling programs are underway and are supported by the pesticide producers. These programs include collection, shredding, granulation, and transformation into secondary products (e.g., plastic pallets and poles).

Due to changes in the society and in our economic environment, there will be many challenges for the crop protection industry in the future. Customer needs and desires are changing, legislation concerning product application and packaging waste is increasing, farmers have needs for safer products, and the public is concerned about contamination of groundwater and food with pesticides.

These changes offer opportunities for the design of new products that have, both biologically more effective and safe active ingredients as well as novel formulations and packagings. Old products, which are less safe, have to be phased out. Industry is also encouraged to pay more attention to packaging waste management and to introduce refillable containers. In addition to developing and producing these new formulation and packaging systems, education and training in application of the new systems will be necessary in order to gain market acceptance. There are substantial opportunities for companies that are able to innovate and who are prepared to take risks to meet the emerging customer needs in order to obtain a competitive advantage in the market place. Much is to be gained from industry working together and with government institutions to establish performance standards in the area of packaging waste management programs.

V. ACKNOWLEDGMENT

We thank our colleagues within Ciba for their contributions and for their support in preparing this manuscript.

REFERENCES

1. **Bailey, J.D.** Safer pesticide packaging and formulations for agricultural and residential applications. *Rev. of Environ. Contam. and Toxicol.*, **129**, 17, 1992.

2. **Dez, G., Lerivrey, J., Schneider, R., and Zurkinden, A.**, Flüssige pestizide Wirkstoffkonzentrate, European Patent Application EP 0449773 A1.

3. **Diepenhorst, P., and Lohuis, H.**, Warum die Formulierung der Wirkstoffe immer wichtiger wird, *Pflanzenschutz Praxis*, **3**, 10, 1991.

4. **Fahnestock A. L.**, Closing in on packaging needs, *Farm Chemicals*, March 1994, 16.

5. Unpublished data. Maritz Marketing Research Inc., Agricultural Division, Fenton, MO, various years.

Chapter 5

Emerging Technology: The Bases For New Generations of Pesticide Formulation

George B. Beestman

CONTENTS

1-8493-7678-5/96/$0.00+$.50

I. INTRODUCTION

Our industry is being asked to protect crops with less pesticide, and expectations are that the public should not see, smell, be exposed to, or be affected by the pesticide in any way. Emerging technologies for formulators to move toward these utopian expectations include new surfactants, new polymers, new processes, and new product forms. Surfactants and polymers will be used to tailor biological efficacy, regulate soil mobility, create new product forms, and combine chemical crop protection chemicals with living biological control agents. As regulatory pressures decrease the total number of adjuvants, new advanced materials will continue to become available for formulators in the future.

II. SURFACTANTS

"Adjuvants" is an encompassing term which is often defined as all materials formulated together with crop protection chemicals or materials which are externally combined with products at the time of application. Current overviews provide a good background of adjuvants used with crop protection products, historical development of adjuvant use in agriculture, and chemical classification of adjuvants.[32,33,34] This section will focus upon surface active adjuvants, i.e., surfactants.

New surfactants for use with crop protection chemicals are not new in the sense of previously unknown chemical structures, but are new in the sense of having gained regulatory approval for their use. Many of these surfactants newly approved for use with agricultural products are well known for formulation of paints, inks, drugs, and other non-agricultural products.

One class of newly approved surfactants are saccharide derivatives.[2,30,50] These more rapidly biodegradable surfactants are effective over a wide temperature range, are salt tolerant, and rapidly lower surface tension of aqueous solutions.

Tristyrylphenol-based surfactants have also recently gained approval of the Environmental Protection Agency.[45] Unique properties are achieved by alkoxylation, which provides a selection of surfactant products from this class.

Organosilicone surfactants are gaining increased use for rapidly lowering surface tension of aqueous solutions.[66] Rapid lowering of surface tension facilitates spreading on leaves, and reduces the tendency of spray droplets to bounce off leaf surfaces. A recent review details use of organosilicone surfactants in agricultural applications.[88]

Organofluorine surfactants have produced the most dramatic lowering of aqueous surface tensions.[54] Synergistic lowering of surface tension is achieved from combinations of organofluorine surfactants with hydrocarbon surfactants.[55]

New agricultural applications can evolve from old established types of surfactants. Ethoxylated alcohols have been used extensively in agricultural applications. Derived from natural sources, the fatty alcohols usually contain fewer than 30 carbon atoms prior to ethoxylation. A synthesis route to produce alcohols with up to 150 carbon atoms opens the door to higher molecular weight alcohol ethoxylates. High carbon number alcohol ethoxylates are used extensively in the cosmetics industry, and for formulation of printing inks.[8] Now, agricultural applications are also being discovered.[18]

Ethylene oxide-propylene oxide block nonionics are an old established class of surfactants available from many suppliers. Solid, paste, or liquid products are available depending upon the alkoxide content. The physical state of a surfactant from this class depends of the ratio of the liquid polypropylene oxide to the solid polyethylene oxide. Generally, surfactants that contain less than 30 wt % ethylene oxide are liquid, and surfactants that contain more than 50 wt % ethylene oxide are solid. However, it has recently been discovered that surfactants from this class that contain 60 wt % ethylene oxide (expected to be solid), are actually liquid.[73] With regulatory approval for their use in agricultural formulations, these recent additions to an old class are likely to provide new formulation properties.

As new knowledge is gained from research, classical models are challenged and new molecular structures will be made. Addition of surfactant to water in excess of the critical micelle concentration will cause formation of surfactant micelles. The classical model of a surfactant micelle is a discrete oil droplet formed by the hydrophobic surfactant tails surrounded by a discrete outer aqueous coating of the dissolved hydrophilic or ionic surfactant heads. Fundamental studies challenge the classical model. Numerous micellar shapes have been shown. Highly interlinked surfactant chains with water-filled cavities and a high degree of water-oil contact are likely.

Synthetic surfactant structures were made that inhibited intramolecular chain-chain association, thereby interrupting self-assembly into micellar structures.[64] These surfactants contain in sequence a long hydrocarbon chain, an ionic group, a rigid stilbene spacer, a second ionic group, and another long hydrocarbon (Figure 1).

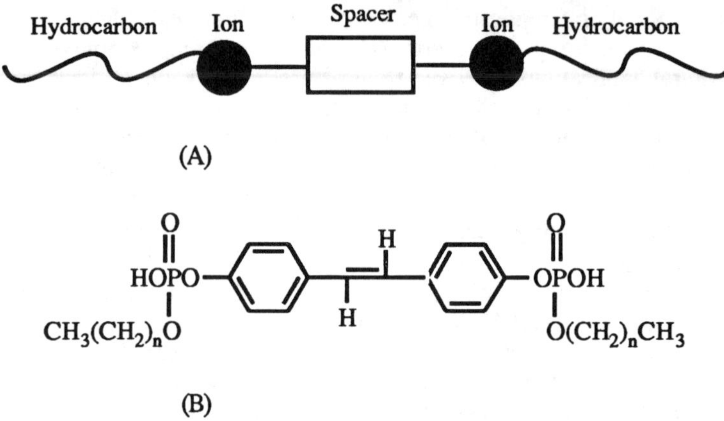

Figure 1. Model (A) and chemical structure (B) of synthetic surfactants which have hindered ability to form chain association micelles.

Detailed studies of phase behavior revealed striking properties of these surfactants. There was a reversal of the classical trend for longer hydrocarbon chains to lower the surfactant critical micelle concentration. The surfactants lie in flat configuration at the oil-water interface. Submicellar structures were formed that have not been fully characterized. As the phase behavior on new synthetic structures are characterized, new structures can be made to provide desired properties.

Tailoring of surfactant structures to produce desired phase behavior, surface mobility, and rheological properties will provide new generations of surfactants to meet the needs of formulators in the future.

III. POLYMERS

A. STRUCTURED POLYMERS

A high degree of suspension stability is achieved from use of polymers to provide steric stabilization. Hydrophobic sections of polymer chains anchor polymers to suspended particles. Pendant water-soluble segments of the adsorbed polymers protrude into the aqueous phase. Particle to particle contact is avoided by resistance of these pendant polymer segments to overlap. Most polymers used to stabilize suspensions of crop protection chemicals are random polymers. Adsorbing segments and water-soluble segments occur randomly along the polymer chains. The tendency of water-soluble polymer segments to move into the aqueous phase weakens adsorption of the hydrophobic segments, leading to destabilization of the dispersions. High molecular weights are required to obtain satisfactory stabilization by random polymers.

A greater degree of stabilization is achieved with structured polymers that contain the hydrophobic and hydrophilic segments as distinct blocks along the polymer chain. These structured block polymers are highly effective stabilizers of pigment particles for inks and paints[61] and crop protection compounds in suspension concentrate formulations.[94] Functional hydrophobic backbones can be ethoxylated, producing water-soluble pendant ethylene oxide chains like teeth of a comb. The comb polymers are also effective stabilizers for industrial dispersion formulations.[95]

Novel synthesis methods produce graft and block polymers in which water-soluble anionic chains are built into a hydrophobic backbone of the polymers.[104] Graft polymers are structured polymers that consist of a continuous adsorbing polymer backbone into which water-soluble segments are contained like teeth in a comb. These structured graft polymers are highly effective stabilizers of suspended particles. Adsorption of a continuous hydrophobic backbone onto a particle surface is stronger and less affected by the tendency of water-soluble portions to move off the particle surface than is the case with random polymers (Figure 2). New agricultural uses for structured polymers are being developed in our laboratory.

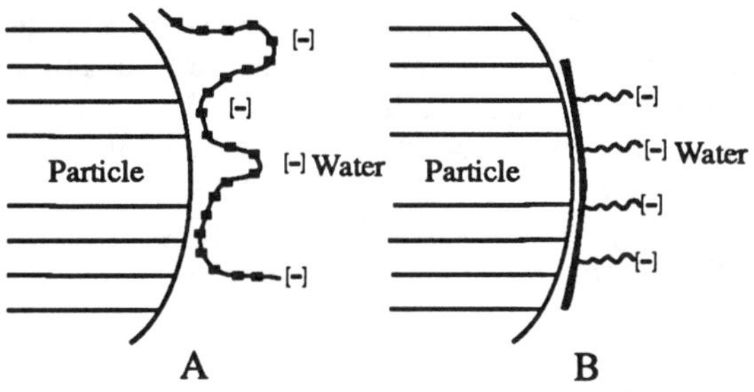

Figure 2. Absorption of random polymer (A) or comb polymer (B) on a suspended particle. Water-soluble segments , of random polymers reduce absorption tendency of hydrophobic absorbed segments .

Highly structured three-dimensional star polymers are being developed.[74,110] The polymer chains emanate from a center point like spokes on a wheel. Unique properties of these polymers are still being discovered. Structured polymers offer the potential for agricultural formulators to chemically construct complex biological and chemical entities.

B. STIMULI SENSITIVE POLYMERS

Polymers that respond to external stimuli such as changes in pH, temperature, light, electric current, and magnetic field are being used to achieve on-demand delivery of active ingredients.[72,85,98] Most of these applications are in nonagricultural areas.

Temperature-triggered release of agricultural chemicals has been proposed based upon selection of polymers with specific glass transition temperature.[57,58] At temperatures below the polymer glass transition temperature, the polymer chains are rigid and only vibrational movement is possible. Entrapped chemicals are released slowly by diffusion through rigid polymer chains. At temperatures above the glass transition temperature, however, polymer chains become mobile and release of the entrapped chemicals is rapid.

Polymer melt point has been used as a trigger for temperature induced release of crop protection chemicals.[89] Hydrocarbon esters of acrylic monomers are polymerized to form meltable side chains on acrylic polymer backbones. When these polymers are added to pesticides prior to microencapsulation, change in release rate from the microcapsules occurs at the temperature at which melting and recrystallizing of the meltable polymer segments occurs.[44] Apparently, enough of the side-chain-melting polymers are incorporated into the microcapsule shellwalls to provide rapid release at temperatures above the melting point of the side chains and lower rate of release at temperatures below the crystallization point of the side chains.

Stimuli-sensitive polymers will continue to play a role in the pursuit of on-demand delivery of crop protection compounds.

C. WATER-SWELLABLE POLYMERS

Cross-linked polymers are available that will absorb hundreds of times their weight of water. Bound water within these linked polymer chains is a carrier for controlled release of active compounds. Drugs dissolved or suspended into bound water can be delivered with varying degrees of control.[84]

Water-absorbing polymers are known to affect biological efficacy of crop protection compounds. Simply adding water-swelling polymers to formulated crop protection compounds has been shown to enhance biological efficacy of the compounds.[76] For direct use in formulated products water-swelling polymers must be developed that are low viscosity and free flowing so that products can be sprayed without plugging of spray nozzles.

Expanded use of water-swellable polymers as controlled release carriers of crop protection compounds requires that discrete sprayable micro-sponges be formulated that do not aggregate in storage.

D. POLYMERS FOR TAILORED DELIVERY OF COMPOUNDS

Natural and synthetic polymers are manipulated to make various types of controlled release formulations.[83] Tocker provides a good review of controlled release technologies and points out the pitfalls to successful application of these technologies for crop protection compounds.[100]

Among natural polymers, starch has been most extensively tested as a controlled release carrier. Chemical and physical modification of starch has been championed by USDA scientists for tailored delivery of crop protection compounds. Harsh processing conditions have limited use of starch as a tailored delivery carrier of crop protection compounds. However, recent discoveries to simplify chemical treatment and improve processing make the use of starch more versatile.

Controlled release has been demonstrated from unmodified starch.[105] Extrusion allows continuous processing of starch with crop protection compounds.[17] Processing conditions can be regulated to produce tailored delivery granules. Porous starch granules with a higher absorption capacity have been produced by controlled enzyme decomposition.[102]

Other natural polymers are proving to be useful carriers for tailored delivery of crop protection compounds. These include hemicellulose from wheat bran,[52] and fresh wood pulp cellulose.[3] Lignins, chitin, lactides, guar, algin, and other natural polymers continue to be processed and modified as carriers of crop protection chemicals. The challenge will be to produce sprayable micro-containers from these polymers.

One approach to making micro-containers is to make porous spheres that are then filled with the crop protection compound. Urea-formaldehyde spheres can be made as discrete entities by careful control of monomer ratio.[68] In the textile industry, yeast cells have been filled with dyes, biocides, or fragrances and then attached to the fibers with cross-linking polymers.[80] Filled pollen grains and spores provide a pulsed release of drugs.[4]

Phospholipids can also form spheres filled with active compounds, but phospholipids are expensive. Use of phospholipids for injection delivery of drugs is well known in the pharmaceutical industry. Filled spheres can also be formed from commonly available dimethyl amides of tall oil fatty acids.[93] These are more cost-effective carriers for agricultural applications.

Both phospholipids and fatty acid vesicles may lack sufficient integrity to withstand transfer, storage, and application forces of agricultural operations. Surface reactions to strengthen the spheres may produce more robust carriers for tailored delivery of crop protection compounds.

E. POLYMER ALLOYS

Polymer combinations, known as alloys, can be created to provide useful properties not available from individual polymers which make up the combination. A useful alloy coating composition was obtained from combining paraffin wax with oxidized polyethylene.[101] Melt combinations of paraffin wax and oxidized polyethylene produced satisfactory slow release compositions which are well suited for application to rice paddies.

Blends of miscible biodegradable polymers are useful carriers of pharmaceutically active agents.[27] Other new and useful combinations of polymers will be created for meeting the diverse needs of agricultural formulators.

F. POLYMERS FOR HIGH CONCENTRATION MICROENCAPSULATION

Microencapsulation became an economically viable technology for tailored delivery of crop protection compounds when directly usable, stable suspensions of microcapsules were produced which contained a high concentration of active compound.[10]

The key feature of high concentration microencapsulation is use of lignosulfonates to produce stable emulsions of a molten organic compound in water. Reaction of monomers to form a plastic film around emulsion droplets produced suspensions that contain a larger volume of organic compound inside of microcapsules than the volume of water suspending the microcapsules. Subsequent patents identify other sulfonated polymers which are useful as emulsifiers for high concentration microencapsulation. Sulfonated polymers produce highly disperse suspensions, consequently, irreversible settling of microcapsules can be a problem once the suspensions are diluted to spray.

In our laboratory, we have discovered that polyvinylpyrrolidone-based copolymers and terpolymers are useful as emulsifiers to produce high concentration suspensions of microcapsules which will readily suspend after dilution of the concentrates into water.

The herbicide alachlor was microencapsulated using lignosulfonate or select polyvinylpyrrolidone copolymers and terpolymers as emulsifiers for high concentration microencapsulation. Formulated microcapsule suspensions were added to water in Nessler tubes and allowed to sediment for three days. The number of inversions of the Nessler tubes required to fully resuspend settled microcapsules was recorded (Table 1).

Table 1

Nessler Tube Inversions Required to Resuspend Settled Microcapsules

Polymer	Number of inversions (3 days)
Vinylpyrrolidone quaternarized dimethylaminoethyl methacrylate	10
Vinylpyrrolidone dimethylaminoethyl methacrylate	12
Vinylpyrrolidone vinylcaprolactam dimethylaminoethyl methacrylate	29
Vinylpyrrolidone quaternarized dimethylaminoethyl methylacrylate, high molecular weight	4
Lignosulfonate	>100

Even after three days, microcapsules prepared in aqueous polyvinylpyrrolidone copolymer and terpolymer solutions were readily resuspended. More highly dispersed microcapsules prepared in an aqueous sulfonated polymer solution settle and pack after dilution.

The search for new polymers that can produce high oil content emulsions that are stable to the shellwall-forming chemical reactions will continue. Future developments will include an understanding of why polymers are useful for high concentration microencapsulation; rational selection of specific polymers for high concentration will be possible.

Microencapsulation reduces leaching to groundwater, improves biological efficacy, and lowers acute toxicity of organic compounds. While these advantages are substantial, an additional advantage of microencapsulation is converting liquid organic compounds to high concentration dry solid form. Dry solid formulations offer storage stability and packaging advantages over liquid products.

In the past it was necessary to first absorb liquids onto inert carriers and then formulate these dry powders to a granular dry flowable formulation. This conversion of liquids to dry powders results in dilute final formulations having less than 50 wt % active compound. Microencapsulation converts liquids to spherical particles with 90 wt % active compound. Removal of water and granulation of a microcapsule suspension can produce dry flowables from liquid active compounds with active ingredient contents in excess of 70 wt %.

IV. DRY SOLID FORMULATIONS

Dry Flowable formulations, i.e., granules which disperse in water back to discrete primary particles were pioneered by Albert and Weed.[1] The primary contribution of this product form was to replace wettable powders and eliminate handling of dusty products. Dry Flowables continue to be a popular product form, and have provided impetus to convert both liquid and solid compounds to Dry Flowable form.

Dry Flowables have been made by spray drying suspensions of microcapsules[91] and powders.[108] Emulsifiable Concentrates and Suspoemulsions have been converted to Dry Flowables with counter-current fluid bed drying.[48] Polymers capable of forming dry solid complexes with liquid organic compounds offer an alternative to microencapsulation for converting liquid compounds to dry granular form.[67]

Removal of water or solvent to obtain Dry Flowable formulations is the most costly aspect of processing. Extensive purification systems are required to handle high volumes of drying air. Water and volatile components must be removed before drying air can be returned to the environment. A new granulation process called heat-activated-binding technology has been developed to avoid the problems associated with purification of drying air.[39] Powders to be granulated by heat-activated-binding are blended with a water-soluble binder as heat is applied, under turbulent conditions, to melt or soften the binding polymer. Dry Flowable formulations are formed as temperature is lowered below the softening point of the binders. A drying step is not required. Useful binding polymers have a unique thermal melting point/freezing point pattern. Heat-activated-binding technology is applicable with common granulating equipment.

Recent extensions of heat-activated-binding technology include stabilized formulations of sulfonylurea herbicides alone,[36] or in combination with water containing acidic compounds.[11]

Tablets are an emerging type of dry formulation, especially useful with compounds that are effective at grams per hectare application rates. Tablet disintegration rate is a primary research focus. A recent review discusses mechanisms of tablet disintegration and identifies materials used to increase rate of disintegration.[46]

Effervescence is used most commonly to obtain rapid dispersion or solution of tablets. Special packaging is required to exclude water which could trigger chemical reaction leading to premature effervescence and evolution of carbon dioxide. Tablets have been produced that contain internal desiccant to ensure tablet stability.[53] Other developments include discovery of new disintegrants to speed dispersion of noneffervescent tablets.

It has been discovered that a range of tablet disintegration rates can be obtained using polyvinylalcohol and vinylalcohol copolymers with up to

10 wt % methyl acrylate or methyl methacrylate.[9] Most rapid tablet disintegration rate was obtained with fully hydrolyzed, heat-treated crystalline polyvinyl alcohol. Blends with amorphous polymers can provide a range of disintegration rates, from immediate release through compositions which provide prolonged release.

Dry solids will continue to be preferred for formulations of crop protection compounds. Technologies will evolve for rapid dispersion into spray solutions, stabilization of compound mixtures, and processing of liquid compounds. Noneffervescing tablets that rapidly disperse to primary particles and do not require moisture-barrier packaging will be developed.

V. LIQUID FORMULATIONS

A. MICROEMULSIONS

Microemulsions are thermodynamically stable, isotropic dispersions of hydrocarbons and water. A common ready-to-use formulation for aerosol homeowner products, microemulsions contain relatively low levels of active ingredient and high levels of surfactant. A recent review describes structure and properties of microemulsions.[70]

Agricultural uses of microemulsion formulations are in specialized applications such as protecting stored grains,[22] but are being developed for crop protection compounds with enhancements such as increased active compound concentration,[37] lower viscosity,[47] and broader temperature stability.[25]

Microemulsions have been used as a carrier of suspended solids in suspension concentrate formulations.[69] Suspensions of finely divided active compound crystals in oil were formulated to produce microemulsions upon dilution. Crystalline solids were non-settling in resultant microemulsions.

Future developments will include low viscosity, high active ingredient compositions, without solvent, that spontaneously disperse in water to microemulsions.

B. GELS

Gels can be water based or solvent based. Solvent based gels are the first liquid formulation to be packaged in water-soluble bags.[49] The gels can be formulated to resist leaking from a pinhole imperfection of a water-soluble bag.

Water based gels have been used as a delivery system of bound water.[6] The gels are effective in an enclosed environment to keep cut flowers fresh or to sustain growing plants.

Water-based gels that are stable formulations of hydrolytically unstable sulfonylureas[63] have been developed by DuPont. Gels containing sulfonylurea and co-pesticides add a new dimension to delivery of compounds that will rapidly degrade.

Expansions of gel technology for agricultural applications could include irrigation, sustained viability of living crop protection systems, and enhanced transport of compounds into plants.

C. MULTIPLE EMULSIONS

A multiple emulsion is, essentially, an emulsion of an emulsion. Used commonly in formulation of drugs, cosmetics, and foods,[71,109] multiple emulsions are now gaining use in agricultural formulations.[96] The oil phase of this W/O/W multiple emulsion prevented a crop protection compound within the inner water core from freely diffusing into the external continuous water phase. The multiple emulsion provided significantly reduced toxicity of the crop protection compound.

Stabilized multiple emulsions could provide controlled osmotic diffusion of compounds to tailor delivery, or may provide multiple compartments for incompatible compounds within a single formulation.

Future developments will include emulsions that are sufficiently robust to tolerate storage and handling forces common in agricultural practice.

D. SUSPENSIONS

Suspensions of crystalline solid compounds in water are a popular form of delivery which provides an alternative to powder formulations. New surfactants continue to be offered and adaptations of suspension formulations continue to be developed.[86]

Chemically stable suspension formulations of hydrolytically unstable sulfonylurea compounds were developed.[51] Both the regulation of suspension pH and complexation were used to create stable aqueous formulations of compounds that are sensitive to hydrolysis.

Water-sensitive compounds have also been formulated by dispersing the active compound into polymer latex particles.[28] The latex particles provide a small uniform particle size for suspension stability along with controlled release of the compound after application.

Future developments will include reduced particle size, increased active compound content, enhanced biological efficacy, and attachment to plants.

E. SUSPOEMULSIONS

Polymer lattices, described above for use in suspensions, are also used in formulations that are a combination of both suspensions and emulsions, or "suspoemulsion" formulations. A good review of recent developments in suspoemulsion formulations is available.[65] These formulations are popular to combine several types of compounds into a single formulation. An emulsion phase containing one or more compounds is combined with a continuous suspension phase that can also contain one or more compounds.

Ethoxylated comb and block type copolymers (Section III. A) used in suspensions are also useful to stabilize the suspension component of suspoemulsions.[97] Alkylglucoside surfactants (Section II) have been successfully utilized in both phases of a suspoemulsion.[29]

Combining of suspensions, emulsions, encapsulated compounds, aqueous phases, and neat compounds into new types of formulations will continue. Broad-spectrum products are being sought to eliminate tank mixing, reduce inventories, and reduce the number of times that a crop must be sprayed. Single active ingredient formulations often cannot meet these requirements. These multicomponent formulations must combine incompatible compounds, must provide maximum biological efficacy, and must regulate the soil mobility of compounds. Suspoemulsions will play a part in providing such improved delivery formulations. Processing technologies are also becoming available that will facilitate the development of these new product forms.

VI. PROCESSING TECHNOLOGIES

Formulation of multicomponent products is relatively simple only if all of the active compounds are chemically, physically, and biologically compatible. Current adjuvants can solve most of the incompatibilities in traditional product forms. However it will become increasingly necessary to physically isolate one or more of the active compounds.

Microencapsulation is a technique that is well suited for isolation of liquid compounds. High concentration microencapsulation facilitates conversion of liquids to dry solid forms (Section III. F). Direct coating of crystalline, dry solid primary particles is a technology still to be developed. Wet chemical batch-type coacervation processes are available, but they require solvent handling and dry solids recovery operations which make them not well suited to coat dry solid primary particles for the purpose of isolating them in a multicomponent, dry flowable form. Novel processing technologies are being developed, however, and direct coating of dry solid primary particles will become a reality.

A unique coating technology utilizes a spinning disc to remove excess coating material from particles that have been immersed in the coating composition.[87] Unlike processes that apply coatings, this process leaves coating on particles. Consequently, difficult to wet particles cannot resist acceptance of coatings; a problem with methods of direct application of the coating material.

This process, known as centrifugal suspension-separation coating, is useful for solid particles as small as 30 μm, but is more applicable for larger particles up to 2 mm (Figure 3).

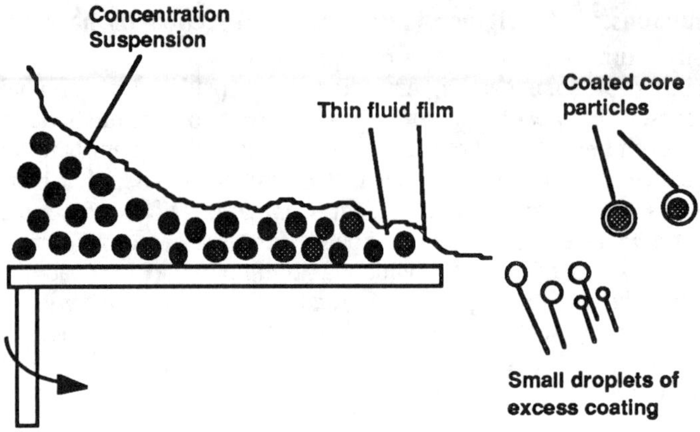

Figure 3. Centrifugal suspension-separation coating process.

The process is also useful to create particles from melt, and for granulation. Particles from melt, as small as 3 to 7 μm median diameter, can be formed. Larger melt particles can be made with a uniform size range. Various compositions can be employed as binders for granulation, including a surface melt of the fine powder active compound itself.

The centrifugal suspension-separation process is useful for applying melt coatings to avoid use of organic solvents. The trend away from solvents will likely continue. Processes using liquid carbon dioxide are also becoming increasingly popular to avoid use of conventional organic solvents.

Unique molecular mixtures of active compound and formulation adjuvants have been made using liquid CO_2 as a carrier solvent.[60] Carbon dioxide in its supercritical state has been successfully used with rapid expansion technology to form loaded microspheres.[23] Rapid expansion of supercritical fluids was developed to create ultrasmall particles and thin films,[75] and has been successfully used to create porous sponges of biodegradable polymers.[24] Since the solvent carrier is a gas after expansion, the technique leads to solvent-free solids in a single step.

The technologies described in this section will become useful tools to create formulations that maximize biological efficacy, minimize the amount of active compounds required, and control mobility of compounds in soils.

VII. TAILORED SOIL MOBILITY

Two general methods of controlled release are matrixing and encapsulation. Liquids and meltable solids are amenable to microencapsulation. Either solids or liquids can be combined with polymers in a matrix or microsponge. Starch matrices (Section III. D) have been most widely tested to control movement of pesticides in soils, but results have been inconsistent. In a recent study, soil mobility of two low-melting-point herbicides was not significantly controlled by starch matrixing.[31] A high-melting-point crystalline solid compound was retained in the upper soil zone by larger starch matrix granules but was too inactivated. Smaller starch matrix granules retained biological efficacy, and may reduce the leaching potential of high-melting-point solids in soils.

In contrast to the results with starch matrices, the microencapsulated low-melting-point compound was retained in the upper soil zone and was biologically efficacious. A complete thin coating around the liquid compound provided a diffusion barrier that retained the compound in the upper soil zone in a biologically effective form. Rapid movement through channels of the starch matrix resulted in unregulated movement through soil.

Thin, uniform diffusional barrier coatings around fine particle crystalline solids, by a direct nonsolvent coating process, are yet to be achieved. Such a coating technology, combined with current capability for selection of a coating material to promote diffusional movement of compound through the coating, would offer a new approach to regulation of soil mobility while retaining biological efficacy.

VIII. MAXIMIZING BIOLOGICAL EFFICACY

Recent studies of biological enhancement of compounds by surfactants identify barriers to pesticide movement,[90] establish relationships between surfactant structure and pesticide efficacy,[56,103] show influence of surfactants to deposition patterns on plant leaves,[69,106] determine interrelationships of absorption of surfactants with pesticides,[79] and elucidate surfactant properties which enhance efficacy of pesticides on crops.[41,43]

All of these studies build upon a growing knowledge base which is essential to guide formulators in judicious selection of surfactants. Yet there is much to be learned about which surfactants to select, and at what concentrations they should be used. An accurate predictor of biological performance of a spray solution is needed to short-circuit exhaustive biological trial and error testing.

Dynamic surface tension measurements may be the needed predictor of biological performance.[42] Initial results established a strong correlation between dynamic surface tension measurements of spray solutions of eight different types of surfactants with biological efficacy of two different classes of postemergence herbicides.

Dynamic surface tension parameters are now known for a number of surfactants. An initial screen can begin with a preselected set of surfactants that covers a prescribed range of dynamic surface tension parameters. Biological response of a new compound with each surfactant will identify the range of dynamic surface tensions that will activate the test compound. An additional search of surfactants within this preferred dynamic surface tension range can then identify surfactants to maximize biological efficacy of the new compound.

Maximizing the biological efficacy of chemical compounds will continue to be a major focus of formulations research. Another focus of formulations research will be preserving the efficacy of living biological pest control products. Formulation needs are a limiting factor in development of biological pest control agents.

IX. BIOLOGICAL PEST CONTROL

All living organisms are subject to diseases, parasites, and predators. Consequently, each of these natural means of control can be exploited to protect crops from pests. Growing interest of companies in biocontrol products ensures that biological control options will be available to growers in the future. A review lists current commercial products available for control of weeds, insects, pathogens, nematodes, and molluscs.[78]

Broad-spectrum biological control of plant diseases could require the deployment of multiple strains of hyper-parasites to combat parasites because parasites are specific both to crops and to soils.[20] Registration may be a barrier to broad-spectrum biological control agents. Discovery and enhancement of sufficient strains for wide geographic control of plant diseases will not lead to commercial fruition if each strain is considered to require an individual registration.[21] Despite inherent specificity, commercial products of biocontrol organisms are utilized in a variety of cropping systems.[92] An overview describes fermentation methods to grow microbial disease control agents and the formulation approaches taken to maximize biological efficacy.[35]

Broad-spectrum biocontrol of insects may not be desirable. The high degree of selectivity of biocontrol organisms ensures that only intended pests will be affected while beneficial insects will suffer no ill effects of the treatment. Once established on plants, insect control bacteria, viruses, and other living agents are highly efficient delivery systems.

Baculoviruses, emerging as viable insect control agents,[59] act slowly so that feeding continues after infection. Genetic engineering technology is being utilized to broaden the spectra of insects controlled and to speed the action of insect control to prevent posttreatment damage prior to insect death.[13] Formulation of biological insect control agents will require that they be maintained in a viable state on the shelf and that they be active in the field environment. Shelf stability is a critical issue and is the focus of considerable formulations research.

Biological weed control agents, likewise, are very host specific. To counter this, highly virulent broad-host-range pathogens are genetically engineered to be host specific for root or crown exudates of target weed species.[81] Several biological weed control agents formulated together could provide significantly broadened weed control spectra.

Biological weed control agents are becoming a bridge to chemical weed control. Synergistic control from the combination of herbicides with biological control agents is known.[19] Commercial products of synergistic combinations are available. Consistency of such products will need to be improved through development of formulations that enhance pest control.[19] Failure of control is frequently not the lack of virulence of the pathogen, but rather the problem is in the formulation and delivery of the pathogen.[111]

Formulation of biological control agents is especially challenging. Formulations must satisfy the requirements of the living agent during storage and must provide conditions required to effect infection in the field environment. Fungi which can attack weeds often require a lengthy dew period to effect infection. However, invert emulsions have been shown to promote infection without a lengthy dew period.[14] Spores formulated in an aqueous suspension were effective without a dew period when an invert emulsion was applied as an overspray.[77]

Many biological control agents cannot be formulated in a liquid, but do perform from solid formulations. Suitable dry solid carriers are being developed. Natural polymers can be pregelled by steam processing, filled with biological control agent, and then dried to provide shelf stable matrices.[12,62,82] Infected insect bodies have been processed directly as a carrier of the virus.[15]

Formulations of biological control agents have been created for non-agricultural application and may be models for agricultural products. Tablets that contain an inner core of dormant live microorganisms and outer layers of nutrients, enzymes, buffers, and barrier polymers provide timed release of agents to decompose organic matter in underwater sediments.[26] Marine bacteria immobilized into water-swelling polymer matrix released antifouling agents which prevented attachment of barnacles onto ships.[38]

Polymeric matrices will continue to be viable systems for delivery of biological control agents. New processing technologies are developing to prepare these matrices. Water-soluble polymers can be formed into beads containing biological control agents by a technique of freezing the beads in nonsolvents and freeze-drying to remove water.[7] The recovered beads can be applied dry, or sprayed by conventional spray operations. In a variation of this process, mixtures are sprayed into cold solvent which can remove water from the beads by diffusion.[40] A microdropper is available for creating small research samples of encapsulated biological control agents.[5]

Genetic engineering enhancements will add new formulation challenges. Plants are being genetically engineered to express pest control agents on demand. Genetically engineered plants generally are grown from callus tissue. Individual transgenic plantlets are cultured, grown to maturity, and seeds are collected for sale to the growers. In the future, transgenic plant material may be processed directly to synthetic seeds. This will require creation of synthetic seed components to deliver viable transgenic plant initiate cell clusters as synthetic seeds. Providing nutrients, matrix for plant initiate material, moisture, air exchange, and seedcoat, all of which function as effectively as the natural seed components, will be challenging.

Current structures of encapsulated plant initiate materials are largely hand made.[99] Zygotic embryo, cell cluster, callus, plantlet, and somatic embryo are embedded in hydrogel then dipped to coat with a waterproof coating into which water-swelling particles have been partially embedded. High volume commercial production needs to be developed. All of the current materials and processes from agricultural and nonagricultural fields will be needed to produce acceptable synthetic seeds in commercial volume.

X. CONCLUSIONS

Enhancement of technologies described in this review, to meet the more stringent temperature and pressure conditions of agricultural practice, will provide new products in the near term. Longer range developments revolve around the union of remote sensing with a constellation of satellites that can cover every spot on the earth every other day with spatial resolution of several meters. On-line translation of crop stress information will alert growers of weed pressure, disease, or insect attack, more quickly and accurately than would be possible by visual inspection of the crop. Highly specific, targeted applications will replace present whole field applications. Treatments may likely be made by unmanned aerial application controlled and directed according to satellite information. Water and nutrients will be provided as needed. Pest control agents, both chemical and biological, will be more preventive than curative. A bumper crop will be anticipated each harvest.

ACKNOWLEDGMENTS

Thanks are expressed to Anna M. Patton for literature review, to Dr. James D. Metzger and Dr. James P. Foster for review of the writing, and to Catherine M. Kershaw for preparation of the manuscript.

REFERENCES

1. **Albert, R. F. and Weed, G. B.**, United States Patent 3,920,442, Nov. 18, 1975.

2. **Aleksejczyk, R. A.**, Alkyl polyglycoside: versatile, biodegradable, surfactants for the agricultural industry, in *Pesticide Formulations and Applications Systems, STP 1146, Vol. 12*, Devisetty, B. N., Chasin, D. G., and Berger, P. D., Eds., American Society for Testing and Materials, Philadelphia, PA, 1993, 22.
3. **Allan, G. G.**, United States Patent 5,252,542, Oct. 12, 1993.
4. **Amer, M. S. and Tawashi, R.**, United States Patent 5,275,819, Jan. 4, 1994.
5. **Arneodo, V., Koka, D., Spenlehauer-Bonthonneau, F., Hundelt, P. and Thies, C.**, Curing of droplets to form microparticles, *Proc. Int. Symp. Contr. Rel. Bioact. Mater.*, 7, 447, 1990.
6. **Avera, F. L.**, International Patent Appl. WO 92/00941, Jan. 23, 1992.
7. **Baker, C. A., Brooks, A. A., Greenley, R. Z., and Henis, J. M.**, United States Patent 5,089,407, Feb. 18, 1992.
8. **Baker, T. J.** and **Woods, J. H.**, United States Patent 5,035,946, July 30, 1991.
9. **Bateman, L. R., DiLuccio, R. C., Stewart, C. A. Jr., Visiola, D. L.,** and **Beach-Coffin, D. P.**, United States Patent 4,990,335, Feb. 5, 1991.
10. **Beestman, G. B.**, and **Deming, J. M.**, Developments in microencapsulation, high concentration, in *Pesticide Formulations and Applications Systems, STP 980*, Vol. 8, Hovde, D. A. and Beestman G. B., Eds., American Society for Testing and Materials, Philadelphia, PA, 1993, 25.
11. **Beestman, G. B.**, International Patent Appl. WO 93/25081, Dec. 23, 1993.
12. **Bok, S. H., Lee, H. W., Son, K. H., Kim, S. U., Lee, J. W., Kim, D. Y., and Kwon, Y. K.**, International Patent Appl. WO 92/20229, Nov. 26, 1992.
13. **Bonning, B. C. and Hammock, B. D.**, Development and potential of genetically engineered viral insecticides, *Biotechnol. Geneti. Eng. Rev.*, 10, 455, 1992.
14. **Boyette, C. D., Quimby, P. C. Jr., Bryson, C. T., Egley, G. H., and Fulgham, F. E.**, Biological control of hemp sesbania (sesbania exaltata) under field conditions with colletotrichum trancatum formulated in an invert emulsion, *Weed Sci.*, 41, 497, 1993.
15. **Buchatskii, L. P. and Kuznetsova, M. A.**, Russian Patent Appl. SU 1,387,221, Sep. 30, 1992.
16. **Bukovac, M. J. and Petracek, P. D.**, Characterizing pesticide and surfactant penetration with isolated plant cuticles, *Pestic. Sci.*, 37, 179, 1993.
17. **Carr, M. E., Doane, W. M., Wing, R. E., and Bagley, E. B.**, United States Patent 5,183,690, Feb. 2, 1993

18. **Chan, J. H., Hasse, K. A., Satre, R. I., and Trusler, J. H.**, United States Patent 5,075,058, Dec. 24, 1991.
19. **Christy, A. L., Herbst, K. A., Kostka, S. J., Mullen, J. P., and Carlson, P. S.**, Synergising weed biocontrol agents with chemical herbicides, in *Pest Control With Enhanced Environmental Safety*, Duke, S. O., Menn, J. J., and Plimmer, J. R., Eds., ACS Symp. Ser. 524, American Chemical Society, Washington, D. C., 1993, chap. 7.
20. **Cook, J. R.**, Making greater use of introduced microorganisms for biological control of plant pathogens, *Ann. Rev. Phytopathol.*, 31, 53, 1993.
21. **Cook, J. R.**, Reflections of a regulated biological control researcher, in Regulations And Guidelines : Critical Issues In Biological Control, Proc. USDA/CSRS Natl. Workshop, 10-12 June, Vienna, VA, 1991, 9.
22. **Dawson, H. B.**, United States Patent 5,037,653, Aug. 6, 1991.
23. **Debendetti, P. G., Tom, J. W., and Yeo, S.**, Supercritical fluids: a new medium for the formation of particles of biomedical interest, in *Proc. 20th Int.Symp. Contr. Rel. Bioact. Mater.*, Roseman, T. J., Peppas, N. A., and Gabelnick, H. L., Eds., The Controlled Release Society, Inc., Washington, D.C., 1993, 141.
24. **DePonti, R., Torricelli, C., Martini, A., and Lardini, E.**, International Patent Appl. WO 91/09079, Jun. 27, 1991.
25. **Derian, P. J., Guerin, G., and Fiard, J. F.**, Microemulsions of pyrethroids : phase diagrams and effectiveness of tristyrylphenol based surfactants, in *Pesticide Formulations and Application Systems Vol. 12, STP 1146*, Devisetty, B. N., Chasin, D. G., and Berger, P. D., Eds., American Society for Testing and Materials, Philadelphia, PA, 1993, 73.
26. **DiTuro, J. W.**, United States Patent 5,275,943, Jan. 4, 1994.
27. **Domb, A. J.**, Degradable polymer blends I. screening of miscible polymers, *J. Polym. Sci. Part A Polym. Chem.*, 31, 1973, 1993.
28. **Dorman, L. C. and Meyers, F. A.**, United States Patent 5,154,749, Oct. 13, 1992.
29. **Fiard, J. F.**, Australian Patent Appl. AU-A-11164/92, Feb. 20, 1992.
30. **Fiard, J. F., Mercier, J. M., and Prevotat, M. L.**, Sucroglycerides: Novel biodegradable surfactants for plant protection formulations, in *Pesticide Formulations and Applications Systems, STP 1146, Vol. 12*, Devisetty, B. N., Chasin, D. G., and Berger, P. D., Eds., American Society for Testing and Materials, Philadelphia, PA, 1993, 33.
31. **Fleming, G. F.**, Effect of Formulation on the Movement of Atrazine, Alachlor, and Metribuzin in a Coarse-Textured Soil., Dissertation, UMI Dissertation Services, Ann Arbor, MI, Order Number 9210802, 1991.

32. **Foy, C. L.**, Adjuvants: terminology, classification, and mode of action, in *Adjuvants and Agrochemicals, Mode of Action and Physiological Activity*, Vol. I, Chow, P. N. P., Grant, C. A., Hinshalwood, A. M., and Simundsson, E., Eds., CRC Press, Inc., Boca Raton, FL, 1989, chap. 1.
33. **Foy, C. L.**, Adjuvants: terminology, classification, and mode of action, in *Adjuvants and Agrochemicals, Mode of Action and Physiological Activity*, Vol. II, Chow, P. N. P., Grant, C. A., Hinshalwood, A. M., and Simundsson, E., Eds., CRC Press, Inc., Boca Raton, FL, 1989, chap. 21.
34. **Foy, C. L.**, Progress and developments in adjuvant use since 1989 in the USA, *Pestic. Sci.*, 38, 65, 1993.
35. **Fravel, D. R. and Lewis, J. A.**, Production, formulation and delivery of beneficial microbes for biocontrol of plant pathogens, in *Pesticide Formulations and Application Systems*, Vol. 11, *American Society for Testing and Materials STP 1112*, Bode, L. E., and Chasin, D. G., Eds., American Society for Testing and Materials, Philadelphia, PA, 1992, 173.
36. **Freeman, R. Q., Sandell, L. S., and Zaucha, T. J.**, International Patent Appl. WO 93/25074, Dec. 23, 1993.
37. **Futcher, I.**, United States Patent 4,995, 900, Feb. 26, 1991.
38. **Gatenholm, P., Kjelleberg, S., and Maki, J. S.**, Immobilization of marine bacteria in hydrogels for controlled release of antifouling agents, *Polym. Mater. Sci. Eng.*, 66, 490, 1992.
39. **Geigle, W. L., Sandell, L. S., and Wysong, R. D.**, International Patent Appl. WO 91/13545, Sep. 19, 1991.
40. **Gombotz, W. R., Healy, M. S., and Brown, L. R.**, United States Patent 5,019,400, May 28, 1991.
41. **Green, J, M., and Green, J. H.**, Surfactant structure and concentration strongly affect rimsulfuron activity, *Weed Technol.* 7, 633, 1993.
42. **Green, J. H. and Green, J. M.**, Dynamic surface tension as a predictor of herbicide enhancement by surface active agents, Brighton Crop Protection Conference - Weeds, 1991, 323.
43. **Green, J. M. and Brown, P. A.**, Influence of surfactant properties on nicosulfuron, DPX-E9636 and thifensulfuron performance on corn, I.U.P.A.C. Conference, Hamburg, 1990.
44. **Greene, L., Phan, L. X., Schmitt, E. E., and Mohr, J. M.**, Side-chain crystallizable polymers for temperature-activated controlled release, in *Polymeric Delivery Systems Properties and Applications*, El-Nokaly, M. A., Piatt, D. M., and Charpentier, B. A., Eds., American Chemical Society, Ser. 520, Washington, D. C., 1993, chap. 17.
45. **Gubelmann-Bonneau, I. V., Mailhe, P. A., and Perrin, M. A.**, Tristyrylphenol surfactants in agricultural formulations: properties and challenges in application., in *Pesticide Formulations and Applications Systems, STP 1234*, Vol. 14, Hall, F. R.,

Berger, P. D., and Collins, H. M., Eds., American Society for Testing and Materials, Philadelphia, PA, 1994.

46. **Guyot-Hermann, A. M.**, Tablet disintegration and disintegrating agents, *S.T.P. Pharma Sci.*, 6, 445, 1992.
47. **Heinrich, R., Haase, D., and Maier, T.**, Canadian Patent Appl. 2,079,092, Mar. 15, 1993.
48. **Heinrich, R., Maier, T., Kocur, J., and Schlicht, R.**, Canadian Patent Appl. 2,060, 745, Aug. 7, 1992.
49. **Hodakowski, L. E., Chen, C. R., Gouge, S. T., and Weber, P. J.**, United States Patent 5,139,152, Aug. 18, 1992.
50. **Hoorne, D., Chasin, D. G., and Rogiers, L. M.**, Novel adjuvants for agrochemical formulations based on sugar ethers, in *Pesticide Formulations and Applications Systems, STP 1146,* Vol. 12, Devisetty, B. N., Chasin, D. G., and Berger, P. D., Eds., American Society for Testing and Materials, Philadelphia, PA, 1993, 3.
51. **Hyson, A. M.**, United States Patent 4,936,900, Jun. 26, 1990.
52. **Ijitsu, T., Shiba, K., Hara, H., and Negishi,Y.**, United States Patent 5,174,998, Dec. 29, 1992.
53. **Jackisch, D. A.**, World Patent, 93/13658, July 22, 1993.
54. **Kissa, E.**, *Fluorinated Surfactants : Synthesis, Properties, and Applications,* Marcel Dekker, New York, 1994, chap. 3.
55. **Kissa, E.**, *Fluorinated Surfactants : Synthesis, Properties, and Applications,* Marcel Dekker, New York, 1994, chap. 8.
56. **Knoche, M. and Bukovac, M. J.**, Interaction of surfactant and leaf surface in glyphosate absorption, *Weed Sci.*, 41, 87, 1993.
57. **Kydonieus, A. F., Decker, S. C., and Shah, K. R.**, Temperature activated controlled release, in *Proc. Intern. Symp. Control. Rel. Bioact. Mater.*, Controlled Release Society, Amsterdam, 1991, 417.
58. **Kydonieus, A. F., Shah, K. R., and Decker, S. C.**, European Patent Appl. EP 0, 497, 626, May 8, 1992.
59. **Leisy, D. J. and Van Beek, N.**, Baculoviruses : possible alternatives to hemical insecticides, *Chemistry & Industry,* 6 April, 250, 1993.
60. **Lindsay, A. D. and Omilinski, B. A.**, International Patent Appl. WO 92/01381, Feb. 6, 1992.
61. **Ma, S. H., Matrick, H., Shor, A. C., and Spinelli, H. J.**, United States Patent 5,085, 698, Feb. 4, 1992.
62. **Marola, J. J., Fravel, D. R., Connick, W. J. Jr., Walker, H. L., and Quimby, P . C. Jr.**, United States Patent 4,724,147, Feb. 9, 1988.
63. **McCollum, W. A., III, Davis, J. S., and Hermansky, C. G.**, International Patent Appl. WO 93/12652, Jul. 8, 1993.
64. **Menger, F. M. and Littau, C. A.**, Gemini surfactants: a new class of self-assembling molecules, *J. Am. Chem. Soc.*, 115, 1083, 1993.

65. **Mulqueen, P. J., Paterson, E. S., and Smith, G. W.**, Recent developments in suspoemulsions, *Pestic. Sci.*, 29, 451, 1990.
66. **Murphy, D. S., Policello, G. A., Goddard, E. D., and Stevens, P. J. G.**, Physical properties of silicone surfactants for agricultural applications, in *Pesticide Formulations and Application Systems, STP 1146, Vol. 13,* Devisetty, B. N., Chasin, D. G., and Berger, P. D., Eds., American Society for Testing and Materials, Philadelphia, PA, 1993, 45.
67. **Narayanan, K. S.**, Polymers for instant dispersions for select family of pesticides, in *Pesticide Formulations and Applications Systems, STP 1234,* Vol. 14, Hall, F. R., Berger, P. D., and Collins, H. M., Eds., American Society for Testing and Materials, Philadelphia, PA, 1994.
68. **Nastke, R., Leonhardt, A., and Neuenachwander, E.**, Canadian Patent Appl. 2,077,884, March 12, 1993.
69. **Nielsen, E.**, United States Patent 5,246,912, Sep. 21, 1993.
70. **Ogino, K. and Abe, M.**, Microemulsion formation with some typical surfactants, in *Surface and Colloid Science,* Matijevic, E., Ed., Plenum Press, New York, 1982.
71. **Ohwaski, T., Machida, R., Ozawa, H., Kawashima, T. H., Takeuchi, H., and Niwa, T.**, Improvement of the stability of water-in-oil-in-water multiple emulsions by the addition of surfactants in the internal aqueous phase of the emulsions, *Int. J. Pharm.*, 93, 61, 1993.
72. **Okano, T., Bae, Y. H., Jacobs, H., and Kim, S. W.**, Thermally on-off switching polymers for drug permeation and release, *J. Control. Rel.*, 11, 255, 1990.
73. **Otten, J. G. and Schoene, K. F.**, United States Patent 5,187,191, Feb. 16, 1993.
74. **Peppas, N. A. and Argade, A. B.**, Dendrimers and star polymers for pharmaceutical and medical applications, in *Proc. Intern. Symp. Control. Rel. Bioact. Mater.,* Roseman, T. J., Pappas, N. A., and Gabelnick, H. L., Eds., Controlled Release Society, Inc., Amterdam, 1993, 143.
75. **Petersen, R. C., Matson, D. W., and Smith, R. D.**, The formation of polymer fibers from the rapid expansion of supercritical fluid solutions, *Pol. Wt. Eng. and Sci.*, 27, 22, 1693, 1987.
76. **Puritch, G. S., McHarg, D., Bradbury, R., and Mason, W.**, United States Patent 5,037,654, Aug. 6, 1991.
77. **Quimby, P. C. Jr., Fulgham, F. E., Boyette, C. D., and Connick, W. J. Jr.**, An invert emulsion replaces dew in biocontrol of sicklepod - a preliminary study, in *Pesticide Formulations and Applications Systems, STP 980, Vol. 8,* Hovde, D. A., and Beestman G. B., Eds., American Society for Testing and Materials, Philadelphia, PA, 1993, 264.

78. **Rodgers, P.**, Potential of biopesticides in agriculture, *Pestic. Sci.*, 39, 117, 1993.
79. **Roggenbuck, F. C., Penner, D., Burow, R. F., and Thomas, B.**, Study of the enhancement of herbicide activity and rainfastness by an organosilicone adjuvant utilizing radiolabelled herbicide and adjuvant, *Pestic. Sci.*, 37, 121, 1993.
80. **Sagar, B., Wales, D., and Nelson, G.**, International Patent Appl. WO91/10772, July 25, 1991.
81. **Sands, D. C. and Miller, R. V.**, Altering the host range of mycoherbicides by genetic manipulation, in *Pest Control With Enhanced Environmental Safety*, Duke, S. O., Menn, J. J., and Plimmer, J. R., Eds., ACS Symp. Ser. 524, American Chemical Society, Washington, D. C., 1993, chap. 8.
82. **Shasha, B. S. and McGuire, M. R.**, United States Patent 5,061,697, Oct. 29, 1991.
83. **Shelley, S.**, Microscopic parcels deliver the goods, *Chemical Engineering*, 45, 1993.
84. **Shih, J. S., Chuang, J. C., and Haldar, R. K.**, United States Patent 5,252,611, Oct. 12, 1993.
85. **Singh, R., Awasthi, S., and Vyas, S. P.**, New development on hydrogels: potential in drug delivery, *Pharmazie*, 48(10), 728, 1993.
86. **Smith, G. W., Mulqueen, P. J., Paterson, E. S., and Heacham, J. C.**, United States Patent 5,321, 049, Jun. 14, 1994.
87. **Sparks, R. E., Jacobs, I. C., and Mason, N. S.**, Centrifugal suspension-separation coating of particles and droplets, in *Polymeric Delivery Systems Properties and Applications*, El-Nokaly, M. A., Piatt, D. M., and Charpentier, B. A., Eds., American Chemical Society, Washington, D. C., *Symp. Ser. 520*, 1993, chap. 9.
88. **Stevens, P. J. G.**, Organosilicone surfactants as adjuvants for agrochemicals, *Pestic. Sci.*, 38, 103, 1993.
89. **Stewart, R. F., Greene, L. C., and Bhaskar, R. K.**, United States Patent 5,120, 349, Jun. 9, 1992.
90. **Stock, D., Edgerton, B. M., Gaskins, R. E., and Holloway, P. J.**, Surfactant-enhanced foliar uptake of some organic compounds : interactions with two model polyoxyethylene aliphatic alcohols, *Pestic. Sci.*, 34, 233, 1992.
91. **Surgant, J. M. and Deming, J. M.**, United States Patent 4,936,901, Jun. 26, 1990.
92. **Sutton, J. C. and Peng, G.**, Manipulation and vectoring of biocontrol organisms to manage foliage and fruit diseases in cropping systems, *Ann. Rev. Phytopathol.*, 31, 473, 1993.

93. **Tabibi, S. E., Sakura, J. D., Mathur, R., Wallach, D. F. H., Schultes, D. T., and Oatrom, J. K.**, The delivery of agricultural fungicides in paucilamellar amphiphile vesicles, in *Pesticide Formulations and Applications Systems, STP 1146,* Vol. 12, Devisetty, B. N., Chasin, D. G., and Berger, P. D., Eds., American Society for Testing and Materials, Philadelphia, PA, 1993, 155.
94. **Tadros, T. F. and Waite, F. A.**, British Patent Appl. GB 2,026,341, Jul. 10, 1979.
95. **Tadros, T. F.**, Industrial applications of dispersions, *Advances in Colloid and Interface Science,* 46, 1, 1993.
96. **Tadros, T. F.**, United States Patent 4,875,927, Oct. 24, 1989.
97. **Tadros, T. F.**, United States Patent 5,139,773, Aug. 18, 1992.
98. **Takagishi, T. and Kono, K.**, Release control of drugs from microcapsules in response to external stimuli, *New Func. Mater.* B, 197, 1993.
99. **Teng, W. L., Liu, Y. J., and Soong, T. S.**, United States Patent 5,250,082, Oct. 5, 1993.
100. **Tocker, S.**, Controlled release formulations of pesticides, *in Proc. 20th Intl. Symp. Contr. Rel. of Bioact. Mater.,* Roseman, T. J., Peppas, N. A., and Gabelnick, H. L., Eds., The Controlled Release Society, Inc., Washington, D.C., 1993, 200.
101. **Tocker, S.**, European Patent Appl. EP 0,529,975, Mar. 3, 1993.
102. **Toka, K. K.**, Japanese Patent Appl., JP05112469-A, 1991.
103. **Wade, B. R., Reichers, D. E., Liebl, R. A., and Wax, L. M.**, The plasma membrane as a barrier to herbicide penetration and site for adjuvant action, *Pestic. Sci.*, 37, 195, 1993.
104. **Webster, O. W.**, Living polymerization methods, *Science*, 251, 887, 1991.
105. **Wing, R. E., Carr, M. E., Doane, W. M., and Schreiber, M. M.**, Controlled release of herbicide from an unmodified starch matrix, in *Polymeric Delivery Systems Properties and Applications ACS Symp. Ser. 520,* El-Nokaly, M. A., Piatt, D. M., and Charpentier, B. A., American Chemical Society, Washington, D. C., 1993, chap. 14.
106. **Winkle, J. R., Wendell, A. R., and Jourdan, G. P.**, Effect of formulation on fungicidal activity, in *Pesticide Formulations and Application Systems: International Aspects,* Vol. 9, *American Society for Testing and Materials, STP 1036,* Hazen, J. L., and Hovde, D. A., Eds., American Society for Testing and Materials, Philadelphia, PA, 1989, 43.
107. **Wirth, W., Storp, S., and Jacobsen, W.**, Mechanisms controlling leaf retention of agricultural spray solutions, *Pestic. Sci.*, 33, 411, 1991.
108. **Yap, W. H.**, United States Patent 5,001,150, Mar. 19, 1991.
109. **Yazan, Y. and Seiller, F. P.**, Multiple emulsions, *Boll. Chim. Farm.*, 132, 6, 187, 1993.

110. **Zhou, G. and Smid, J.**, Amphiphilic star polymers from tri- or tetraisocyanates and poly(ethylene glycol) derivatives, *Polym. Prepr.*, 32(3), 613, 1991.
111. **Zorner, P. S., Evans, S. L., and Savage, S. D.**, Perspective on providing a realistic technical foundation for the commercialization of bioherbicides, in *Pest Control With Enhanced Environmental Safety*, Duke, S. O., Menn, J. J., and Plimmer, J. R., Eds., ACS Symp. Ser. 524, American Chemical Society, Washington, D. C., 1993, chap. 6.

Chapter 6

Water Dispensible Granules - Past, Present, and Future

Jeremy N. Drummond

CONTENTS

ABSTRACT

Formulation research into water dispersible granules has increased greatly in recent years because of the regulatory pressures put on solvent based formulations and the concerns about operator exposure during the use of wettable powders.

Initially the popular methods of manufacture were pan granulation and spray drying. In recent years, attention has become more focused on fluid bed

1-8493-7678-5/96/$0.00+$.50

granulation and extrusion. Each of these methods will be described and will be compared and contrasted. The underlying physicochemical factors influencing each method of granulation will also be discussed.

The problems of solidifying low melting active ingredients and increasing the biological activity of active ingredients formulated as water dispersible granules with low toxicity using globally acceptable inert ingredients remain formidable challenges to the formulation chemist.

I. INTRODUCTION

Water dispersible granules are defined by the International Group of National Associations of Manufacturers of Agrochemical Products (GIFAP) as "a formulation consisting of granules to be applied after disintegration and dispersion in water."[2] They are designated by the code "WG" and are often known as dry flowables. They are designed to be diluted in spray tanks, but applying them as dry granules to the soil has also been investigated.[7] Water soluble granules are regarded as a separate category by GIFAP[2] and are abbreviated as "SG". However, they are similar enough to WGs that they will also be covered in this chapter but may not always be distinguished from WGs.

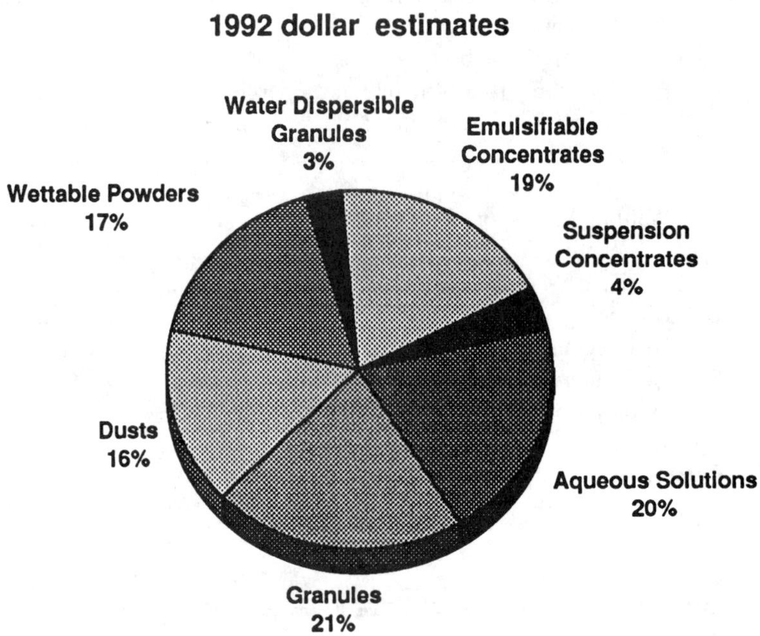

Figure 1. World pesticide market by formulation type

In the early days of pesticide application most of the formulations were wettable powders (WPs) or dusts (DPs). After 1945, emulsifiable concentrates (ECs) increased in popularity and processes moved towards liquid products. This trend was further enhanced by the introduction of suspension concentrates or flowables (SCs). However, the pendulum is now starting to swing back to dry products with the introduction of many different water dispersible granule products. Water dispersible granules are currently only a small portion of the market (Figure 1), but their proportion will grow at the expense of the other formulations, in particular wettable powders.[23]

The move to water dispersible granules is led by the regulatory pressures being put on the solvents that were traditionally used to make emulsifiable concentrates, and because of concerns over operator exposure during the application of wettable powders. The pressure for change has come, therfore, from developed countries, especially the United States, and as yet has had no significant impact on the developing world which accounts for 30% of the world market.

The continuous reduction in dose rates, that has occurred over the whole history of pesticide application, has also positively influenced the trend towards water dispersible granules. Granulation adds at least one step to product manufacture. The extra costs involved can be borne more easily by the highly active pesticides, which have dose rates per hectare measured in grams instead of kilograms.

Water dispersible granules are easier to handle than wettable powders because they are both dust free and exhibit improved flowability (hence the name dry flowables). They are also easy to measure by volume.

WGs can be packaged in plastic or cardboard containers. The Dutch covenant[12] on packaging residue of 0.01% can usually be met without any requirement to rinse out the package. This gives an advantage over suspension concentrates which, when dried out, can be difficult to dislodge. Some water dispersible granules are packaged in water soluble bags (e.g. Tell®[*] and Scepter®[†]) which ensures that there is little chance of residue remaining in the outside package and minimizing contact with the operator. Although this is a convenient solution to the packaging problem, it must overcome the farmers' dislike of water soluble bags and must justify the extra manufacturing expense in addition to the granulation costs. It must also be remembered that dispersion characteristics of granules from the confines of a water soluble bag may be different from those poured in loose. With their flexibility, water dispersible granules offer the possibility of innovative package design, which future legislation may demand, although they do not lend themselves to direct injection systems.

---

[*]Registered Trademark of Ciba Agriculture, Basel, Swiss.

[†]Registered Trademark of American Cyanamid Co., Wayne, New Jersey.

The relatively high manufacturing costs of water dispersible granules can often be reduced by having a high loading of the active ingredient. This allowed for the pioneering commercialization of triazine WG formulations in the U.S. However, the active ingredient concentration very much depends on the physical properties of the active ingredient and in some cases its toxicity classification.

Acute toxicity, particular via dermal routes, can often be reduced by reformulating solvent based formulations as water dispersible granules. The European CPL transport classifications are also more favorable with regard to acute toxicity for solid products over liquids (with tablets being an exception), but no distinction is made in the U.S.

II. METHODS OF MANUFACTURE

The method of manufacture is crucial in determining the formulation requirements and the final properties and morphology of a water dispersible granule. The different methods for producing water dispersible granules can be summarized as follows:

A. Pan
B. Spray Drying
C. Fluid Bed Granulation
D. Extrusion
E. High Shear Agglomeration

When selecting a particular process the volume of a single product must justify the initial capital investment cost or the process should be sufficiently flexible to allow several products to be made on a single line.

In the past, the popularity of methods of manufacture could be divided regionally: The U.S. adopted water dispersible granules made by pan granulation, Europe tended towards spray drying or fluid bed granulation, and Japan produced granules by extrusion for application to rice paddies in which they were designed to sink and where quick dispersion was undesirable. Recent global trends have been towards hybrid fluidized-bed spray drying, continuous fluid bed granulation, and extrusion.

A. PAN GRANULATION

A typical pan granulator is shown in Figure 2. It consists of a wide shallow flat-faced pan in which the powder and binder solution is mixed.[11,28] The pan can normally be rotated at various speeds and various angles. A scraper is usually used to keep oversize granules to a minimum. Powder is charged into the pan and water is sprayed on the powder at various points as the pan rotates. The water first wets the powder and binds individual particles together through capillary forces. As the moisture content increases all the intergranular spaces are filled by the binder solution. Once there is sufficient moisture present and the primary particles start to become coated, the granules start to roll and increase in size by ball growth. Deaeration and densification also occur at this time.

Alternatively, the powder can be prewetted in a continuous mixer prior to addition to the pan. This reduces dust and can reduce the breadth of the granule size distribution.

The path the particles take around the pan depends on their stage of growth. Nucleating particles travel around the outer edge of the pan (Path 1, Figure 3) while large particles take the shortest route and are removed from the pan by gravitational forces (Path 3, Figure 3). Growing particles follow an intermediate route (Path 2, Figure 3).

Pan granulators are relatively simple pieces of equipment, but there are many variables which can affect the quality of the granules produced. On a laboratory scale the process is usually carried out in batches but a production process is run continuously with the finished product spilling over the edge of the pan. The relationship between the depth and the diameter of the pan is important in determining the size range of the

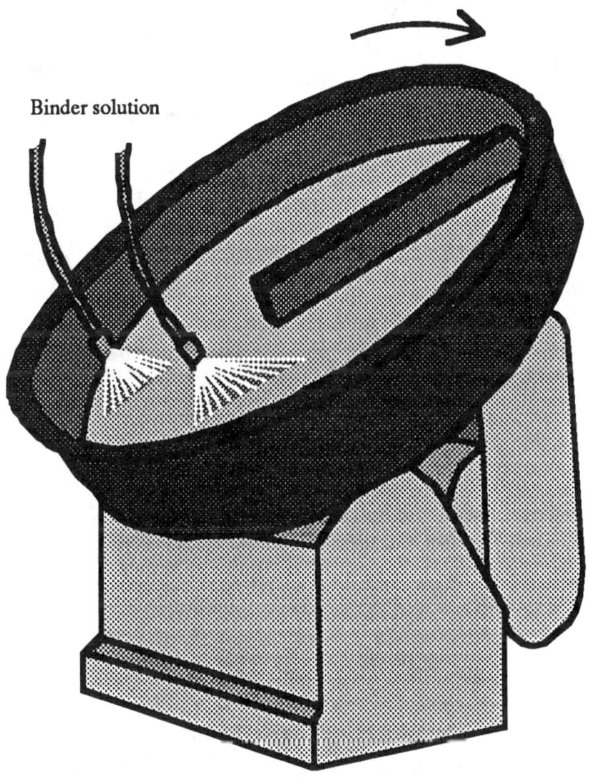

Figure 2. Pan granulator

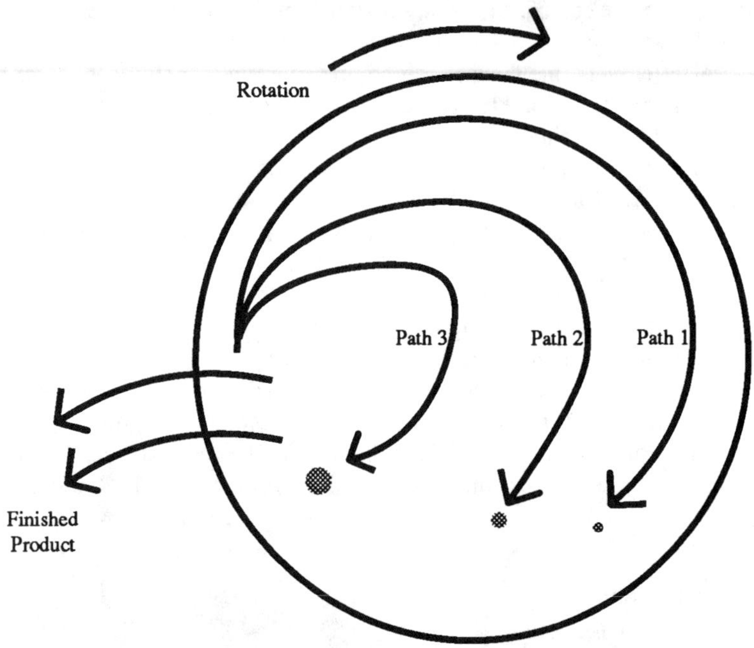

Figure 3. Movement of granules in pan granulator

granules. Increasing the speed of the pan densifies the granules and reduces their size. Increasing the angle of the pan also results in smaller granules.

The amount of binder solution added is very critical and affects both the size range and density of the pellets. The moisture range for optimum production is invariably narrow.

The position of the powder and binder solution feeds are also very important. If the binder solution is sprayed over the nucleating particles (Path 1, Figure 3) the pellets will be relatively small in size. By spraying large particles (i.e., over Path 3, Figure 3) ball growth is promoted and the size range of the product increases. The position of the powder feed has a similar effect.

The WGs produced by pan granulation tend to have a relatively large particle size which may lead to a low bulk density and poor disintegration characteristics. The granules are usually soft because the agitation is mild and the feed powder may require milling. Optimizing the process of pan granulation is still something of an art form, and achieving a high throughput of granules with the desired charateristics remains a challenge at each stage of the scale-up. Even when optimized, the amount of off-size recycling required is usually high (10 to 25%).

Densification and granule strength can be improved through increased agitation using either counter-current batch granulators, such as those made by Erich, or the high-speed granulators commonly used in the pharmaceutical industry. These have the disadvantage of being batch units, and easy discharge to a drier must be engineered. But with their increased shear they are less sensitive to the method of water addition and end point additions can be accurately judged based on motor power consumption.

B. SPRAY DRYING

Spray drying technology was first developed in the food and detergent industry. A solution or suspension is sprayed into the top of a chamber along with hot air (Figure 4) and each droplet dries to form a single particle.[21] The process is continuous and has the added advantage that granules can be made from a suspension where particle size can be reduced through wet milling. This is more efficient than dry milling and avoids any potential explosion hazards, although the risks in the spray chamber must be considered.

The particles have an eggshell-like appearance, and although they are relatively fragile, their small size and superb flowability reduce friability. With one particle being formed per droplet, the final particle size distribution is invariably small. Therefore, the products are often dusty and it can be difficult to separate the dust from the product. The dust can be difficult to contain on a manufacturing scale and may be a potential hazard. They are very quick to disperse and can bloom in the spray tank like an

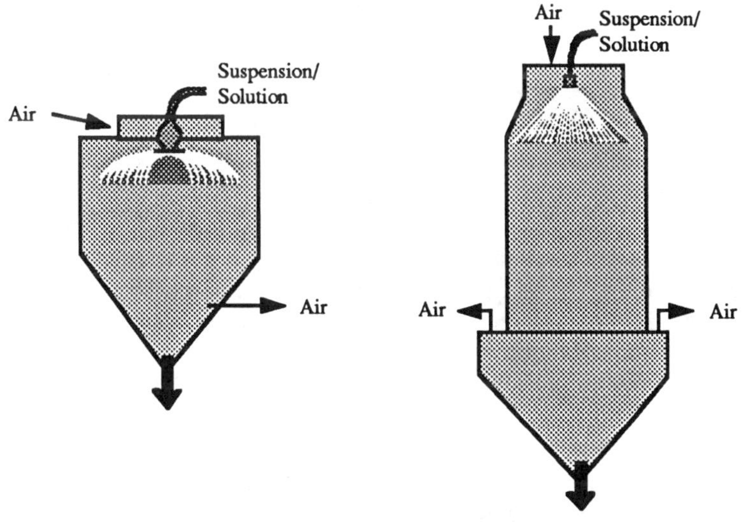

Figure 4. Schematic diagrams of different spray drier designs

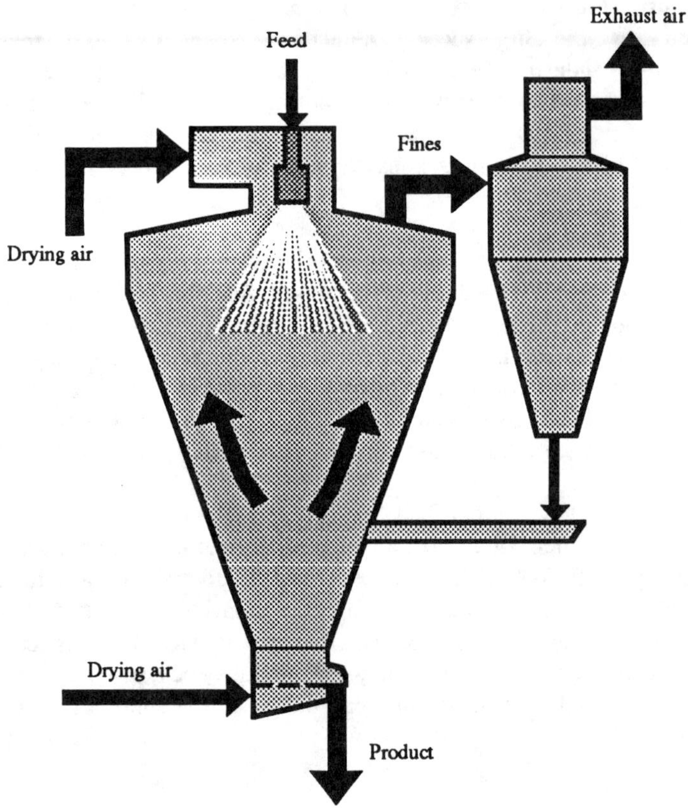

Figure 5. Fluid bed spray drier

emulsifiable concentrate.

Ingredient segregation can occur during spray drying. Segregation is dependent upon the relative water solubilities of the ingredients and the drying conditions. If wetters and dispersants become concentrated on the surface of the granules the result can be incomplete dispersion after an initial rapid disintegration.

Particle size can be increased by increasing the droplet size distribution of the spray. Drops are either formed by a rotary atomizer or a pressure nozzle (which gives a larger particle size) and the spray pattern influences the shape of the chamber (Figure 4). For larger droplets it is necessary to have larger chambers to prevent wall deposition. However, even with larger droplets the product is never dust free because of the wide particle size distribution generated, and fines are usually separated using a cyclone.

Manufacturing plants have to be very large in order to achieve suitable throughputs.

Agglomeration of the spray-dried particles can be carried out in a fluid bed drier after the spray chamber to increase the particle size. Alternatively, a fluid bed can be used as a final drying step to improve the throughput of the spray drier.

The process parameters associated with spray drying need to be optimized at each scale of operation. It is not possible to dry large droplets on a laboratory scale and therefore the particle size achievable with full scale manufacturing can not be reproduced. This must be remembered during formulation development since particle size variability influences the quality of the granules. Furthermore, yields on a laboratory and pilot scale are generally poor. Formulations are thus difficult to optimize when there is only limited technical material available.

Inlet air and the spray droplets normally enter at the top of the chamber together (co-currently) and the product and air outlets are at the bottom. However, various combinations can be used so that, for example, the feed is sprayed into the chamber in the opposite direction to the inlet air (counter-currently). Each modification has a different effect on the granules' properties.

By introducing warm air at both the bottom of the spray drier and around the atomizing nozzles, the advantages of spray drying and fluid bed granulation can be combined. The equipment required is smaller than the spray drier needed to achieve the same granule size distribution because the spray dried particles agglomerate.

Niro's fluid spray drier[21] (Figure 5), for example, is used commercially to produce water dispersible granules. Air is introduced around the spray nozzle at the top of the conical chamber as well as at the bottom. The warm air from the base throws dust particles back up into the spray drying region where they agglomerate with the spray droplets. The air introduced at the base also acts as a classifier. Particles have an open structure, typically do not exceed 0.4 mm, and have good dispersion characteristics.

C. FLUID BED GRANULATION

Fluid bed granulation has traditionally been a batch process in which binder solution is sprayed on to a fluidized powder (Figure 6).[13,15,21] Granulation is initiated by two or more primary particles bonded together by the capillary forces associated with liquid bridges (nuclei). They continue to grow by agglomeration of nuclei and layering mechanisms. At the same time, the spray solution is being evaporated and binders are concentrated early in the granulation process.

As with all granulation processes, the quantity of spray solution is crucial in determining the particle size distribution and density of the final product. The primary granule size is also dependent on the droplet size distribution of the spray. Droplet size is related to the viscosity and surface

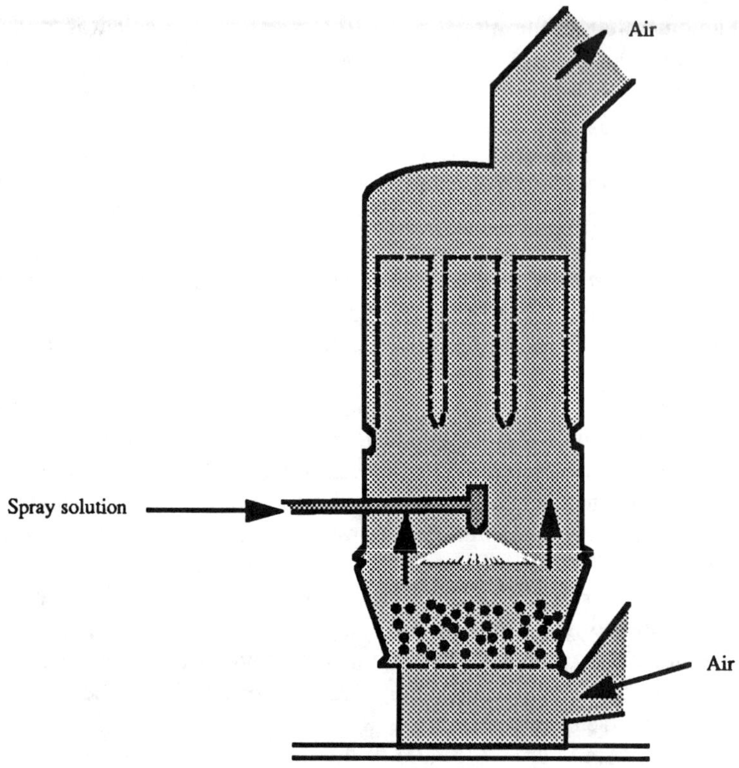

Figure 6. Schematic fluid bed drier

tension of the solution, its flow rate, and the type of nozzle.[26] However, the conditions in the drier can be altered during the process to promote growth (by spraying) or drying (by increasing the temperature and air flows).

The position of the spray nozzle is also important. The nozzle should usually be at a height above the powder bed which gives maximum coverage. This prevents local wetting and leads to maximum uniformity of particle size.

Granules from a fluid bed have an open structure which aids dispersion. They are typically in the size range 0.1 to 1 mm and are roughly spherical, but lack the uniformity of spray-dried or pan granules.

A big advantage of fluid bed granulation is that the granulation and drying can be carried out in the same unit. A disadvantage is that these units are typically not continuous and any particle size reduction of the powder prior to granulation must be done by dry milling.

Explosion hazards are also a concern during the initial fluidization of the powder. However, closed systems have been developed if inertion is necessary, and vacuum fluid bed drying systems are now available.[18] They also make the use of solvents feasible. Fluid bed driers can also be used to coat water dispersible granules if desired.

Some or all of the ingredients for a fluid bed granulation can be introduced as a suspension or solution through a spray nozzle. The process is usually initiated by spraying onto seed particles. However, with the right configuration and start-up conditions, it has been claimed that nuclei are not required.[27] If the active ingredient is added as a liquid suspension it can be wet-milled prior to granulation if necessary. The relationship between the feed process variants and granule structure are summarized in Figure 7.

Workers at Hoechst (AgrEvo) have reported the granulation of suspoemulsions (SEs) containing active ingredients dissolved in a high boiling point solvent using a fluid bed drier.[24] They claim that the biological activity of the emulsified active ingredients is equivalent to an EC formulation, even after granulation.

More recently, continuous fluid bed granulation equipment has been

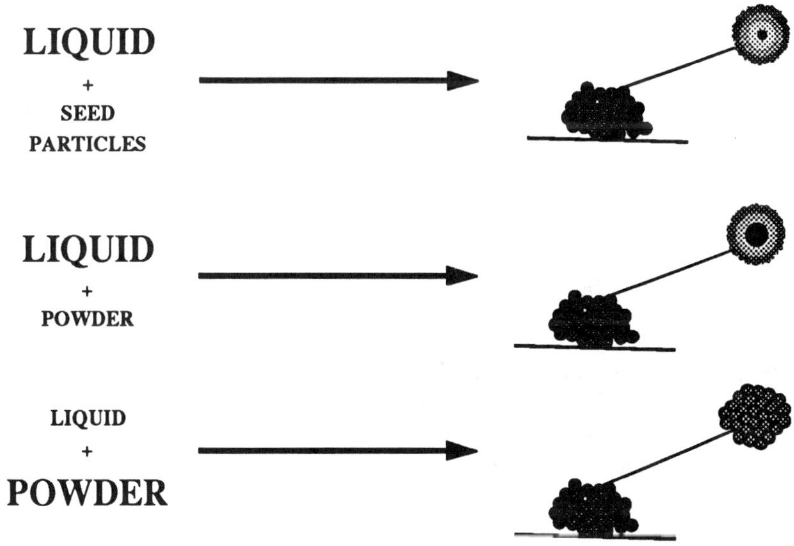

Figure 7. The effect of feed process variants on granular structure

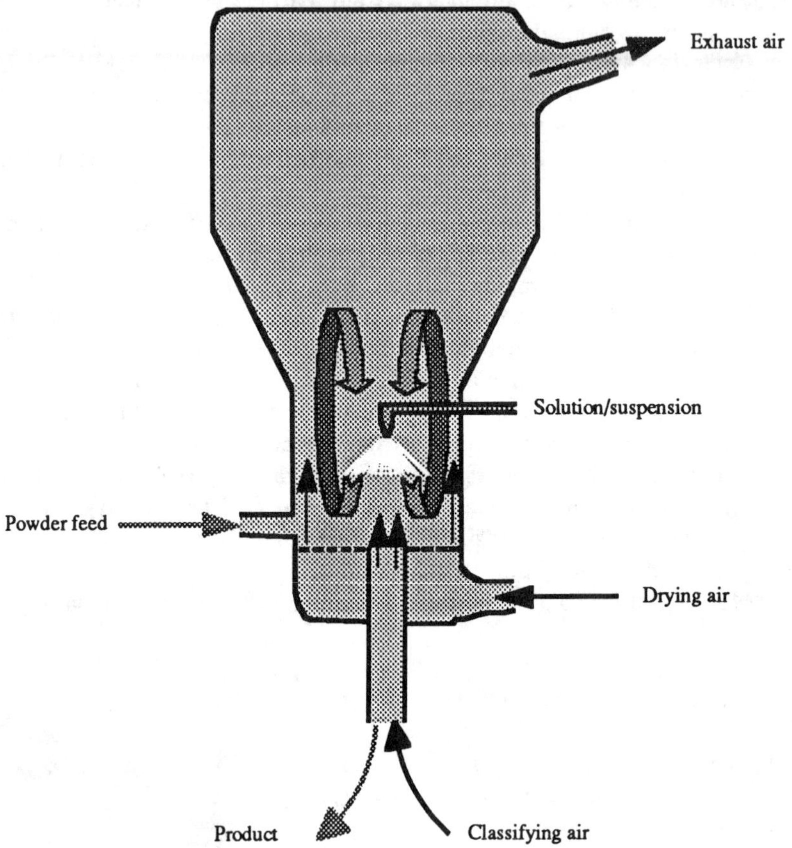

Figure 8. Schematic of Glatt AGT process

developed. For example, Glatt's AGT process* (Figure 8) is an interesting new development. The fluid bed classifies the particles within the chamber, which gives granules with a very narrow size distribution. However, the production of rapidly dispersing granules from this process has yet to be proven.

*Further information available from Glatt Air Techniques Inc., 20 Spear Road, Ramsey, NJ,07446

D. EXTRUSION

Producing water dispersible granules by extrusion has become of increasing interest to agrochemical companies on both sides of the Atlantic in recent years. Much of this is due to the development of low pressure extrusion equipment*. The cylindrical granules are relatively large, typically with a radius of 0.8 to 1.0 mm. They are relatively dense and so disintegration rate is an important parameter to investigate during formulation development.

The process has a high throughput rate, is high yielding, and with the addition of a back feed mixer it can be made continuous. The granules have a uniform aesthetic appearance and can be made essentially dust free if the formulation has good friability characteristics. It is relatively simple to produce large quantities of material on a small scale and formulations can be scaled up with a fair degree of confidence.

There are essentially two types of extruders commonly used in the industry to make water dispersible granules. Twin screw extruders with radial discharge have a higher back pressure than the basket type extruders (Figure 9). In general, the lower the back pressure of the extruder the more sensitive it is to the nature of the formulation and the water requirements. All the extruders used cylindrical screens 1 mm in depth, and 1-mm-diameter dies are typically used.

The dry powder, which may have to be dry milled, is blended with the binder solution. The wet cake must be thoroughly wetted and mixed prior to extrusion, and must still be easy to handle on a manufacturing scale. Extrusion requires that the material is able to undergo plastic deformation.

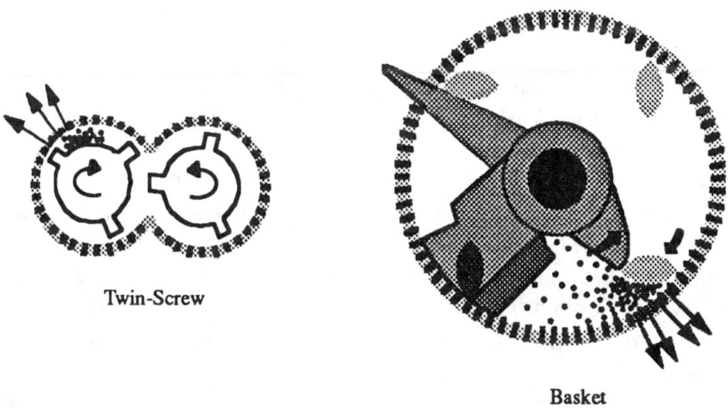

Figure 9. Schematic cross-sections through different extruder designs

*Further information on this quipment is available from Fuji Paudal, Japan and Niro-Fielder, Southampton, UK.

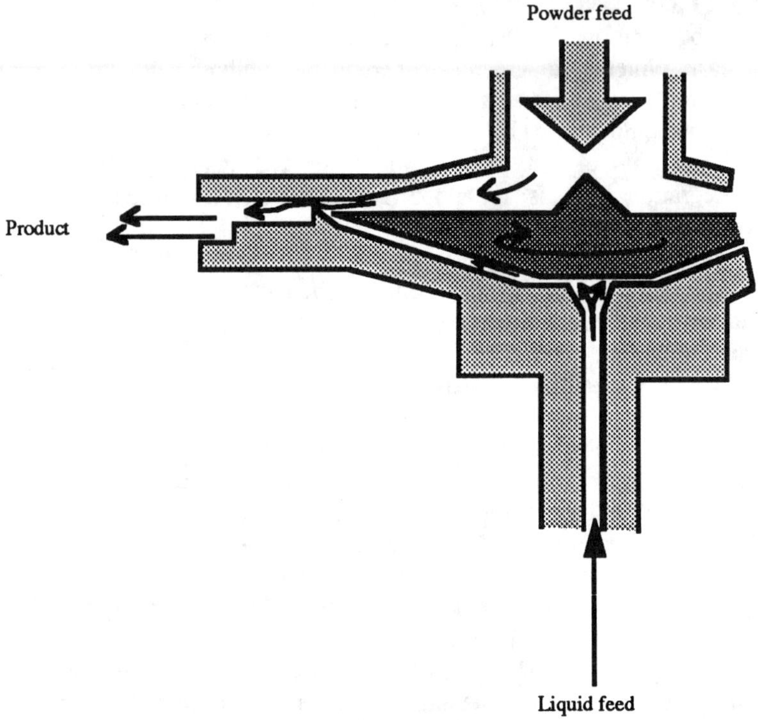

Figure 10. Schematic of Nica Turbine Mixer

The lubricant is usually water although hydro-alcoholic solutions or alcohols can be used. Surfactants and polymers can also be used to increase the plasticity of the extrudate.[14]

The movement of water within the wet mass is crucial.[20,9] Prior to compression the water should be held intergranularly so that the wet mass is still powdery. Under compression, the water is forced out of the intergranular pores and acts as a lubricant as the extrudate passed through the dies. Once through the die the water must move back into the extrudate so that different strands do not stick together and intergranular spaces are maintained.

The extrudate strands can be conveyed directly to a fluid bed drier where the agitation will break them into uniform lengths. Alternatively they can be made spherical prior to drying.[20,9,17] The latter process is practiced in the pharmaceutical industry but it adds an extra manufacturing step, puts higher demands on the formulation, and tends to overdensify the product.

E. HIGH SHEAR AGGLOMERATION

High shear agglomeration is a continuous method of wetting and agglomerating powder under high shear conditions. It has been used to make water dispersible granules, but it has proved less popular than the methods of manufacture previously discussed. Equipment of this type is made by Shugi and Niro-Fielder* (Figure 10). The equipment can be used as either a granulator feeding into a fluid bed drier, or as a high throughput continuous mixer prior to extrusion. The resultant granules have an open structure, but particle size distribution and quality can be difficult to control. Furthermore, the high throughput of these units results in high levels of product wastage during formulation development and production start ups.

III. FORMULATION REQUIREMENTS FOR WATER DISPERSIBLE GRANULES

The formulation of a water dispersible granule cannot be optimized until a method of manufacture has been selected. However, water dispersible granules are typically based on the ingredients listed below:

Active ingredient
Wetting agent
Dispersant
Filler
Binder
Disintegrant
Antifoam
Adjuvant

WGs can be as simple as an active ingredient, wetter, and dispersant, but it is becoming common for them to have a much more sophisticated combination of ingredients. The granulation of microencapsulated active ingredients has even been claimed.[3]

A. ACTIVE INGREDIENT

The properties of the active ingredient are of course fundamental to any formulation. The properties particularly pertinent to water dispersible granules are

Melting point	Hydrolytic stability
Particle size and shape distribution	Adhesive properties
Water solubility	Explosivity
Plasticity	Toxicity

*Further information is available from Schugi B.V., Lelystad, Netherlands and Niro-Fielder, Southampton, U.K.

1. Melting Point

Liquid active ingredients or those with low melting points are major challenges to formulate as a water dispersible granules. They can be solidified by being either absorbed on fillers[22,16] or coprecipitated/comelted with polymers.[4,10,19] The presence of high boiling point solvents have also been claimed to aid in the formulation of low melting active ingredients as water dispersible granules. Storage tests are crucial because of the potential for actives of this type to migrate within the granule over time.

2. Particle Size And Shape Distribution

For granules with a high active ingredient concentration, the hardness of the granules will depend on the particle size distribution of the technical material. As a consequence, particle size reduction usually leads to longer disintegration times but better suspensibility. The level of dispersant in the formulation must be sufficient to overcome increases in particle surface area on particle size reduction.

Particle shape affects both the adhesive properties and suspensibility of particles. However, particle morphology can not usually be influenced by the formulation chemist and there are no convenient quantitative methods of analysis.

3. Water Solubility

A highly water soluble active ingredient will be easier to disperse. However, the saturation of the granulation solution will effect evaporation and lubrication properties during the manufacturing process. There are many commercial examples of water soluble granules, including glyphosate, N-(phosphonomethyl)glycine, and pirimicarb, 2-dimethylamino-5,6-dimethylpyrimadin-4-yl dimethylcarbamate.

4. Plasticity

This is an important property for extrusion. For granules with high active ingredient concentrations, the correct plastic compression properties of the wetted powder is required to ensure high throughputs and prevent clumping of the wet extrudate.[14,20] The plasticity can be modified by fillers, binders, and dispersants. It can be quantified using a capillary rheometer, but as yet this technique has not been reported to have been applied to water dispersible granules.

5. Hydrolytic Stability

Water dispersible granules are good candidate formulations for active ingredients susceptible to hydrolysis (e.g., sulfhonylureas) where water-based formulations are not feasible. However, it may not be possible to spray dry the active ingredient or even use water as the granulation solution if the hydrolysis is rapid.

6. Crystal Surface Structure And Energies

These parameter play a significant part in determining the adhesive forces between the active ingredient and fillers, and also the cohesive forces between active ingredient particles; these forces will influence the strength and dispersibility of the granules. The surface energies are affected by the adsorption of dispersants and wetters as well as by the process used to synthesize the technical material. The molecular structure of the crystal surfaces can be elucidated by X-ray diffraction and the Miller indices. This information can give unexpected insights into granulation problems.

7. Explosivity

This is an important safety consideration when deciding upon a method of manufacture. It is dependent on particle size and it can be effectively prevented by inertion provided that this can be implemented within cost constraints. Burst panels can also be incorporated into equipment designs.

8. Toxicity

Both the dermal and oral acute toxicity of an active ingredient can often be reduced, compared to solvent based products, by formulating it as a water dispersible granule. The acute toxicity of the active ingredient may also limit the active ingredient content in order to influence the hazard classification of the final formulation.

B. FILLERS

Fillers, such as kaolin, are typically used to maintain a consistent concentration of the active ingredient in water dispersible granules.

Adsorbent fillers, such a diatomaceous earth and silica, are also used to adsorb low melting or liquid active ingredients. However, silicas are hydrophobic and they tend to inhibit disintegration, although calcined precipitated silica has been claimed to give improved performance.[22] Diatomaceous earth is hydrophilic and aids the ingress of water into the granule, but the diatoms are extremely abrasive. Wet or dry milling of diatomaceous earth suspensions or powders prior to granulation can cause unacceptable wear on the equipment. Furthermore, most grades of diatomaceous earth also have relatively large particle size distributions which can lead to poor suspensibility.

Poor disintergration can also be due to redistribution of the active ingredient away from the filler on storage. It has been claimed that this can be improved by premixing the active with only a portion of the filler prior to milling and then adding the remainder prior to granulation.[16] However, even this method gave suspensibility and wet sieve analyses that were poor after storage.

C. WETTING AGENTS AND DISPERSANTS

The selection of the wetting agents and dispersants is a crucial part of the development of water dispersible granules as they usually affect all significant physical properties.

1. Wetting Agents

Initial wetting of granules on dilution is not usually a problem unless their particle size is very small. However, wetting agents aid the wicking of water and hence dispersion. They also improve the extrusion of water dispersible granules by aiding the movement of water. However, wetting agents can increase the amount of foam generated on dilution and may thus necessitate the use of an antifoam. Alkylnaphthalene sulfhonates and alkylsulfhates are examples of commonly used wetting agents.

2. Dispersants

Dispersants often have a binding effect as well as stabilizing the suspension after dilution. They also affect the disintegration and resuspensibility of the granules and, in the case of some lignin sulfhonates, can dramatically change their color. Dispersant selection has traditionally been done empirically based on experience and favoritism. Zeta potentials can now be measured easily. This allows the electrostatic component of the dispersion stabilization forces to be quantified which can aid selection. Lignin sulfhonates and naphthalene formaldehyde condensates are the most commonly used dispersants.

D. BINDERS

If the formulation has poor adhesive properties, the hardness and friability can be greatly improved by the addition of a binder. These are typically water soluble polymers like polyvinylpyrrolidone or some starches. They increase the adhesion between particles by film forming and bridging mechanisms. They may also aid the processing by reducing the water requirement during granulation and by increasing flexibility.[14,6]

Thermoplastic binders can also be used to produce granules under non-aqueous conditions. The binder can be a surfactant or a plasticized water soluble polymer. These have been used to granulate agrochemicals in mixers,[8] fluid bed driers, and by extrusion.[26] Alternatively, the active ingredient can be coprecipitated or comelted with a binder.[4,10,19]

E. DISINTEGRANTS

Disintegrants are required if adhesion forces prevent quick disintegration. Water soluble inorganic salts are often used. They act as a wick on dilution. They then quickly dissolve, leaving spaces for water ingress between the water insoluble particles. However, many are hydroscopic and sublimation on storage can be a problem. Water soluble ingredients can also disturb water movement during extrusion. Super-disintegrants, such as cross-linked polyvinylpyrrolidone, which is commonly used in pharmaceutical tablets, have also been used.[8,25] Even at low concentrations they absorb the water quickly, swell and break the granules apart. Granules can also be made effervescent, which speeds up disintegration, but this restricts manufacture to nonaqueous methods.

F. ADJUVANTS

The most serious drawback of water dispersible granules is that they can lack the biological activity of solvent based formulations because the active ingredient is particulate. Furthermore, water dispersible granules made from dry milled active ingredients may not offer sufficient cost-performance benefits as compared to suspension concentrates because of the larger particle size distribution of the dispersion.

Claims have been made that these problems can be overcome by modifying active ingredients by such processes as comelting or co-precipitation with surfactant or polymers. Comelts with wax or thermoplastic resins can extend the activity of volatile insecticides.[1] It has been claimed that coprecipitates of polyvinylpyrrolidone can control the dispersion of organotin insecticides in water resulting in biological activity equivalent to a liquid formulation.[10]

Whatever the physical form of the active ingredient, it may be desirable to incorporate an adjuvant to boost its biological performance. The adjuvant may either simply be a solid surfactant[4] or more sophisticated adjuvants can be used as granulation fluids.[5] This is presently an intense area of formulation research because it combines two highly desirable goals: achieving maximum biological performance and formulating the active ingredient as a solid.

IV. QUALITY OF WATER DISPERSIBLE GRANULES

Water dispersible granules must be of good quality and, in particular, virtually dust free under commercial use conditions in order to justify the granulation step when a wettable powder is a viable alternative. Early commercial examples of water dispersible granules were not received well by farmers because they were dusty.

A list of important physical properties for assessing the quality of water dispersible granules are displayed in Table 1. CIPAC, the Collaborative International Pesticides Council, has been developing a set of standard methods for assessing the physical properties of water dispersible granules on behalf of the FAO and the relevant method numbers are listed in Table 1*. Regulatory authorities are starting to demand that these tests be carried

*Copies of CIPAC methods are available from Black Bear Press, King's Hedges Road, Cambridge CB4 2PQ, U.K. or the FAO Plant Protection Officer, FAO Plant Protection Services, Via delle Terme die Caracalla, I-00100 Rome, Italy.

Table 1 Methods for Assessing the Physical Properties of Water Dispersible Granules

Method	CIPAC Method No.	Relevant performance characteristic
Suspensibility	MT 168	Spray tank performance
Wet sieve analysis	MT 167	Spray tank performance
Dispersibility	MT 174	Spray tank performance
Particle size of dispersion		Spray tank performance
Resuspensibility		Spray tank performance
Wetting time	MT 53.3	Spray tank performance
Foam	MT 47.2	Spray tank performance
pH	MT 75	Spray tank performance
Bulk density	MT 169	Transportation and packaging
Hardness/Friability		Transportation and packaging
Dustiness	MT 171	Operator handling
Flowability	MT 172	Operator handling
Sieve analysis	MT 170	Quality control

out under GLP protocols, but many are not suitable for formulation development.

The perceived quality of a water dispersible granule formulation is very dependent upon its use rate. This must always be considered when setting specifications. With low-dose active ingredients the specification requirements for granules, on a weight for weight basis, can be less stringent than for active ingredients applied in kilograms per hectare.

The following comments are pertinent about each method:

1. Suspensibility (MT168)

This typically involves analyzing the lower 10% of a dilute suspension after settling for half an hour. Gravimetric and chemical analysis may lead to quite different results. The results are very dependent on the use rate, and the highest concentration should thus be used as a worst case scenario.

2. Wet sieve analysis (MT167)

A 75 micron sieve is usually used to mimic spray nozzle filters. CIPAC stipulates 10 g of formulation, but this should generally be increased to 50 or 100 g during development for most use rates.

3. Dispersibility (MT174)

The CIPAC test for this property is cumbersome. Most companies measure disintegration times by visually observing the granule break-up under specified conditions.

4. Particle size of dispersion

This method is very useful for comparing the particle size distribution of the diluted granules with those of the formulation ingredients prior to granulation. It may also correlate with biological activity.

5. Resuspensibility

It is important that spray tank suspensions do not form dilatant sediments when left for as long as 24 hours without agitation. However, if there are areas in the spray tank which experience only very mild agitation, then redispersion will not occur after the product has sedimented, regardless of how well it resuspends under laboratory conditions.

6. Wetting time (MT53.3)

This is generally only important for products with very low particle size distributions, for example those made by spray drying. The test should use relatively high weights of formulation to mimic addition to an induction tank.

7. Foam (MT47.2)

This property is very dependent on formulation use rate and is best tested in a small or full size spray tank.

8. pH (MT75)

The pH of a 10% dispersion must be measured as part of the regulatory submission but it is also important if the chemical stability of the active ingredient is pH dependent.

9. Bulk density (MT169)

This is typically a tap density method. It is generally used to set packaging specifications.

10. Hardness and friability

There are several different methods used to assess hardness and friability. The methods are based on either the force required to fracture individual granules or dust formation after agitation in the presence of steel or plastic balls. The methods usually show good consistency, but determining a realistic specification is often difficult.

11. Dustiness (MT171)

The CIPAC method requires 30 g of formulation to be dropped into a box through which an air stream at a specified flow rate is passing. The dust levels are measured gravimetrically or by light scattering. This test is very equipment dependent and weights greater than 30 g should be used if the water dispersible granules will be applied at over 1 kg ha^{-1}.

12. Flowability (MT172)

The flowability of water dispersible granules is superior to a wettable powder. Therefore, this test is rarely important for water dispersible granules.

13. Dry Sieve analysis (MT170)

Granule size distribution affects many of the properties of the product. Therefore, it is important to do a sieve analysis in order to make valid comparisons between formulation recipes.

Additional tests will be required for water dispersible granules packaged in water soluble bags.

Even having carried out the above-mentioned tests, a formulation should be tested in a full-size spray rig. This requires the production of a considerable amount of material, but it is only under these conditions that the development specifications can be judged with confidence.

V. CONCLUSIONS

Water dispersible granules are the agrochemical formulations of the future because of the many advantages they have over many of the current commercial formulation types. However, the technical challenges in developing them and the cost of the processing equipment mean that their global growth will be driven by regional economic development and tougher environmental legislation.

The selection of a method to manufacture water dispersible granules is crucial prior to formulation development. In the future, extrusion and fluidized-bed spray drying are likely to be the most popular methods.

The problems of solidifying low melting active ingredients and increasing the biological activity of active ingredients formulated as water dispersible granules with low toxicity, using globally acceptable inert ingredients, are formidable challenges to the formulation chemist.

The methods for assessing the quality of water dispersible granules are still evolving. They are very dependent upon dose rate, and their suitability has to be evaluated on a case by case basis during formulation development.

REFERENCES

1. **Brown, D. J., and Marrs, G. J.**, Insecticidal Compositions, European Patent 369,612, 1989.
2. Catalogue of Pesticide Formulation Types and International Coding System, Technical Monograph No. 2, International Group of National Associations of Manufacturers of Agrochemical Products (GIFAP), Brussels, Belgium, 1984.
3. **Deming, J. M., and Surgant, J. M.**, Water Dispersible Granules of Herbicides, US Patent US 4,936,901, 1990.
4. **Djafar, R. R., and Benke, A. H.**, Method of Preparation and Use of Solid Phytoactive Compositions, European Patent 256,608, 1987.
5. **Fu, E., Narayanan, K.S., Hall, F.R., and Downer, R.A.**, Water Soluble and Water Dispersible Granules with Spreader-Sticker Incorporated, in Pesticide Formulations and Application Systems Vol. 14, ASTM STP 1234, Eds. Berger, P.D., Collins, H.M. and Hall, F.R., American Society for Testing and Materials, Philadelphia, 1994.
6. **Fu, E., Chaudhuri, R. K., and Narayanan, K. S.,** Graft and Co-Polymers of Vinyl Pyrrolidone for Improved Water Dispersible Granules, in Pesticide Formulations and Application Systems Vol. 13, ASTM STP 1183, Berger, P.D., Devisetty, B.N. and Hall, F.R., Eds., American Society for Testing and Materials, Philadelphia, p404 - 412, 1993.
7. **Gandrud, D. E. and Haugen, N. L.,** Dry Application of Dry Flowable Formulations, Pesticide Formulations and Application Systems, ASTM Spec. Tech. Publication, 158, 1985.
8. **Geigle, W., Sandell, L.S., and Wysong, R.D.**, Water Dispersible and Water Soluble Pesticide Granules from Heat-Activated Binders, WIPO Patent WO 91/13546, 1991.
9. **Harrison, P. J., Newton, J. M. and Rowe, R. C.**, Flow Defects in Wet Powder Mass Extrusion, J. Pharm. Pharmacol., 37, 81, 1985.
10. **Hill, A. C., Reid, T. J., and Steer, B. D.**, A Solid Pesticide Formulation, European Patent 413,402, 1990.
11. **Holley, C. A.,** Disc Pelletizing: Theory and Practice, Ferro-Tech, 467 Eureka Road, Wyandotte, MI48192, 1979.
12. Implementation of Dutch Covenant on Leftover Crop Protectants and Used Packaging, Dutch Inst. of Ag. Eng. (IMAG), Oct 12, 1988.
13. **Jones, D. M.**, Factors to consider in Fluid-Bed Processing, Pharmaceutical Technology, April, p50, 1985.
14. **Kuchikata, M., Richardson, R. O., and Sato, T.**, Improved Glyphosate Formulations, European Patent 448,538 1991.

15. **Lin, K. C.**, Development of Solid Pesticide Formulations by Fluidized-Bed Technology, ACS Symposium series, Pesticide Formulations: Innovations and Developments, Chap. 20, 1988.
16. **Lloyd, J. M.**, Water Dispersible Granules of Low Melting Point Pesticides, WIPO Patent WO 93/14632, 1993.
17. **Lloyd, M. J. and Stuart, G. A.**, Preparation of Water Dispersible Granules, WIPO Patent WO 8,900,079, 1989.
18. **Luy, B., Hirschfeld, P., and Leuenberger, H.**, Granulation and Drying in Vacuum Fluid Bed Systems, Drugs Made in Germany, 32, 68, 1989.
19. **Narayanan, K. S.**, Polymers for Instant Dispersions, in Pesticide Formulations and Application Systems Vol. 14, ASTM STP 1234, Berger, P.D., Collins, H.M. and Hall, F.R., Eds., American Society for Testing and Materials, Philadelphia, 1994.
20. **Newton, J. M.**, Extrusion and Extruders, in Encyclopedia of Pharmaceutical Technology, Swarbrick, J., and Boylan, J.C., Eds., Marcel Dekker, New York, 1987.
21. **Mortensen, S.**, Granulation and Agglomeration by Fluidized Bed and Spray Drying Technology, F-270, Niro A/S, Copenhagen, Denmark.
22. **Ogawa, M., Ohtsubo, T., and Tsuda, S.**, Water Dispersible Granules of Low Melting Active Ingredients, UK Patent GB 2,234,678, 1991.
23. **Pritchard, D**, personal communication, 1994.
24. **Rochling, H., Schumacher, H., and Baumgartner, J.**, Water Dispersible Granules from Suspoemulsions, European Patent 541056 (1992).
25. **Sandell, L. S.**, Water Dispersible Granular Agricultural Compositions Made by Heat Extrusion, European Patent, 501,798, 1992.
26. **Schaefer, T. and Woerts, O.**, Control of Fluidised Bed Granulation,
    Part I Arch. Pharm. Chem. Sci. Ed., 5, 51, 1977.
    Part II Arch. Pharm. Chem. Sci. Ed., 5, 178, 1977.
    Part III Arch. Pharm. Chem. Sci. Ed., 6, 1, 1978.
    Part IV Arch. Pharm. Chem. Sci. Ed., 6, 14, 1978.
    Part V Arch. Pharm. Chem. Sci. Ed., 6, 69, 1978.
27. **Schlicht, R., Rochling, H., and Albrecht, K.**, Process for Preparation of Water Dispersible Granules, Canadian Patent 2,012,660, 1990.
28. Water Dispersible Granule Development, Form 332, Ferro-Tech, 467 Eureka Road, Wyandotte, MI48192, 1987.

Chapter 7

Microencapsulation Technology and Future Trends

Alan J. Stern and David Z. Becher

CONTENTS

1-8493-7678-5/96/$0.00+$.50

I. INTRODUCTION

Microencapsulation is the process of creating small particles consisting of a core containing one or more materials surrounded by a barrier layer. This process has become the basis of the carbonless copy paper industry and is important in other areas such as pharmaceuticals and flavorings. Despite a number of successful applications of this technology to pesticide formulations, it has not been used very widely as a formulation method for agricultural chemicals. This review will address the potential for the application of microencapsulation in the agricultural chemical industry.

In this review, the benefits and problems associated with microencapsulated agrochemicals are discussed. The emphasis will be placed on the general types of encapsulation technology that are available and their potential utility. The literature on microencapsulation is vast and has been extensively reviewed.[21,24,28,45] The broad classes of encapsulation technology are discussed, but a detailed review of the microencapsulation patent literature will not be included. The final section discusses the current trends in agrochemical encapsulation and what the future might hold for this technology.

In this review, microencapsulation will be distinguished from both "matrixing" and "macroencapsulation." Although matrixing and macroencapsulation can be used to address some of the same objectives as microencapsulation, they differ greatly in the technologies they employ and the products they produce.

Matrixing is the dispersing of the compound to be formulated in a second (usually solid) inert material having some degree of barrier properties. It differs from microencapsulation in that the active material is in the form of many small particles dispersed more or less evenly throughout a continuous phase which is the barrier material, and in that the active material may be exposed at the surface of the particle (Figure 1). The USDA's starch encapsulation technology is a good example of a matrix.[47,49,50] It should also be noted that micro- or macroencapsulation can include capsules with multiple particles of active material, but they should be located entirely in the interior of the capsule and be surrounded by a continuous layer of the barrier material.

Macroencapsulation differs from microencapsulation primarily in the size of the capsules generated. For the purpose of this review, a process that generates capsules of hundreds or thousands of micrometers is considered macroencapsulation. Wurster[18] coating and orifice drop formation methods[46] are examples of macroencapsulation processes. Materials in this size range can be very useful, but because of the difficulty of suspending such large particles in a spray solution, macrocapsules are not easily formulated for use as agricultural sprays.

Figure 1. Matrixing versus microencapsulation.

II. BENEFITS OF ENCAPSULATION

Microencapsulation can produce formulations with many desirable attributes. These include reduced solvent usage, reduced acute toxicity, modified biological activity, reduced degradation of unstable active ingredients, reduced volatility, reduced leaching, and protection of the active ingredients from other incompatible formulation components.

A. REDUCTION OF SOLVENT

Many of the active ingredients used in agricultural formulations have low melting points, which makes it difficult to formulate them as suspension concentrates or water dispersible granules. They are, therefore, often sold as emulsifiable concentrates. These products may contain large amounts of organic solvent to prevent the formulations from crystallizing during storage. Microencapsulation provides a method of making stable liquid formulations of such materials in which the solvent content is reduced or eliminated. This provides formulations with improved characteristics in the areas of flammability, air and water pollution, and odor, as well as reduced raw material costs.

B. REDUCED ACUTE TOXICITY

Encapsulation can have a substantial impact on the mammalian toxicology of the pesticide being formulated. This can be of great importance when the pesticide has a relatively high inherent toxicity. For example, Penncap

M®* (encapsulated methyl parathion) was found to be 12 times less toxic by dermal absorption than the emulsifiable concentrate formulation, and at least 6 times less toxic orally, depending on the test species.[19,25,29] This reduction is very significant, since the toxicity of methyl parathion makes handling it very dangerous.

C. SLOWER DEGRADATION

An active agent inside a shell wall is not as exposed to the chemical, photochemical, and microbial forces known to degrade pesticides. For example, Penncap M® was found to be more resistant to photodegradation than the emulsifiable concentrate. In tests on glass plates the half life of the encapsulated product was increased by about three or four times. Control of *Heliothis zea* was similarly extended.[29]

D. IMPROVED CONSERVATION TILLAGE EFFICACY

Encapsulation can also be used to modify the biological activity of a pesticide. For example, it has been reported that Micro-Tech®** (encapsulated alachlor) is more effective under no-till and conservation tillage systems than Lasso EC®**, the comparable emulsifiable concentrate.[38,48] It has been proposed that this is due to the greater ability of the encapsulated product to penetrate the trash layer without being adsorbed or lost to volatility. However, this explanation is not universally accepted.[32]

E. REDUCTION OF VOLATILE LOSS

Encapsulation can also reduce the rate of evaporation of a volatile pesticide. This can lead to activity modifications such as reduced off site injury caused by volatility and in some cases extended residual control because less pesticide is lost to the atmosphere. In one example, the vapor concentration of diazinon in a simulated carpeted room was compared for Knox Out 2 FM* and a standard diazinon 4 EC. The vapor concentration from the encapsulated product was both initially lower and varied much less over time.[29]

F. REDUCTION OF LEACHING

The use of various types of entrapment to reduce the leaching of pesticides is well established in the literature. This can both increase the activity of the pesticide by maintaining the concentration in the area where it is useful and improve the environmental characteristics of the product.[29] Encapsulated

---

* Penncap M is a registered trademark of Elf Atochem North America, Inc., Philadelphia, PA, 19102.

** Micro-Tech and Lasso EC are registered trademarks of Monsanto Agricultural Group, St. Louis, MO 63167.

* Knox-Out 2FM is a registered trademark of Elf Atochem North America, Philadelphia, PA 19102.

alachlor was found to leach less than the emulsifiable concentrate and, therefore, presumably maintains a greater amount of active material in the area where its biological effect is desired.[15]

G. FORMULATION ODOR

In some cases, encapsulation can mitigate formulation odor. If the odor in question is that of the active ingredient, entrapping it within the shell wall may reduce it. If the odor is due to the organic solvents used in the formulation, replacing them with water by microencapsulation will eliminate the problem.

H. STABILITY

Encapsulation has the potential to protect an unstable active ingredient from the other components of a formulation. This may allow the preparation of formulations containing ingredients which would otherwise be incompatible. In the case of biological pesticides such as pathogenic microorganisms, encapsulation and matrixing have proven very successful at maintaining the viability of such organisms prior to use.[8,39]

I. PHYTOTOXICITY

Finally, some pesticides can have their phytotoxicity to desirable species reduced by encapsulation. Penncap M®, for example, is less phytotoxic to plants on which it was used than the emulsifiable concentrate.[9,13]

III. SHELL WALL PROPERTIES

It is obvious that the performance of a microencapsulated formulation will depend on the physical characteristics of the encapsulating polymer. Unfortunately, the small size of these particles makes these properties very difficult to measure directly. Some of the critical properties can, however, be estimated. The critical properties of a shell wall include its thickness, its flexibility, its stability to chemical and microbial breakdown, its resistance to diffusion of the active ingredient, and the total surface area over which diffusion can occur.

Of these, the easiest to estimate are the shell wall thickness and total surface area. The surface area of a typical microencapsulated product is large and the polymer layer which separates the active from the surrounding medium is consequently normally quite thin. The total surface area, in square meters per liter of solution, of the particles in a suspension of spheres of fixed diameter is easily calculated to be

$$\frac{6000*(F_M)}{D}$$

where F_M is the volume fraction of the suspension that is microcapsules, and D is the microcapsule diameter in micrometers.

For example, a theoretical formulation with a loading of 50 volume per cent microcapsules containing particles with a diameter of 5 µm would have a total surface area of the particles of 600 m^2/l. For a real suspension, of course, the particle size is a distribution and the calculations are more complicated. However, the magnitudes of the values that would be obtained are about the same.

An estimate of the thickness of the polymer layer requires some additional assumptions. As a limiting case, since it gives the thickest value, it can be assumed that the polymer layer is homogeneous and of constant thickness and that all of the polymer is a simple shell surrounding the core. In this case, the thickness of the shell (R_S) is given by:

$$R_S = R_T*(1-(\%Core)^{1/3})$$

where R_T is the particle radius and %Core is the percentage of the particle that is not shell wall material. For example, for a particle with 10% shell wall material the calculated shell wall thickness is 3.45% of the particle radius or 0.086 µm for a particle with a 5 µm diameter. This is clearly extremely thin and highlights one of the critical areas for effective microencapsulation of pesticides.

In order to get a 0.5 µm thick shell, the percentage of core material must either be reduced to about 50% or the particle diameter increased to about 60 µm. Since both of these alternatives are undesirable in most applications, it will usually be necessary to obtain the desired results from extremely thin polymer membranes. This means that the choice of an appropriate shell wall and method of manufacture can be extremely critical.

The other properties of the shell wall must usually be estimated from the properties of bulk materials of analogous composition. The difficulty of doing this varies considerably with the nature of the material used. If the shell wall consists of a highly cross-linked polyurea, for example, it may be almost impossible to duplicate the polymer in a bulk polymerization. On the other hand, the physical properties of a polymer which is deposited by evaporating a solvent may be essentially unchanged from the bulk and easily determined.

While there is as yet no good way to predict the effect of varying shell wall composition on formulation performance, there are a number of properties that are obviously important. These include the degree of cross-linking of the polymer used, its chemical composition, and the standard polymer bulk properties such as flexibility and strength. If a bulk sample of the polymer used can be obtained, it may be possible to correlate its properties with the formulation performance. Otherwise, the formulator must do the best he can with estimates based on the known chemical properties of the starting materials and measurements of the extent of reaction.

IV. METHODS OF ENCAPSULATION

Many different encapsulation processes have been used in various industries and even more have been proposed.[21,24,45] Both chemical and physical methods of creating a protective barrier have been used. Improved methods for specific applications are constantly being developed. All the processes have in common, however, the production of an inert barrier, or shell wall, around a core containing the compound(s) of interest.

Some processes such as coacervation, *in situ* polymerization, and interfacial polymerization have been widely applied in fields other than pesticides, such as carbonless copy paper, food, drugs, and cosmetics. Many others have been technical but not commercial successes. However, microencapsulation has not been as widely used for pesticide formulation as the potential advantages of the method, described above, would suggest. The major encapsulation methods applicable to pesticides and their strengths and weaknesses are discussed below.

A. INTERFACIAL POLYMERIZATION

Interfacial polymerization[36] is a widely used technique for making condensation polymers. It is distinguished from other methods of polymer synthesis by the fact that the reaction takes place at the interface between two non-miscible phases rather than in the bulk of a single phase.[37]

The basic procedure for encapsulating a material using the most common form of this process is quite simple. The first monomer is dissolved in the core material. The resulting solution is then dispersed in the continuous phase (usually water), which normally contains one or more dispersing agents. The second monomer is then added to the resulting emulsion. The shell wall forming reaction occurs at the oil/water interface of the emulsion droplets. The resulting suspension of microcapsules can then be further formulated to produce the final product.

Interfacial polymerization has a number of advantages over bulk reactions methods. It generally produces high molecular weight polymers. It can be carried out at or near ambient temperatures, since large viscosity increases are not encountered during the reactions. It is less critically sensitive to the exact stoichiometry of the reactants, than bulk polymerizations.

There are, however, limitations on both the materials that can be encapsulated and the monomers that can be used. The greatest limitation is that the material to be encapsulated must exist as a liquid dispersed in the continuous phase, which is usually water. It must either be a liquid that is essentially insoluble in the continuous phase or a suspension in a suitable solvent under the encapsulation conditions. In addition, there must be no significant reaction between the monomers and the material to be encapsulated. The most important limitation on the shell wall materials is that one monomer must be soluble in each phase. In addition, the reaction of the monomers with each other must be significantly faster than any side reactions with the solvents or other ingredients.

When interfacial polymerization is performed at the interface between two bulk phases, the area of the interface is small relative to the amount of the monomer reactants. Under these conditions, the polymerization will normally consume only a small percentage of the monomers, because once a polymer skin forms at the oil/water interface, further reaction becomes very slow. When interfacial polymerization is used for microencapsulation, however, the large interfacial area relative to the amount of the monomers allows a higher degree of conversion of monomer into polymer. It should not, however, be assumed that the addition of stoichiometric amounts of the reactants insures that the reactions between the monomers will be complete.

For pesticides, an oil-in-water emulsion is the most usual form for the two-phase system, with the reaction taking place at the surface of the oil drops. This produces a suspension of pesticide particles in water, which can then be converted into a water based formulation. However, other two-phase systems such as water-in-oil emulsions have also been used.[5]

The most commonly used oil-soluble monomers are polyfunctional isocyanates and acid chlorides. Polyfunctional amines are preferred as the water-soluble monomers because they dissolve readily in water and react with the oil-soluble monomers more rapidly than the water. These monomer combinations produce a polyurea if an isocyanate is used or a polyamide if an acid chloride is chosen.

By selecting suitable other monomers, it is possible to form shell walls of other classes of polymers, such as poly(sulfonamide), polyurethane, and even polyepoxide. Monomers can also be mixed to generate mixed polymers. Well-known examples of this are Elf Atochem's insecticide products, which are encapsulated in a mixed polyamide/polyurea shell wall. Even within a single chemical type of shell wall it is possible to make polymers that vary greatly in their physical properties by the appropriate selection of monomers. Polymer shell walls can be produced which will give a variety of release characteristics for a given core material.[25]

The result of this process is a suspension of microcapsules in water. If the water is removed, for example by spray drying the suspension, a dry formulation can be obtained. An example of such a product is Monsanto's Partner®* herbicide.[10]

---

* Partner is a registered trademark of Monsanto Agricultural Group, St. Louis, MO 63167.

Interfacial polymerization has been the most commercially successful method of encapsulation for pesticides. This is primarily because it has a number of advantages over the other traditional encapsulation methods. It is relatively inexpensive, the reactions are rapid, and it is possible to produce material on a large scale. It is also possible to produce more concentrated pesticide formulations than with many other encapsulation methods and the wide range of starting materials available make it possible to optimize the shell wall properties for a particular active ingredient. Its primary disadvantages, which it shares with most other methods of encapsulation, are higher manufacturing costs, when compared to nonencapsulated formulation types and in some cases reduction of postemergent or knockdown activity.[3,4,6,35]

B. COMPLEX COACERVATION

In complex coacervation a shell wall is formed around a water-immiscible active when an anionic, water soluble polymer is reacted with a cationic material (which may or may not be a second polymer).[1,22] The resulting coacervate is insoluble in water and tends to separate from the solution as a second phase. If a dispersed phase is present in the solution the coacervate will tend to coat the suspended particles creating a protective shell.

The best known example of coacervation is the reaction of gelatin with gum arabic. The basic method for producing a formulation using this system is as follows. A suspension or emulsion of the core material is prepared in a warm, dilute gelatin solution. To this is added a dilute gum arabic solution and the pH lowered to less than 4.5, causing a coacervate to form around the dispersed core material, which acts as nuclei. The system is then cooled, causing the coacervate, which is a 1:1 mixture of gelatin and gum arabic, to gel. For the shell walls to form properly, this material must completely coat the particles of core material. The final step is the hardening (cross-linking) of the soft shell walls by adding an aldehyde, typically formaldehyde or glutaraldehyde, raising the pH to 9, and warming the mixture. Following preparation, capsules made by this process may be isolated and dried by several methods if desired.

This process has received much attention in the microencapsulation literature and has been tested in a wide range of applications, including agricultural chemicals. Capsules made by complex coacervation have been commercial successes in several other industries, but due to the complexity and cost of the process, coacervation has not been found to be of value for agricultural formulations. The primary advantage of coacervation, which is the source of the interest, is that any core material which can be dispersed in a liquid phase can potentially be coated. This is a significant advantage over interfacial polymerization, which is limited to liquid or liquefiable core materials.

This method has, however, several significant disadvantages that have discouraged its application in agriculture. The most significant is that coacervation is normally carried out at low concentrations. Unless the process is done in relatively dilute solutions, it is common to obtain aggregates rather than individually encapsulated particles. The need to use an aldehyde cross-linker with the associated toxicological concerns is another drawback of this method. Another limitation of the commonly used technology is that most coacervation research has been concentrated on a small number of naturally occurring polymers, such as gelatin, gum arabic, alginates etc., which has limited the range of shell wall characteristics available. The technique should, however, be applicable to many synthetic polymers having the appropriate ionic characteristics. Finally, the technical difficulties in developing a process that efficiently coats the core material with coacervate on an industrial scale are considerable.

C. *IN SITU* POLYMERIZATION

In-situ encapsulation is a term that is used to describe two very different encapsulation processes. Although in both processes a water-insoluble pesticide is dispersed in an aqueous phase, the processes differ in both the type of shell wall produced and the phase in which the shell wall polymerization occurs. In one method, the shell wall forms from the polymerization of a polyamine and an aldehyde in the aqueous phase. This is commonly known as the melamine/formaldehyde or urea/formaldehyde encapsulation, after the ingredients normally used. In the other class of *in situ* encapsulation, the shell wall polymer forms from *within* the dispersed pesticide-containing oil phase, usually by the hydrolysis of an isocyanate.

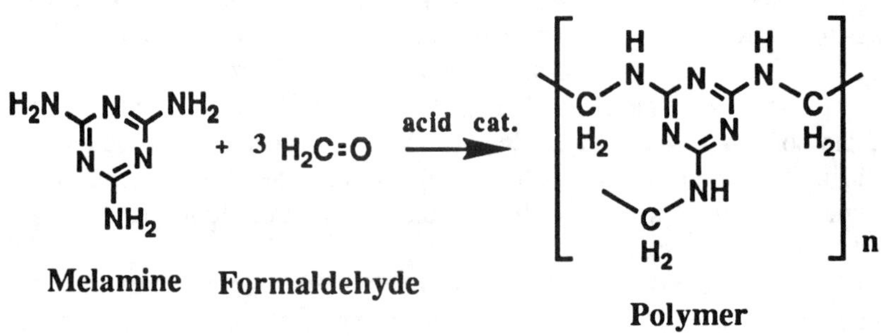

Figure 2. Melamine/formaldehyde polymerization.

1. Polyamine Formaldehyde Shell Walls

In this process, a low molecular weight melamine/formaldehyde or urea/formaldehyde prepolymer is first dissolved in water. The water insoluble material to be encapsulated is then emulsified (or dispersed) into this solution. The pH is then lowered to around 3.5, and the mixture is heated to 50°C for several hours causing the prepolymers to further polymerize into an insoluble shell wall around the active material (see Figure 2). The use of the prepolymers avoids the need to add free aldehyde to the reaction. The main disadvantages of this method are the long reaction time needed for shell wall formation and the limited number of polymers that can be made by this process. The material to be encapsulated must be stable at low pH and not must not contain any functional groups that will react with the amine or aldehyde.[17,23,44]

2. Isocyanate Hydrolysis

In the other process commonly referred to as *in situ* polymerization, an oil-soluble monomer or prepolymer is converted to a polymer shell by reaction with the water from the aqueous phase. In the commonest version of this process, as in interfacial polymerization, the shell wall is a polyurea polymer. Instead of reacting the isocyanate with an aqueous amine, some of the polyisocyanate starting material in the oil phase is hydrolyzed by the surrounding water to an amine. This amine then reacts with the remaining isocyanate groups. This produces a polyurea somewhat similar to those produced by the reaction of polyisocyanate and polyamines by interfacial polymerization. The hydrolysis reaction is slower than the reaction between the isocyanate and amine, so the formation of excessive amounts of amine is not normally a problem. Figure 3 shows an example of the reactions involved.[40]

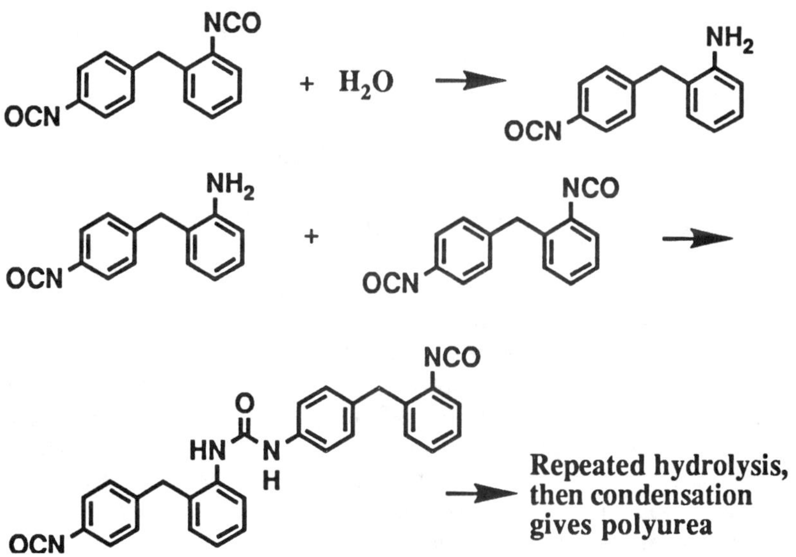

Figure 3. *In situ* polymerization (isocyanate hydrolysis).

The major advantage of this process of *in situ* polymerization is that the amine does not have to be introduced in the aqueous phase. This avoids potential problems with variable amine concentrations during the reaction. There are, however, a number of problems that can be encountered with this method. They include the generation of large amounts of carbon dioxide from the amine hydrolysis reaction. This can lead to processing problems and may also cause the shell walls to be porous and have poor integrity. Since the hydrolysis reaction is slower than the reaction between amine and isocyanate, the time required to create the shell wall may be longer then that for a similar interfacial polymerization. Finally, the range of polymers that can be prepared is more limited than those that can be made by interfacial polymerization, since only the isocyanate monomer can be varied. This can, however, be at least partly overcome by the use of mixtures of isocyanate monomers.

D. SOLVENT EVAPORATION

This process is one that is often proposed for pharmaceuticals, but has not yet seen significant applications in agriculture. Unlike the preceding processes in which the shell wall polymer is formed by reaction of low molecular weight monomers, in this process the shell wall is made from a preexisting polymer which is deposited around the active ingredient. To make the encapsulated product, the active ingredient to be encapsulated and the polymer from which the shell wall is to be formed are dissolved in a volatile water-insoluble solvent. The polymer and the active ingredient must both be soluble in the solvent, but not in each other. The solution of active ingredient and polymer is emulsified in water containing a suitable surfactant. The solvent is then removed from the stirred mixture by heat, or reduced pressure, or both. As the solvent evaporates, the polymer separates from the solution, forming a continuous layer at the surface of the emulsion droplets.

The primary advantage of this process is that a wide range of preexisting polymers with well-defined physical properties can be used to form the shell wall. The appropriate solubilities of the polymer in the volatile solvent, core material, and continuous phase are the only limitations on which polymers can be used. The cost and hazard involved in evaporating and recycling the highly volatile solvents (often chlorinated) that must be used in this process are its primary drawback. The requirement that the polymer be soluble in the volatile solvent also limits this approach to polymers with little or no cross-linking. It is possible, however, to perform a hardening reaction, similar to that used for coacervated systems, after stripping the solvent, if desired. Because of the solvent volumes and costs involved, the process has been more appropriate to the lower volumes and higher values of pharmaceutical products, than those of typical pesticides.

E. SPRAY ENCAPSULATION

This is one of the oldest encapsulation techniques and has many variations. Several related methods that produce capsules using air rather than a liquid as the continuous phase are combined in this general class. These methods are particularly useful if the desired product is a dry formulation rather than an aqueous suspension, since removing the aqueous phase is unnecessary. In its simplest form, this technique consists of spray drying an emulsion of a core material dispersed in an aqueous solution of the shell wall material. When the water is removed, the dissolved polymer remains behind, coating the core material. With the simpler versions of this process, the particles that are produced are usually matrices which contain several core particles rather than true microcapsules. The capsules produced by these methods also tend to be larger than those made by other methods discussed here, since it is difficult to produce very fine drops with ordinary spray equipment. The average particle size of the resulting powder is usually in the range of about 10 to 150 μm, and more often at the upper end of the range. It can be difficult, therefore, to produce sprayable products having handling properties comparable to conventional formulations by these methods.

Many improvements have, however, been made on this basic concept. Through nozzle engineering it is possible to generate true microcapsules containing a single coated core. Southwest Research Institute (SWRI), for example, has developed a technology using a specially designed concentric orifice device that works on a centrifugal extrusion principle. The material to be encapsulated moves through the inner orifice, while the shell wall or coating material moves simultaneously through the outer orifice. The device is rotated, causing a stream of material to be ejected from the nozzles by centrifugal force. As the stream leaves the device, it spontaneously breaks into droplets composed of the core material surrounded by the coating material. While the capsules are in flight, the shell wall is formed by evaporation of solvent from the particle. The capsules fall into collectors below the spinning orifice generator. They can then be posttreated chemically or physically to harden the shell walls, if necessary. The capsules made by this technology are usually large (125 to 3000 μm) and are consequently difficult to suspend and spray. This technology has, however, apparently been successfully used in at least one commercial product.[11,27]

Other "drop formation" encapsulation techniques based on spinning disk technology are promoted by SWRI and Washington University Technology Associates (WUTA). In these methods, the core material (dispersed phase) and wall material (continuous phase) are fed onto the center of a disk spinning at a high rate. Small droplets of core plus coating fly off the perimeter of the disk, forming capsules. Shell wall formation can be completed by drying, cooling, or curing the droplets using methods appropriate to the shell wall material. These methods have the advantage of encapsulating solids as well as liquids and can achieve high loadings, but are somewhat limited in terms of producing small, easily suspended capsules for agricultural spraying.

These spray encapsulation methods are potentially useful, but have so far seen only limited application to agricultural products. They have the advantage of producing dry capsules that can be incorporated into nonaqueous systems and that a wide range of active ingredients and coating materials are compatible with the technology. The primary shortcomings from the point of view of applications to agriculture are the large particle size and the relatively low manufacturing capacity of most equipment. To produce reproducible and homogeneous drops it has been necessary to use spray equipment that produces a relatively small number of relatively large drops. Improvements in this technology could greatly increase its potential utility.

Table 1. Commercialized Microencapsulated Pesticides*

Active Ingredient	Trade Name	Shell Wall Polymer	Company
Herbicides			
Alachlor	Micro-Tech	Polyurea	Monsanto
Alachlor	Bullet	Polyurea	Monsanto
Alachlor	Partner	Polyurea	Monsanto
EPTC	Capsolane	Polyurea	ICI
Insecticides			
Carbaryl	SLAM	Gelatin	BASF
Chlorpyrifos	Duraguard		Whitmire
Chlorpyrifos	Empire 20		Dow Elanco
Chlorpyrifos	Kayatack MC	Polyurea	Nippon Kayaku
Diazinon	Knox-Out 2FM	Polyamide/polyurea	Elf Atochem
Fenitrothion	Lumbert MC		Sumitomo
Fenitrothion	Kareit MC	Polyurethane	Sumitomo
Fonofos	Dyfonate MS	Polyurea	ICI
Parathion	Penncap-E	Polyamide/polyurea	Elf Atochem
Parathion-methyl	Penncap-M	Polyamide/polyurea	Elf Atochem
Parathion-methyl	Fulkil		Rhone-Poulenc
Permethrin	Penncapthrin 200	Polyamide/polyurea	Elf Atochem
Pirimiphos methyl	Actellic M20	Polyurea	ICI
Pyrethrin	Micro-Sect	Polyurea	3M
Pyrethrin	Sectrol	Polyurea	3M
PGR			
Chlormequat	Cap-Cyc	Urea/formaldehyde	3M
Rodenticide			
Warfarin	Tox-Hyd		Haaco Inc.

V. CURRENT COMMERCIAL PRODUCTS

Microencapsulated pesticides have not achieved the importance that the numerous advantages attributed to them would suggest. Table 1, adapted from information from Southwest Research Institute with the addition of material from other sources, lists some commercial products that are believed

* Capsolane, Dyfonate MS, and Actellic M20 are registered trademarks of Zeneca, London, England; SLAM is a registered trademark of Micro Flo Co., Lakeland, FL 33807; Duraguard is a registered trademark of Whitmire Research Laboratories, St. Louis, MO 63122; Empire 20 is a registred trademark of DowElanco, Indianopolis, IN 46268; Kayatack MC is a registered trademark of Nippon Kayaku Co., Ltd. Tokyo, Japan; Lumbert MC and Kareit MC are registered trademarks of Sumitomo Corp., Tokyo, Japan; Penncap-E and Penncapthrin 200 are registered trademarks of Elf Atochem Philadelphia, PA 19102; Fulkil is a registered trademark of Rhone-Poulenc Agrochemie, Lyon, France; Micro-Sect, Sectrol, and Cap-Cyc are registered trademarks of 3M, St. Paul, MN; Tox-Hyd is a registered trademark of Haaco, Inc., Madison, WI 53707.

to be encapsulated. Although all of these products are pesticides, it appears that some of them are not products intended primarily for the agricultural market, but rather home and garden or commercial lawn care products. Since there is in general no requirement to disclose formulation information, there probably are other encapsulated products on the market.

Little information is publicly available on the market of most of these products. Encapsulated alachlor (Micro-Tech®, Bullet®*, and Partner®) has probably been by far the largest-volume encapsulated pesticide, accounting for about 20 million kg/year of encapsulated alachlor at its peak. Most of the other products are intended for special applications and have relatively low sales volumes.

VI. FUTURE TRENDS IN MICROENCAPSULATION

The future of microencapsulation will be driven by two kinds of factors, changes in the industry and improvements in the technology. Both are critical to the increased use of microencapsulation in agricultural products. The changes in the industry have the potential to make the existing microencapsulation technologies more attractive in comparison with more traditional formulation types. Concurrently, improvements in the technology present the opportunity to overcome the shortcoming of the existing technologies and increase their usefulness. The use of microencapsulation will also feel the effects of industry trends that will impact all new technologies. The competitive drive to do everything better and faster will both help and hinder new technology. The ability of new technology to deliver improved performance will be desired, but this will compete against greater speed to market and lower risk of failure of well established technologies. Which trend will prove the stronger only time will tell.

A. INDUSTRIAL AND REGULATORY TRENDS

1. Regulatory, Environmental, and Safety

Current environmental and safety policies favor water based products over solvent-containing ones, because they release less volatile material to the environment and present a reduced fire hazard. This gives microencapsulated products a potential regulatory advantage over some more traditional types of formulations.

The U.S. Environmental Protection Agency (EPA) or other regulatory agencies are likely to require extensive testing before they will approve a new microencapsulated version of an existing unencapsulated product. Since the formulation may have different release characteristics, the residue and dissipa-

---

* Bullet is a registered trademark of the Monsanto Agricultural Group, St. Louis, Mo.

tion studies done for the old formulations may not be adequate. Despite the new formulation's many environmental benefits and reduced user exposure, it will probably be necessary to provide data to show that the crop and soil residues remain within acceptable levels. Also, it may be necessary to demonstrate that the capsules do not adversely affect local bee populations. This issue first arose with Penncap M®, one of the earliest encapsulated pesticides. After the product was introduced it was discovered that the capsules could be inadvertently picked up by bees along with pollen they were gathering and carried back to the hive, destroying large numbers of bees.

On the positive side, the US EPA has recently granted tolerance exemption to "crosslinked polyurea-type encapsulating polymer" applied to growing crops.[14] This exemption was requested by ICI Americas Inc. who manufacture several microencapsulated pesticides. This removes one of the major impediments to the development of microencapsulated products, since it removes the necessity of getting separate approvals for every new shell wall. This exemption was based on the EPA "polymer exemption guidelines". They concluded that because of their molecular weight the crosslinked polyureas are nontoxic and will not be absorbed or metabolized by the crop to which they are applied. Similar reasoning can be applied to many other possible shell wall polymers which, therefore, should be approved quickly when requested. Overall the cost and effort of introducing a microencapsulated product should be no greater than that for introducing any other new type of formulation.

2. Cost Effectiveness

Historically, microencapsulation has been perceived as more expensive than other formulation types. This is generally not due to the cost of the shell wall material, but rather to manufacturing costs: capital, engineering, and processing. This has made it very difficult to justify microencapsulating low cost, lower margin pesticides. Only the existence of a significant product enhancement could justify the use of this formulation type. However, Monsanto has been able to sell Micro-Tech® and Bullet® herbicides at prices competitive with their "conventional" solvent-containing counterparts. This demonstrates that under the right conditions, microencapsulation can be a cost competitive formulation method.

However, the products most suited to encapsulation from a cost perspective are new, proprietary lower-use-rate actives. Because profit margins are typically higher on proprietary compounds, and because low-use-rate compounds can more easily adsorb higher formulation costs, the potential for the use of encapsulation is greatly increased. As rates are reduced, the cost of encapsulating a given amount of active is spread over more and more acres. With the higher cost and value of these actives it is also more critical to obtain the maximum value from the formulations. This should drive a tendency to use more complicated and expensive types of formulation technology.

3. Adoption of No-Till Practices

The use of conservation tillage methods is increasing in the United States and around the world, especially where soil erosion is a concern. However, some pesticides seem to be less effective than expected when used in conservation tillage practices. It has been proposed that the crop residues left on the soil surface after the previous harvest hinder soil active products from reaching the soil surface. Microencapsulation has been shown to improve the performance of alachlor under conservation tillage practices. If this is also true for other pesticides, the industry may see a trend toward increasing use of microencapsulated herbicides in the conservation tillage market. Because soil erosion is a concern worldwide, conservation tillage practices should be widely adopted, and this should increase the use of microencapsulation.

4. Use as Enabling Technology

One frequently overlooked advantage to microencapsulation is that it is an enabling technology for producing formulations that would otherwise be impossible. For example, low melting or liquid actives such as alachlor and acetochlor may be formulated as a suspension concentrate or a water dispersible granule. Microencapsulation can also permit the formulation of package mixes that would not be possible by other methods. For example, mixtures of actives that are chemically incompatible can be made by using microencapsulation to keep them separate in the formulation.

B. NEW TECHNOLOGY

The other potential driver for microencapsulation is the introduction of new technology that would make a step change in its viability as a formulation method. The literature is full of new patents and discoveries in the area of encapsulation, but from a commercial point of view, the number of products being introduced remains small.

The rapid changes occurring in the industry and the improvements being made in the technology, however, suggest that the trend noted above is not likely to continue. While most of the new patents are minor improvements or applications of old technologies to new actives, some new and interesting improvements are being discovered.

A complete listing of all of the new patents in microencapsulation does not appear useful. General microencapsulation patents routinely claim the use of pesticides for technologies, which, at least currently, have little potential for use. Many other patents contain narrow claims for specific uses or minor improvements. While it is impossible to predict which of the new ideas will lead to commercial results, a discussion of the major areas of improvement and some interesting work being done on them are presented below.

1. Interfacial Polymerization

Interfacial polymerization is the most important encapsulation technology for pesticides. It is also, however, an area in which continual improvements are being made. This technology was originally developed in the carbonless copy paper industry and this field is still an important source of new ideas. Great possibilities for rapid improvements have been opened by the EPA's policy of rapidly granting exemptions from tolerance to wide classes of polymers. This has the potential to greatly increase the range of shell walls that are commonly used. For example, Jabs et al. disclose microcapsules with improved shell walls made from aromatic di-isocyanates functionalized with an eight to twenty carbon alkyl or alkoxyl group.[20] These monomers have much more solubility in aliphatic solvents than the purely aromatic isocyanates commonly used. The poor solubility of the conventional monomers in some actives has been a significant problem.

Another example of the potential to widely increase the range of shell walls used is disclosed by Meinard and Taranta.[30] The monomers used in this work were phthalic acid diesters (the oil-soluble component) and amino plastic resins such as polymethylated formaldehyde melamine resin (the aqueous soluble component). The reaction is acid catalyzed, and takes about 2.5 hrs at 65 °C. In current form, this method suffers from a low active concentration in the capsule suspension which makes it desirable to recover the capsules by filtration and drying.

Another idea with potential to give new modes of controlled release is the inclusion of labile groups in the shell wall. Dauth et al. for example, claim capsules containing azo groups in the shell wall.[12] It is proposed that this will make the release dependent on the presence of light or heat. The primary initial use for this technology is in thermal copying, but possible applications to controlled release in other fields may exist.

The other major potential area for improvement in this technology is the physical properties of the formulations. Beestman, for example, claims that the use of co- or terpolymers of vinyl pyrrolidone as dispersants in interfacial polymerization gives formulations with improved properties.[7] It is claimed that the resulting formulations do not form hard-packed deposits when the diluted formulations are allowed to settle out in the spray tank.

2. *In situ* Polymerization

This is probably the second most used technology for pesticides. Many of the comments made above for interfacial polymerization also apply to this technology. There have been a number of interesting advances in this area. For example, Scher and Rodson have disclosed a novel twist on the standard urea-formaldehyde *in situ* polymerization method of encapsulation.[41-43] In this case, the urea-formaldehyde prepolymer is 50 to 98% etherified with a C-4 to C-10 alcohol, making it soluble in the oil phase. Because of this, the polymerization takes place inside the dispersed oil droplet rather than in the continuous aqueous phase. This has several advantages including the ability to produce a higher concentration of capsules in the formulation.

Zsifkovits et al. describe a method in which shell walls or micromatrices are formed by the *in situ* polymerization of droplets containing oil-soluble silicon containing monomers or prepolymers.[51] Capsule formation is slow, taking from one to four days, depending on the choice of catalyst and reaction conditions. A variety of organic side chains on the silicon atom are disclosed. The method works for a range of pesticides. It is suspected that for a given active ingredient, release rates could be greatly affected by the choice of the alkyl group on silicon.

3. Spray Encapsulation

The biggest issues for this technology have always been the cost and availability of equipment and ability to make sufficiently small particles. The increased use of spray drying to make WDG formulations may address the first question and improvements in spraying technology the second. Whether this technology has a future as a significant method of producing pesticide formulations will depend on these factors.

Misselbrook et al. have disclosed a method of encapsulating herbicides like trifluralin through a spray-drying process.[33,34] In the case of trifluralin, the patent claims that a specific, low melting, more biologically active polymorphic crystalline form of trifluralin can be formed as the molten trifluralin cools during the spray-drying process, and that this polymorphic form is preserved in the microcapsules. Each capsule comprises a multitude of trifluralin particles, separated by a water soluble film-forming polymer such as polyvinyl alcohol. It was claimed that this would allow the production of a stable dry trifluralin WDG.

4 Other Technologies

The other technologies discussed above, complex coacervation and solvent evaporation, appear less likely to be immediately useful for typical pesticides. Low-use rate compounds may, however, cause increased interest in these methods. In addition to the methods that have been investigated for years, new types of encapsulation technologies are constantly being proposed for pesticides. New types of pesticides such as biological products or living biological systems will require new approaches.

The encapsulation of living organisms is an area of great current interest. Controlled-release formulations of microbial pesticides have been reviewed by Connick.[9] A related area is the encapsulation of pesticides (usually biologically derived agents) in cells. Mycogen Corporation has described a technology for producing biotoxins such as B.t. encapsulated in killed cells.[16]

Of course, microbes can be encapsulated in other matrixes as well. Baker et al. describe a method for encapsulating biological materials such as microbes in nonionic polymer beads.[2] The process requires suspending the agent to be encapsulated in water containing a dissolved polymer, then spraying the mixture into very cold water immiscible solvent to freeze the water. The ice capsules are then recovered and dried to remove all unbound water. Encapsulation of *Bacillus thuringiensis, Pseudomonas* sp, and *Altrernaria cassiae* is claimed.

The future of these technologies is dependent on the long term future of biopesticides, which are facing potential competition from resistant plants developed by biotechnology. If living organisms and biologically derived materials obtain a significant part of the pesticide market, their encapsulation could become an important area of research, since they can be difficult to formulate in more conventional formulation types.

Landec Corporation is promoting a technology for encapsulating pesticides in polymers that undergo a change in crystallinity at a particular temperature.[31] This technology has considerable potential against pests such as some soil insects whose activity is temperature dependent. The potential exists to develop formulations that release their active ingredient only when the target organisms are active. The primary difficulty that must be overcome is the development of formulations that have satisfactory storage stability at temperatures above the release temperature.

The production of very small capsules (less than 1 μm) is an area of potential interest. There are number of possible ways of doing this. Various technologies exist for making very fine emulsions, for example, that could be adapted to the existing emulsion based encapsulation technologies such as interfacial polymerization. Other methods have also been proposed. Devissaguet et al., for example, produced nanocapsules (capsules less than 500 nm in this case), by a method they call the "in liquid drying technique".[13] The shell wall material in this method may be a synthetic or natural polymer, such as polylactic acid. One advantage claimed for this invention, which uses a kind of spontaneous emulsification, is that very little energy is needed to form the capsules.

The major question about very small capsules is whether a sufficiently strong shell can be formed without using a large amount of wall material. Even for 10 μm particles, the shell walls are typically very thin. Reducing the particle size by a factor of ten or more is likely to produce capsules with shell walls with a thickness of less than 10 nm. On the other hand, most of the suspension and resuspension problems that can be encountered with microencapsulation would be eliminated with smaller capsules. This is an area that appears worth further investigation.

VII. CONCLUSION

Microencapsulation is a technology whose potential in agriculture is still mostly unrealized. The current trends to lower use rates, conservation tillage, and solvent reduction are likely to encourage its use in the future. However, even the most advanced of these technologies still need a great deal of improvement to achieve their full potential.

New technologies and improvements in the old ones are constantly being introduced into the literature, but the actual use of this technology remains small. Up to this time, microencapsulation has not been able to compete successfully with the older more established formulation approaches. Another factor for the future is the competition from other new technologies, such as matrixing and the linking of pesticides to polymers. Overall the future of microencapsulation as an important pesticide formulation method is still undecided.

REFERENCES

1 Bakan, J. A., Microencapsulation using coacervation/phase separation techniques, *Controlled Release Technologies: Methods, Theory, and Applications*, Volume 2, Kydonieus, A. F., Ed., CRC Press, Inc., Boca Raton, FL, 1980, chapter 4.
2 Baker, C. A., Brooks, A. A., Greenley, R. Z., and Henis, J. M., US Patent 5,089,407, February 18, 1992.
3 Becher, D. Z., and Magin, R. W., US Patent 4,563,212, Jan. 7, 1986.
4 Beestman, G. B., and Deming, J. M., US Patent 4,280, 833, July 28, 1981.
5 Beestman, G. B., US Patent 4,534,783, Aug. 13, 1985.
6 Beestman, G. B., US Patent 4,640,709, Feb. 3, 1987.
7 Beestman, G. B., WO 941319, June 6, 1994.
8 Connick Jr., W. J., Formulation of living biological control agents with alginate, in *Pesticides Formulations Innovations and Developments*, Cross, B. and Scher, H. B., Eds., American Chemical Society, Washington, DC, 1988, 241.
9 Connick Jr., W. J., Microbial pesticide controlled-release formulations, *in Controlled Delivery of Crop-Protection Agents*, Wilkins, R. M., Ed., Taylor and Francis, Inc., Bristol PA, 1990, chap. 12.
10 *Crop Protection Chemicals Reference*, 10th edition, Chemical and Pharmaceutical Press, New York, 1994, 1268.
11 *Crop Protection Chemicals Reference*, 11th edition, Chemical and Pharmaceutical Press, New York, New York, 1995, 350.
12 Dauth, J., Nuyken, O., and Pekruhn, W., WO 9112883, September 5, 1991.
13 Devissaguet, J., Fessi, H., and Puisieux, F., US Patent 5,049,322, September 17, 1991.
14 Federal Register, Vol 58, No. 138, 58 FR 38977.
15 Flemming, G. F., Wax, L. M., Simmons, F. W., and Felsot, A. S., Movement of alachlor and metribuzin from controlled release formulations in a sandy soil, *Weed Sci.*, 40, 606, 1992.
16 Gaertner, F., Cellular delivery systems for insecticidal proteins: living and non-living microorganisms, *in Controlled Delivery of Crop-Protection Agents*, Wilkins, R. M. ,Ed., Taylor and Francis, Inc., Bristol PA, 1990, chap. 13.
17 Golden, R., US Patent 4,157,983, Jun. 12, 1979.
18 Hall, H. S., and Pondell, R. E., The Wurster process, *Controlled Release Technologies: Methods, Theory, and Applications*, Volume 2, Kydonieus, A. E. Ed., CRC Press, Inc., Boca Raton, FL, 1980, 133.
19 Ivy, E. E., Penncap M: an improved methyl parathion formulation, *J. Econ. Entomo.* Vol. 65(2), 473, 1972.
20 Jabs, G., Nehen, U., and Scholl, H. J., US Patent 4,847,152, July 11, 1989.
21 Kondo, A., *Microcapsule Processing and Technology*, Marcel Dekker, Inc., New York, 1979.

22 Kondo, A., Microencapsulation utilizing phase separation from an aqueous solution system, *Microcapsule Processing and Technology*, Marcel Dekker, Inc., New York, 1979, chap. 8.
23 Ibid. page 50.
24 Kydonieus, A. F., *Controlled Release Technologies: Methods, Theory, and Applications*, Volume 2, CRC Press, Inc., Boca Raton, FL, 1980.
25 Lowell, J. R., Culver, W. H., and DeSavigny, C. B., Effects of wall parameters on the release of active ingredients from microencapsulated insecticides, *Controlled Release Pesticides*, Scher, H. B. Ed., ACS Symp. Series 53, 1977, 145.
26 Lowell, J. R. and Murnighan, J. J., data presented to the 172nd ACS. National Meeting, San Francisco, CA, August, 1976.
27 McMahon, W. A., Lew, C. W., and Branly, K. L., US Patent 5,292,533, March 8, 1994
28 Marrs, G. J. and Scher, H. B., Development and uses of microencapsulation, *in Controlled Delivery of Crop-Protection Agents*, Wilkins, R. M., Ed., Taylor and Francis, Inc., Bristol PA, 1990, chap. 4.
29 Meghir, S., Microencapsulated insecticides, *Med. Fac. Landbouww. Rijksuniv.* Gent 45(3), 513, 1980.
30 Meinard, C., and Taranta, C., US Patent 5,051,306, September 24, 1991.
31 Meyers, P. A., Greene C. L., and Springer, J. T., The use of Intelimer®[†] Microcapsules to Control the Release Rate of Agricultural Products and Reduce Leaching, in *Pesticide Formulations and Application Systems: 13th Volume, ASTM 1184*, Paul D. Berger, Bala N. Devisetty, and Franklin R. Hall, Eds.., American Society for Testing and Materials, Philadelphia, 1993.
32 Michael, D. J., and Wyse, D. L., The influence of herbicide formulation on weed control in four tillage systems, *Weed Sci.*, 37, 239, 1989.
33 Misselbrook, J., Hoff, Jr. E. F., Bergman, E., McKinney, L. J., and Lefiles, J. H., US 5,073,191, December 17, 1991.
34 Misselbrook, J., Hoff, Jr. E. F., Bergman, E., McKinney, L. J., and Lefiles, J. H., US 5,160,530, November 3, 1992.
35 Monsanto, Unpublished data.
36 Morgan, P. W., Interfacial polymerization a versatile method of polymer preparation, *SPE J.*, June 1959, 485.
37 Odian, G., *Principles of Polymerization*, McGraw-Hill, New York, 1970, 91.
38 Peterson, B. B., Shea, P. J., and Wicks, G. A., Acetanilide activity and dissipation as influenced by formulation and wheat stubble, *Weed Sci.*, 36, 243, 1988.
39 Raun, E. S., and Jackson, R. D., Encapsulation as a technique for formulating microbial and chemical insecticides, *J. Econ. Entomol.*, 59(3), 620, 1966.
40 Scher, H. B., US Patent 4,285,720, Aug. 25, 1981.
41 Scher, H. B., and Rodson, M., US Patent 4,956,192, September 11, 1990.
42 Scher, H. B., and Rodson, M., US Patent 5,160,529, November 3, 1992.
43 Scher, H. B., and Rodson, M., US Patent 5,332,584, July 26, 1994.
44 Schibler, L., US Patent 3,594,328, July 20, 1971.
45 Sliwka, W., Microencapsulation, *Angew. Chem. Internat. Edit.*, 14(8), 539, 1975.
46 Somerville, G. R., and Goodwin, J. T., Microencapsulation using physical methods, *Controlled Release Technologies: Methods, Theory, and Applications*, Volume 2, Kydonieus, A. E. Ed., CRC Press, Inc., Boca Raton, FL, 1980, 155.
47 Trimmnell, D., and Shasha, B. S., Autoencapsulation: a new method for entrapping pesticides within starch, *J. Controlled Release*, 7, 25, 1988.
48 Wilson, H. P., Hines, T. E., Hatzios, K. K., and Doub, J. P., Efficacy comparisons of alachlor and metolachlor formulations in the field, *Weed Tech.*, 2, 24, 1988.
49 Wing, R. E., Maiti, S., and Doane, W. M., Amylose content of starch controls the release of encapsulated bioactive agents, *J. Controlled Release*, 7, 33, (1988),
50 Wing, R. E., Maiti, S., and Doane, W. M., Factors affecting release of butylate from calcium ion-modified starch-borate matrices, *J. Controlled Release*, 5, 79, 1987.
51 Zsifkovits, W., Gruning, B., Kollmeier, H., Schaefer, D., and Weitemeyer, C., US Patent 4,931,362, June 05, 1990.

† INTELIMER is a registered trademark of Landec Corporation, Menlo Park, California.

Chapter 8

Macro- and Microemulsion Technology and Trends

Kolazi S. Nayayanan

CONTENTS

1-8493-7678-5/96/$0.00+$.50

ABSTRACT

This paper provides brief theories for the formation and structure of emulsions, microemulsions, and dispersions. A few examples are provided to illustrate the principles involved. Thermodynamic and kinetic stabilities and factors affecting them are discussed from a practical standpoint.

Modern methods of analysis and tools available are discussed, with particular reference to formulations for agricultural active ingredients. Availability and use of safe and effective inert formulation ingredients is reviewed. More recent formulation techniques such as computer-aided experimental design and optimization procedures are reviewed.

Some of the recent trends and considerations in formulation development of liquid concentrates and factors that affect reformulation of existing active ingredients are discussed. A projection of the future trend regarding the status of liquid concentrates in the agricultural industry is presented.

I. INTRODUCTION

Out of hundreds of different active ingredients currently used in agricultural formulations for herbicides, insecticides, fungicides, and growth regulators, about 90% of the active ingredients are hydrophobic complex organic compounds (practically insoluble in water).[40,115] Liquid formulations, wherever possible, are the preferred mode, because of their increased biological activity.[47]

It is also the current practice and is increasingly important to use multiple active ingredients with different biological spectra of activity in a single formulation for effective pest control. Multiple active ingredients in a single formulation would ensure the proper dosage by the user, as compared with mixing separate formulations on the application site. It is also economical, convenient to use, and will reduce the liability for the manufacturer from overdose or insufficient dose of the component active ingredients.

There is also a trend to reduce the use of toxic solvents and toxic inert ingredients, especially those included in EPA's List 1 and 2, from all formulations.[59,116] Further, in these days of environmental consciousness, there is a strong driving force to provide formulations using safe proven ingredients that not only provide additional safety to the user, but also enhance the biological activity of the active ingredients.[28]

Thus, there is a constant demand for safe and effective formulations, not necessarily from new active ingredients. Many existing active ingredients are reformulated today to meet the industry's demand. The formulator's task is becoming increasingly challenging.

Availability and use of safe and effective inert ingredients[24] such as solvents (EPA approved), newer, safer emulsifiers, polymers, thickeners, rheology modifiers, better formulation techniques for stability evaluation, and the use of computer-aided experimental design and statistical analysis for optimization[64] have become common practice in formulation development.

We shall examine the nature of emulsions/microemulsions and dispersions and their applications in agricultural formulations.

II. EMULSIONS, MICROEMULSIONS, AND TYPES OF DISPERSED SYSTEMS

A. DESCRIPTION

A comprehensive description of an emulsion will include dispersion of one isotropic Newtonian liquid in another in the form of spherical droplets.[6,30,107] In many cases, a third phase is recognized which plays a significant role.[321,42] Stability of such systems is accomplished by modifying interfacial compositions and interfacial properties by the use of emulsifiers. Emulsifiers are generally polydisperse systems. A single emulsifier is seldom used to produce stable emulsions. Often, mixed emulsifiers are used along with cosolvents and additional stabilizers. The functions of these components will be evident after reviewing the causes and remedies of destabilization of an emulsion system.

Emulsifiers are amphiphiles with polar and non-polar segments in the molecule. Emulsifier molecules orient at the interface and cause reduction in the surface/interfacial tension. Reduction of the interfacial tension alone is not a sufficient element for stability.[54] Very low interfacial tension may lead to instability if the interfacial film does not have sufficient thickness. Emulsions and methods of emulsification are reviewed in several books and monographs.[6,20,52,86,107,112] Emulsion systems are divided into three types based on the particle/droplet size distribution: (1) macroemulsion with particle size of the dispersed phase > 400 nm (opaque); (2) microemulsions that are often transparent with particles < 100 nm; and (3) miniemulsions which may appear blue-white with intermediate particle size of 100 to 400 nm. (see Table 1) Many systems are also described as multiple emulsions in which the dispersed particles are themselves emulsions.[56]

Figure 1(a) and 1(b) represent o/w and w/o systems. Figure 2(a) and (b) represent w/o/w and o/w/o multiple emulsions.[57]

The appearance of the dispersed system depends upon particle size distribution of the dispersed phase, difference in the refractive index of the two phases, natural color of the components, and presence of solids. The particle size can be reduced by (a) use of increasing amounts of the emulsifier, (b) improving hydrophilic-lypophilic balance (HLB), or (c) preparing the emulsion by phase inversion method or preparing at a temperature close to the PIT followed by cooling and by improved agitation. (See discussions later for formulations of emulsions.)

Table 1 summarizes the relation between particle size of the dispersed phase and appearance of the dispersed system.

Table 1
Effect of Particle Size of Dispersed Phase on Emulsion Appearance

Particle size	Appearance
Macroglobules	Two phases may be distinguished
Greater than 1 μm	Milky-white emulsion
1-0.1 μm	Blue-white emulsion, especially a thin layer
0.1-0.05 μm	Gray semitransparent, dries bright
0.05 μm and smaller	Transparent, dries bright

From Jones, S., Kirk Othmer Encyclopedia of Chemical Technology, Vol. 8, Grayson, M., Ed., John Wiley & Sons, New York, 1979, 900. With permission.

The type of emulsion (o/w or w/o) can be determined by various methods.[98] Usually the component having the larger (> 10) volume ratio (ratio of the volume of the dispersed phase and volume of the dispersion medium) will form the outer phase. When the volume ratio is < 3 other factors such as type of emulsifier, and order of addition are to be considered in determining the outer phase. Gross properties of the emulsion will be closer to the properties of the outer phase. For example:

- An emulsion can be readily diluted with its outer phase (dispersion medium) i.e., o/w emulsion disperse readily in water and vice versa for w/o emulsion.
- O/W emulsions would show higher electrical conductance, whereas w/o emulsions will not conduct electricity.

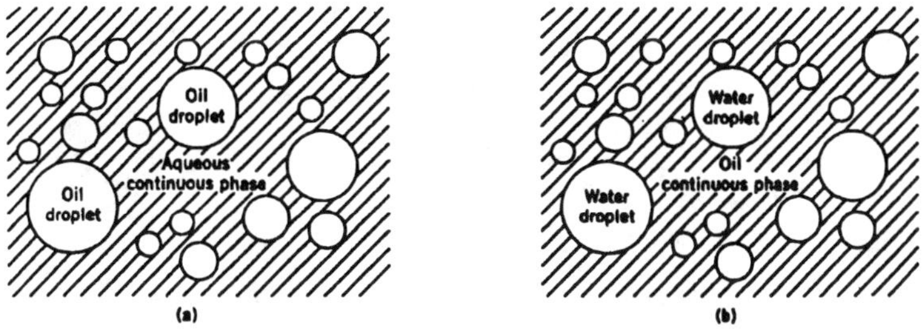

Figure 1 In an oil-in-water (o/w) emulsion (a), macroscopic oil droplets are dispersed in water. In a water-in-oil (w/o)emulsion (b), the situation is reversed. (From Friberg, S. E. and Jones, S., Othmer, K. Enclyclopedia of Chemical Technology, Vol. 9, Kroschwitz, J. I., Ed., John Wiley & Sons, New York, 1993, 393, With permission.)

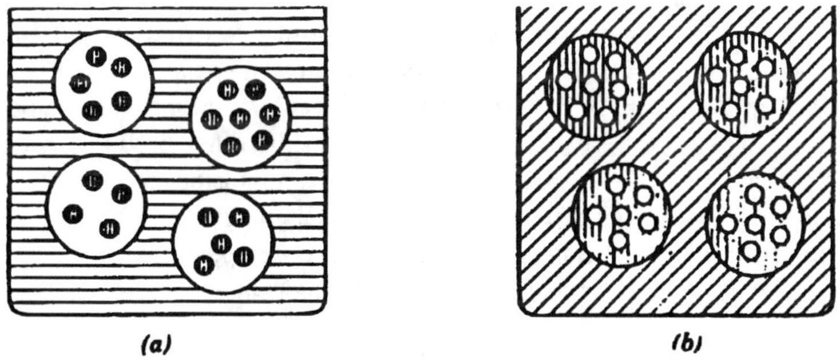

Figure 2 Multiple emulsions. (a) w/o/w emulsions: □ , inner water phase; □ inner oil phase; □ outer water phase. (b) o/w/o emulsion: □ , inner oil phase; □ inner water phase; □ outer oil phase. (From Rosen, M. J., Surfactants and Interfacial Phenomena, 2nd ed., John Wiley & Sons, New York, 1989, chap. 8. With permission.)

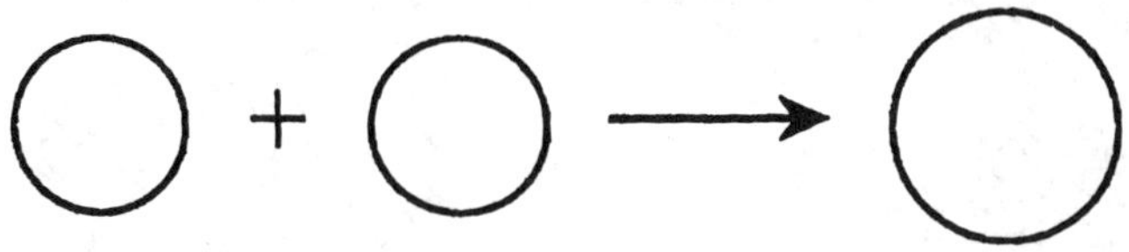

Figure 3 Coalescence of two droplets, forming a bigger droplet.

- O/W emulsion will be colored by water-soluble dyes, whereas w/o emulsions will be colored by oil-soluble dyes.
- If the refractive indices of the two phases are known, examination under the optical microscope could determine the nature of the dispersed droplet. A droplet will appear brighter on focusing upward if it has a refractive index greater than that of the dispersion medium, and darker if the droplet has the refractive index smaller than the dispersion medium.
- When an o/w emulsion is placed on a filter paper, it will spread immediately. A dried filter paper previously treated with a solution of hydrated $CoCl_2$, will appear pink when an o/w emulsion is placed on it, whereas a w/o emulsion will maintain the blue color because of poor wetting.

B. FORMATION

To prepare an emulsion system consisting of two or more immiscible liquids, one of the liquid phases is broken up into tiny particles in order to disperse in the other. Interfacial tension between two immiscible liquids is always positive and dispersion of one phase in the other produces a larger increase in the surface area. Formation of an emulsion or dispersion is accompanied by a large increase in the interfacial energy given by the change in the Helmoltz free energy (ΔF) at constant temperature, volume, and composition (in the absence of a stabilizing component).

$$\Delta F = W = \gamma \, dA \tag{1}$$

where W is the work done to increase the surface area by dA at a constant interfacial tension γ. Thus, the emulsion produced by dispersing one liquid in another is thermodynamically unstable compared to the two bulk phases separated by a minimum interfacial area. Presence of emulsifying agents at the interface will stabilize this basically unstable system for a sufficient length of time required for its application. The role of emulsifying agents will be discussed in detail (see Sections II.C and III.).

The emulsifying agent reduces the interfacial tension γ, thus reducing the amount of work required (Equation 1). The emulsifier will also reduce the rate of coalescence of the dispersed phase due to film formation by building mechanical, steric, or electrical barriers to the approach of two liquid droplets.

Use of high shear mixers, colloid mills, homogenizers, and sonic/ ultrasonic treatment are well described in the literature[102,123] for preparing concentrated, viscous emulsions.

C. STABILITY OF EMULSIONS

The term stability, for all practical purposes, refers to resistance of dispersed droplets in an emulsion to coalesce. Figure 3 represents the coagulation of two droplets, which is a spontaneous process, as it accompanies a decrease in total area.

The Helmholtz free energy change for coagulation is negative and hence the coagulation process is spontaneous. For an emulsion droplet containing a well-designed adsorbed emulsifier(s) at the interface, the net change in the free energy on coalescence can be made positive at least during the duration of use, if the system has a high negative free energy of adsorption for the 3rd component(s) [x] (the emulsifier or stabilizing components, i.e., polymeric films). The reduction in the area accompanying the coalescence process, will provide less room for the adsorbed stabilizing species [x] (which was already at the interface spontaneously) and desorption will require energy (see Section III.E).

Rising or settling of the droplets causes 'creaming' because of the difference in density of the dispersed phase and continuous medium and is not usually considered as instability. The steps involved in breaking of the emulsion are shown below:

$$\text{Flocculation} \rightarrow \text{Coalescence} \rightarrow \text{Separation}$$

Figure 4 summarizes the various steps involved in separation of an emulsion system. Flocculation, although leading to eventual coagulation and breaking of the emulsion, is not often a serious problem for many practical purposes.

The stability of the emulsion can be measured by the rate of coalescence (R) of the droplets.[12] Rate can be measured by counting the number of droplets in a unit volume of the emulsion as a function of time. Several methods are described in the literature for measuring R such as: (a) direct counting using an optical microscope, generation of size distribution by counting ~ 400 droplets;[51] (b) use of hemocytometer cell or a coulter centrifugal photo sendimentometer are described;[29,39,107] and (c) for counting submicron particles which are not visible under an optical microscope, methods based on quasi-elastic light scattering are necessary.[41] Commercial instruments with developed software are available.[50] The particle

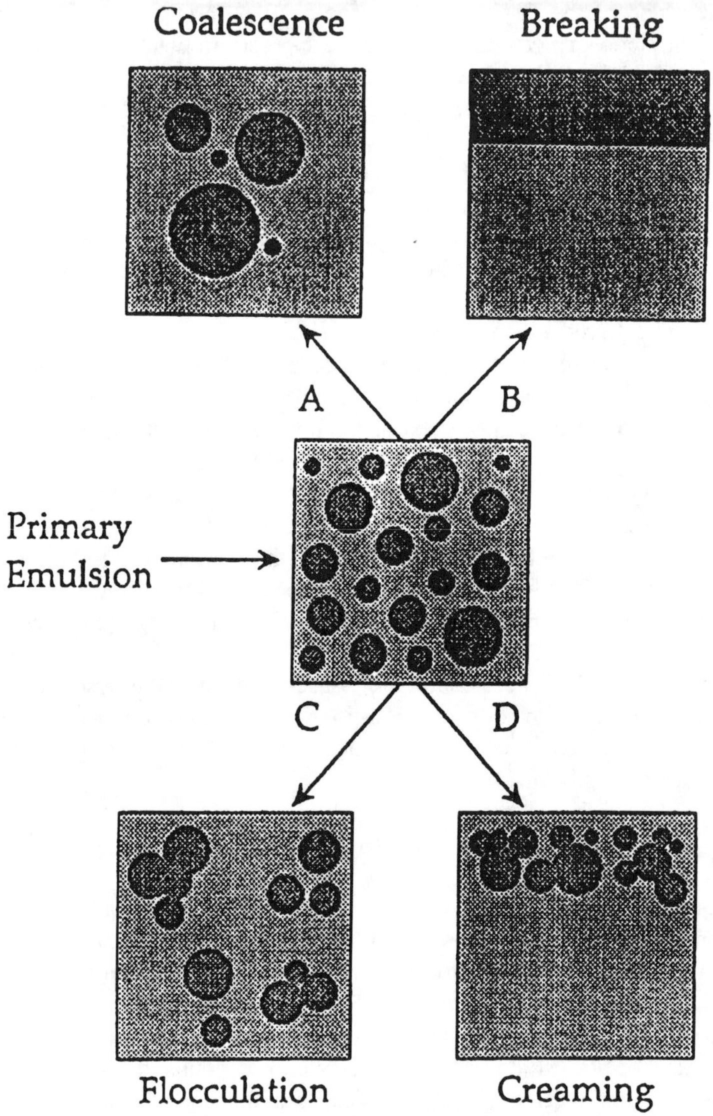

Figure 4 Effects of aggregation of droplets. (From Saroja, Surfactant and Emulsion Stability, PhD Thesis, Clarkson University, A Bell and Howell Company, Michigan, 1993, 48. With permission)

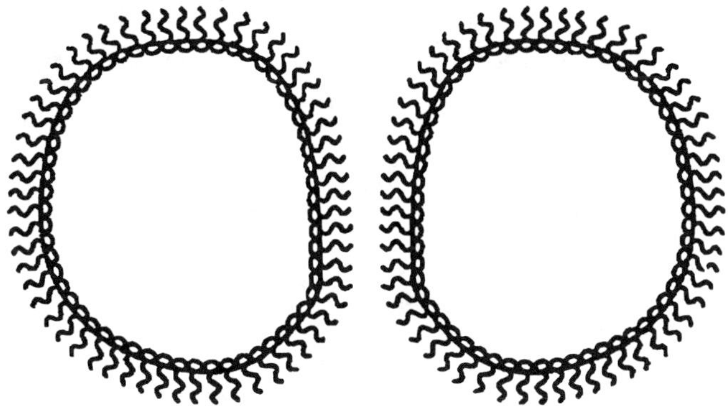

Figure 5 Film formed by the emulsifier in a flocculated droplet. (From Kirk Othmer Encyclopedia of Chemical Technology, Vol. 9, Kroschwitz, J. I., Ed., John Wiley & Sons, 1993, 393. With permission.)

size distribution can be measured in a microemulsion system (in a dilute aqueous system within certain viscosity range (< 5cp). The method is based on an induced Doppler effect produced by a laser beam. The frequency spectrum is transformed into geometric dimension by Fourier transform analysis of data.[50,77]

In flocculation, two droplets are attached to each other, but are separated by a thin film of liquid. When more droplets are attached, an aggregate is formed producing a cluster of droplets. During this process the emulsifier molecules (and other stabilizing components) remain at the surface of the individual droplets (Figure 5).

1. Rate of Coalescence (R)

The rate of coalescence of an emulsion droplet, leading eventually to breakage of the emulsion, depends upon a number of factors:

- Physical nature of the interfacial films
- Existence of an electrical or steric barrier
- Size distribution of the droplet
- Viscosity of the medium
- Phase volume
- Temperature

2. Nature of the Interfacial Film

The Brownian motion of the droplets will cause collision between them, and during such collisions the interfacial film between the droplets (formed by the surfactant molecules and other stabilizing molecules) can rupture, causing

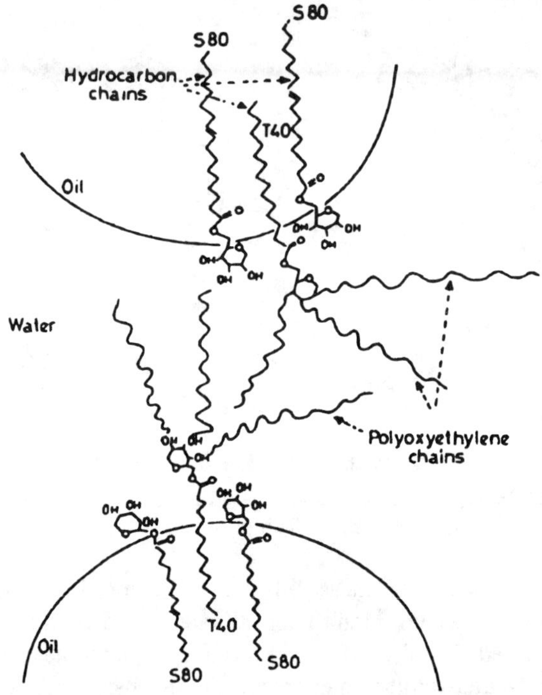

Figure 6 Complex formation between Span 80 and Tween 40 at o/w interface. (From Boyd, J., Parkinson, C., and Serman, P., J. Colloid Interface Sci., 41, 359, 1972. With permission.)

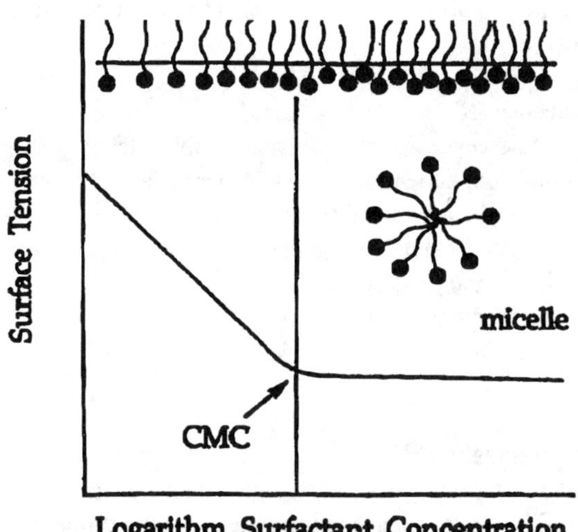

Figure 7 Monolayer and micelle formation.

the two droplets to coalesce to form a larger droplet. If this process continues the dispersed phase (oil) will eventually separate. Therefore the interfacial film should have a high mechanical strength. This can be achieved by providing strong lateral intermolecular forces between the surfactant molecule and the solvent molecules at the interface. The surfactant molecules should adsorb strongly at the interface. Highly purified surfactant is generally ineffective, as it forms interfacial films that are not closely packed. Mixed surfactants are used as a rule. A combination of water-soluble and oil-soluble surfactants will increase the lateral interaction and will form films of high mechanical strength.[27]

An example of a combination of oil-soluble and water-soluble surfactant used in many applications is a mixture of Span 80 [sorbitol ester (stearate)] and Tween 40 [polyethoxylated sorbitol ester]. The hydrophilic groups from Tween 40 extend further into the aqueous phase compared to using one type of emulsifier alone. Penetration of the polyethoxylate into the aqueous phase also causes the hydrophobic groups of both Span 80 and Tween 40 to approach closer in the oil droplet. The result is formation of a strong interfacial film.[12]

Criteria for oil solubility is a long hydrophobic chain with a small hydrophilic head. For example, small quantities of lauryl alcohol produce a close packed film with sodium lauryl sulfate (SDS).[122] Addition of NaCl would also increase stability of SDS emulsions, as NaCl will compress the electrical double layer, causing the hydrophobic, dodecyl group to compress and close pack.[121] Area occupied by the surfactant molecules is a measure of the association number and nature of packing of the interfacial film in a monolayer.

> For a w/o emulsion, the film that surrounds the droplet has to be particularly strong, as the internal phase, which is the water droplet, carries no charge and has no electrical barrier to offer stability against coalescence. Thus, for w/o emulsions, the main stabilizing factor is the mechanical strength of the interfacial film. Figure 6 illustrates the complex formation and the strong interfacial film.

3. Molecular Dimensions and Orientations of Surfactant Molecules

The area per molecule can be calculated from Equation 2 from the surface excess Γ . The surface excess Γ can be obtained from the slope of surface tension/interfacial tension (γ) and log C using Equation 3,

$$a = 10^{16}/N\Gamma \tag{2}$$

$$\Gamma = -\ 1/2.303\ RT\ [d\gamma/d\log C]_T \tag{3}$$

where C is the concentration of the surfactant, a is the area in Å^2 and N is Avogadro's number.

Figure 7 shows the relationship between surface tension and log C, and regions of monolayer and micelle formation.[103]

Figure 8 shows the relative orientations and area per molecule for a few representative surface active molecules. The relatively large spreading of tri-*p*-cresyl phosphate is worth noting.[93]

Figure 9 shows various modes of surfactant action for reducing the surface/interfacial energies.

4. Electrical or Steric Barrier

Presence of electrical charge on the dispersed droplet (especially in o/w emulsions) can build an electrical potential that can produce an energy barrier for close approach of two droplets. Presence of added electrolytes can shrink the electrical double layer (reduce the Debye radius) and can increase the stability. Direct evidence for such an effect and measure of such stability is an increased value of the zeta potential and the corresponding stability observed by added electrolytes for o/w emulsions.[87]

The close packing of the interfacial film can also be accomplished by interaction of the groups in the surfactant molecules. Dehydration of hydrated $[EO]_x$ chains on close approach of o/w emulsion is one such example. In a w/o emulsion, the long hydrophobic chain can penetrate into the oil phase. Such interaction can result in an energy barrier via steric repulsion. Adsorption of external molecules or segments of polymers can also yield a steric barrier. (See Section II.C.10).

5. Size Distribution

An emulsion with a narrow particle size distribution is more stable than one with a broader size distribution even though both emulsions have identical average particle size. Larger particles have less interfacial surface per unit volume, and in a macroemulsion, larger particles are thermodynamically more stable than smaller ones and they grow at the expense of the smaller ones, eventually causing separation of oil.

The sedimentation velocity is dependent on the particle size. The sedimentation velocity is given by Equation 4 from balancing the buoyancy force and viscous drag for perfect elastic spheres.

$$f_b = 4\Pi a^3 \Delta d/3 = f_d = 6\Pi\eta a v \qquad (4)$$

where f_b = buoyancy force, $\Pi = 3.143$, a = radius of the droplet, η = viscosity of the continuous phase, Δd is the density gradient, f_d = drag force, and v is

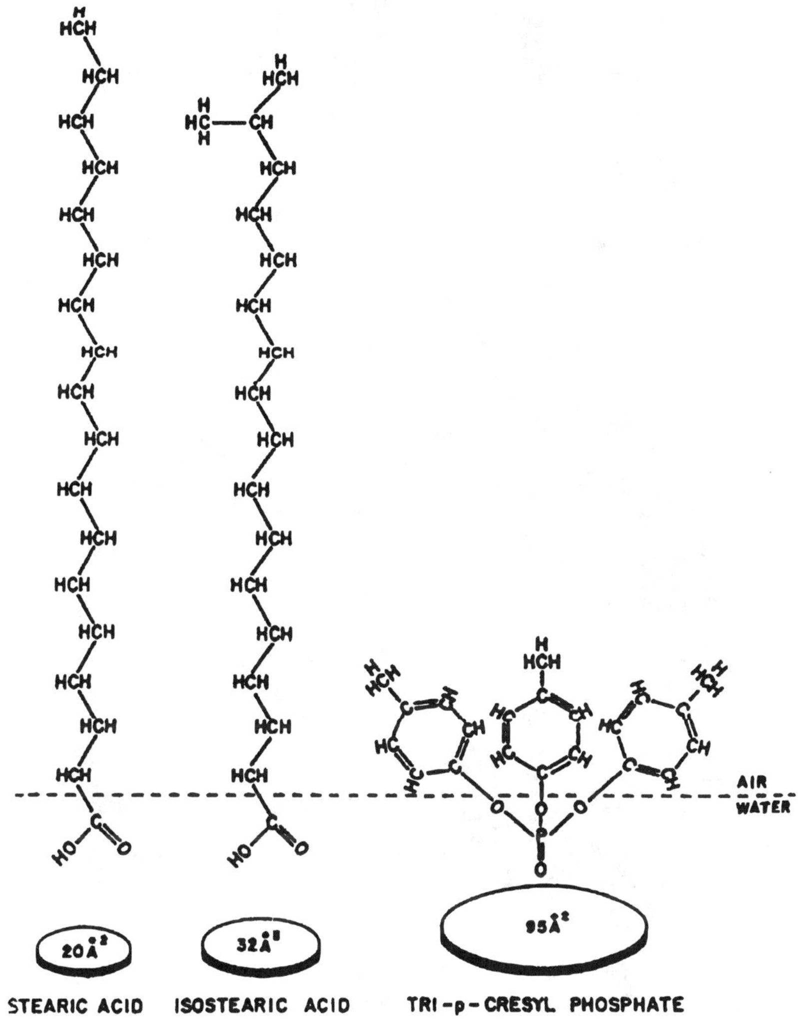

Figure 8 Structures and orientations of stearic acid, isoatearic acid, and tri-p-cresyl phosphate in air-water interface. (From Ries, H. E., Jr., Sci. Am., 204, 152, 1961. With permission.)

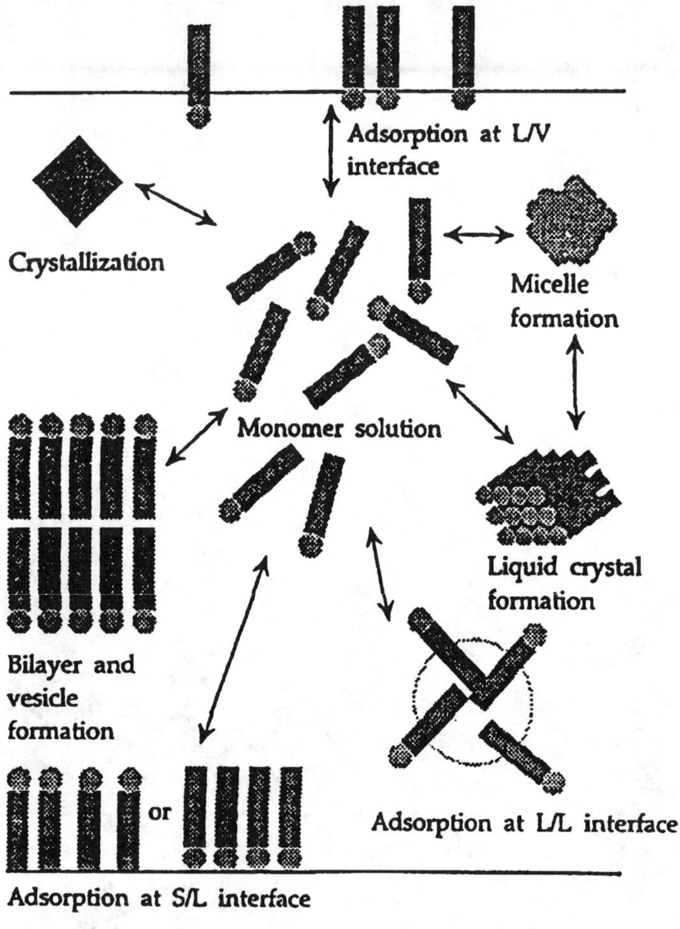

Figure 9 Modes of surfactant action for reduction of surface/interfacial energies. (From Myers, D., Surfactant Science and Technology. Reprinted with permission from VCH Publishers, New York, 1988.)

the sedimentation velocity. Solving Equation 4 for v we obtain:

$$v = 2a^2\Delta d/9\eta \quad (4a)$$

This Equation predicts that larger droplets sediment faster. For example, a 10-μm droplet will sediment 10,000 times faster than a 0.1-μm droplet, which will lead to enhanced flocculation.

6. Phase Volume

If more and more dispersed phase is added to an emulsion, the volume of the dispersed phase increases, causing expansion of the interfacial film. If the volume of the dispersed phase increases beyond certain limits, the emulsion can invert, if the area of the new interface for the inverted emulsion is significantly smaller. Inversion may not occur if the emulsifier stabilizes only one type of emulsion.

7. Viscosity of the Continuous Phase (Dispersion Medium)

Increase in viscosity of the continuous phase generally stabilizes the emulsion droplets. The viscosity is inversely proportional to the diffusion coefficient:

$$D = kT/6\Pi\, \eta\, a \quad (5)$$

where D is the diffusion coefficient, k is the Boltzman constant, η is the viscosity, and a is the radius of the internal phase droplet. The higher the viscosity of the medium, the lower is the diffusion constant and the lower will be the velocity of collision and collision frequency.

An increase in viscosity from 1 cp to 10 cp for an o/w emulsion will reduce the flocculation rate by 10^4. Such change for an unprotected emulsion will only increase the $t_{1/2}$ by a few hours. The key factor for increasing $t_{1/2}$, i.e., time taken for 1/2 the number of droplets to agglomerate, is to increase the energy barrier. Increase in viscosity can be accomplished by using certain polymeric materials or by formation of liquid crystals. Formation of liquid crystals can produce a mesophase, resulting in a great increase in the stability of emulsions.[32]

8. Temperature

Change in temperature can cause variation of several parameters such as interfacial tension (usually lower at higher temperature), viscosity of the film, relative solubility of the emulsifier in the two phases, and water solubility of many ethoxylates (less at high temperature). Lower solubility of the emulsifier can cause higher stability, as emulsifiers are more effective when they are least soluble, yielding elastic films. Beyond a certain temperature, the emulsifier becomes more oil soluble and inversion of o/w to w/o can occur at higher temperature. As temperature increases, Brownian motion and vapor pressure increase and there will be larger movement of molecules across the interface, causing general instability [see PIT method of preparation of emulsions and three phase formation]

9. Quantitative Measure of Coalescence and Energy Barrier

The rate of coalescence for diffusion-controlled coalescence of spherical particles via collisions in a dispersed system is given by Smoluchowski,[19] in the absence of an energy barrier;

$$- dn/dt = 4\Pi\, Drn^2 \qquad (6)$$

where D = diffusion coefficient, r = collision radius i.e., the closest distance between two spherical droplets that cause coalescence, n = number of droplets per unit volume (cm^3). If we assume an energy barrier E , Equation 6 is modified as Equation 7:

$$- dn/dt = 4\Pi\, Drn^2 .\ e\text{-}[E/kT] \qquad (7)$$

where k is the Boltzman constant. Equation (7) assumes that every collision results in agglomeration. Only binary collisions are considered and space variation of the energy barrier is ignored in this treatment. Integrating Equation (7) we get Equation (8)

$$1/n = \{4\Pi\, D\, r.\ e\text{-}[E/kT]\}t + \text{constant} \qquad (8)$$

Combining Equations 5 and 8, we obtain Equation 9:

$$1/n = [\{4\ kT.\ e\text{-}(E/kT)\}/3\ \eta\ ]\, t + \text{constant} \qquad (9)$$

Thus, E can be calculated from the slope graphically by plotting 1/n versus time, or measuring n as a function of time, n being the number of particles in a unit volume.

Defining a mean volume for a particle by Equation 10, where V is

$$V_m = V/n \qquad (10)$$

the volume fraction of the dispersed phase, and combining Equations 9 and 10 we have Equation 11

$$V_m = [\{4\ V\ kT.\ e\text{-}(E/kT)\}/3\ \eta\ ]\, t + \text{constant} \qquad (11)$$

Differentiating Equation 11, we obtain Equation 12

$$- dV_m /dt = \{4 V kT/3 \eta\} . e_{\text{-}}\text{-}(E/kT) \qquad (12)$$

Summing the effect for all particles, Equation 12 can be rewritten as Equation 13, where N is the total number of droplets/particles in the system

$$- d(\Sigma V_m /dt = \{4 V kT/3 \eta\} . e_{\text{-}}\text{-}(\Sigma E/NkT) \qquad (13)$$

$$\Sigma E = E_{Total} \qquad (14)$$

Equation 13 gives a quantitative measure of the volume distribution rate and can be expressed in a simplified form by Equation 15:

$$- d\Sigma V_m /dt = A e_{\text{-}}\text{-}(E_{Total}/kT) \qquad (15)$$

A is a constant for a given system, E_{Total} is the total energy barrier that contains both mechanical (steric) and electrical barriers to coalescence.

Thus, using Equations 7, 13 or 15, we can relate the stability of emulsions either in terms of sedimentation rate or more usefully, in terms of $t_{1/2}$ (time taken for half the population of droplets to agglomerate) as a function of the size of the energy barrier. The $t_{1/2}$ can be calculated by integrating Equation 7 with limits for time, and number of particles between t = 0, n = n to t = $t_{1/2}$, n = n/2. Table 2 summarizes the half life calculated for different values of E.

Thus, building an energy barrier of 20 kT units can provide a $t_{1/2}$ of ~ 4 years, which would be sufficient for many applications.

10. Energy Barrier

DLVO theory is useful in interpreting stability particularly for o/w emulsions.[22,119] Figure 10 represents the attractive (Van der Waals) potential energy (E_A), the repulsive electrical energy (E_R), and the total energy (E_A + E_R) as a function of inter droplet distance r. As can be seen, the total energy (dotted line) has a maximum which corresponds to the energy barrier discussed earlier.

This energy barrier essentially arises from the electrical double layer, and can be calculated for certain model cases from the surface potentials. An estimated value of 40 mV in the zeta potential will be sufficient to stabilize an o/w emulsion containing < 0.1 *M* monovalent ions or < 0.01 *M* divalent ions.

Table 2
Sedimentation Rates, $t_{1/2}$ as a Function of Energy Barrier, E_{Total} in Units of kT

E_{Total}, kT	$t_{1/2}$
0	Seconds
5	40 Seconds
10	1.5 Hours
20	4 Years
50	4 X 10^{13} Years

From Friberg, S.E., and Jones, S., Kirk Othmer Encyclopedia of Chemical Techonolgy, Vol. 9, John Wiley & Sons, New York, 1993, 393. With permission

Presence of excess salts in the continuous water phase can cause a reduction in the double layer potential and energy maxima or energy barrier. This leads to loss of emulsion stability. Addition of salt would have negligible effect in the Van der Waals potential. Thus, o/w emulsions are generally unstable in salt solutions, while w/o emulsions are less affected by the presence of salts.

11. Stabilization by Polymers

Use of polymers can produce excellent stabilization, and can in some cases replace the use of surfactants.[9,38,67,68,104] Figure 11 shows adsorption of polymer on a surface. The polymer adsorption produces different segments of adsorption such as loops, tails, and trains. A single loop can provide 20 kT of steric energy barrier.[16,67]

Stabilization by a certain polymer depends upon the extent of adsorption on the surface and sometimes there is a narrow range of absorption energy that affords protection, as shown in Figure 12.[16]

The relationship between the fraction of molecules and total adsorbed energy identifies the region affording protection from flocculation. At low energy region A (Figure 12), there is weak adsorption of the polymer and surface coverage is poor. At high energy region B, the polymer absorbs too strongly (too many trains). Thus it remains flat on the surface and affords no protection.

Sometimes complication can arise if the 'tail' or 'loop' from two protected droplets entangle and interact and can shrink or collapse the protection.[104] Block polymers are often the polymers of choice for offering steric protection.

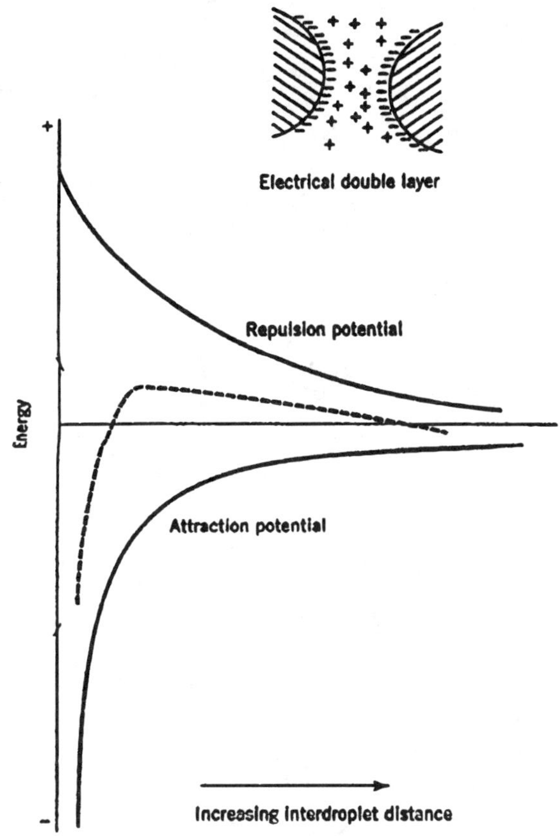

Figure 10 The von der Waals' potential between droplets is increasingly negative with reduced inter-droplet distance, whereas the electrical double layer potential is increasingly positive. The resulting potential (dotted line) may have a maxima. (From Friberg S. E., and Jones, S., Kirk Othmer Encyclopedia of Chemical Technology, Vol. 9, John Wiley & Sons, New York, 1993, 393. With permission)

D. PREPARATION OF EMULSIONS

Several methods are used for the preparation of emulsions.

1. Phase Inversion

This method is generally useful for the preparation of concentrated emulsions. To prepare an o/w emulsion, the emulsifier is added in the oil phase that contains all the oil-soluble ingredients. The aqueous phase is prepared separately. As increasing amount of the aqueous phase is added to the oil phase containing the emulsifier while the emulsion is passed through a

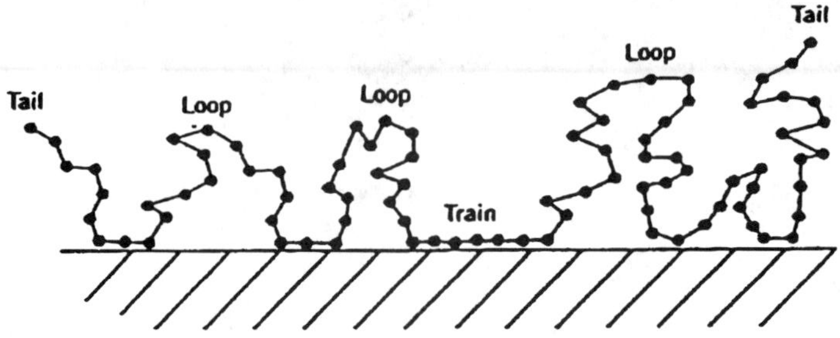

Figure 11 A polymer is adsorbed to a surface in loops, tails, and trains. (From Friberg S. E., and Jones, S., Kirk Othmer Encyclopedia of Chemical Technology, Vol. 9, John Wiley & Sons, New York, 1993, 393. With permission)

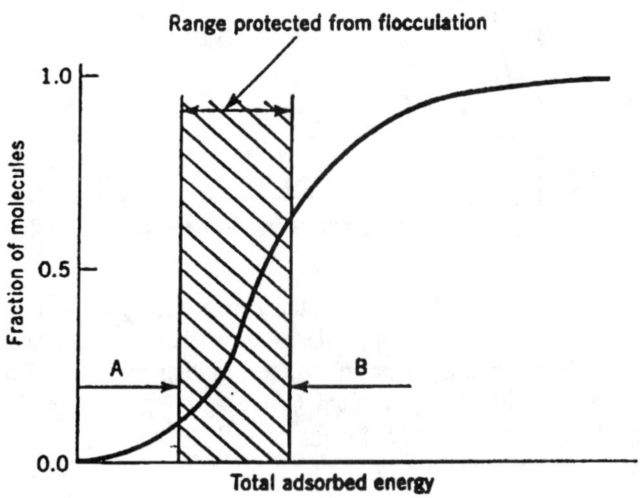

Figure 12 Calculation of the adsorption of a polymer at an interface show pronounced sensitivity to the total adsorption energy, leaving out a limited range in which the polymer can serve as a stabilizer. (From Clayfield, E. J. and Jumb, E. C., reprinted with permission from Macromolecules, 1, 133, 1968, copyright American Chemical Society.)

colloid mill. A w/o emulsion with an appearance of an ointment is produced at 60 to 70% dilution (i.e. 30 to 40% aqueous phase). This thick emulsion is transferred into an oil tank and water is added to the desired level with mild stirring, when the original w/o emulsion inverts to an o/w emulsion. The emulsifier used above should be oil soluble and should have predominantly hydrophilic groups.

2. Phase Inversion Temperature (PIT) Method

The presence or formation of a third phase in the emulsification process has been recognized for nonionic surfactants (e.g., $(EO)_x$ alkyl lauryl ethers).[108] If the solubility of the emulsifier has opposing temperature dependence in the aqueous and oil phase, one can make use of the third phase for easy emulsification. For example, at a given temperature (Figure 13a) the emulsifier is completely soluble in water. At high temperature (Figure 13c) the emulsifier is completely oil soluble. Somewhere in the intermediate temperature range, a three-phase system appears where the middle phase is a surfactant-rich phase. This intermediate temperature (Figure 13b) is a good tool to choose the appropriate emulsifier.

An equal amount of oil and aqueous phase with all components are mixed with a known quantity (~ 5%) of several emulsifiers to be screened. For preparing the o/w emulsion, the samples are thermostated at ~ 55°C to equilibrate. Those emulsifiers producing a three-phase system are the most effective candidates. All components are mixed and the temperature is raised toward PIT (phase inversion temperature). As the temperature is raised, the interfacial tension becomes continuously lower which enables the formation of small droplets. Once the fine emulsion is obtained, the system is cooled rapidly and additional stabilizing agents such as anionic or cationic emulsifiers can be added.

3. Intermittent Mixing or Stirring

While mixing the emulsions, especially on dilution of a concentrate, intermittent mixing is often better than continuous mixing. The rest intervals during the intermittent stirring allows the stabilizers to migrate to new interfaces created during the process. Continuous agitation promotes coalescence by collisions of unstabilized droplets.

4. Electric Emulsification

In this technique, the internal phase is ejected through a fine capillary held at a high electrical potential. A fine spray of droplets emerges from the top and is dispersed into the continuous phase.

5. *In Situ* Emulsification

The two components of the emulsifier are separately dissolved in the oil and aqueous phase. The emulsifiers are generated at the interface during mixing. For example, a fatty acid can be dissolved in an oil; the base is dissolved in the aqueous phase. During mixing the interface would be rich in the salt of the fatty acid, which is the emulsifier in this case. This technique is often used to create instant interfaces for microencapsulation or for emulsion polymerization.

In practice, any one method or combinations of the above can be adopted depending on a particular application.

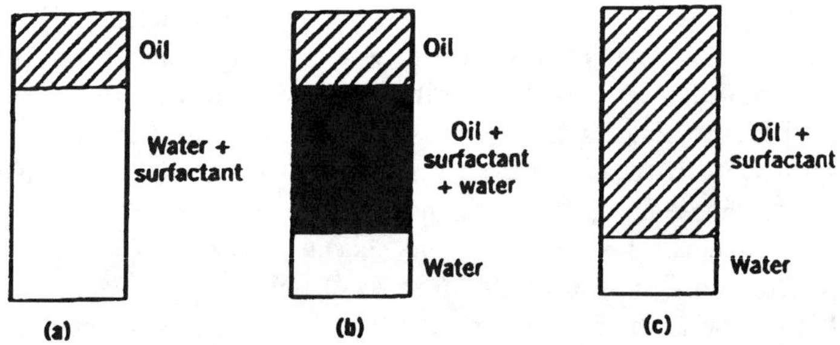

Figure 13 In a system of water and hydrocarbon a nonionic emulsifier with a poly(ethylene)glycol chain as the nonpolar part dissolves in the aqueous phase at low temperatures (a) and in the oil phase at high temperatures (c). At intermediate temperatures (b) three isotropic liquid phases may be found. (From Friberg, S. E. and Jones, S., Kirk Othmer Encyclopedia of Chemical Technology, Vol. 9, Kroschwitz, J. I, Ed., John Wiley & Sons, New York, 1993. With permission.)

E. THERMODYNAMICS, STABILITY OF MICROEMULSIONS AND MINIEMULSIONS

Characteristics of microemulsions and miniemulsions were discussed earlier. Table 3 summarizes the essential differences between macro and microemulsions. Formation of microemulsions is a spontaneous process and therefore a microemulsion is thermodynamically stable. Factors that favor the formation of microemulsions are: low interfacial tension and large free energy of adsorption of the surface active components (emulsifiers, polymers, cosolvents, hydrotropes).

Table 3
Comparison of Micro-and Macroemulsions

Macroemulsion	Microemulsion
Prepared from ECs by dilution with water just prior to use.	Can be prepared from an ME concentrate by dilution with water or can be presented as a stable, prediluted formulation.
Sequence of adding ingredients is important.	Sequence of ingredient additions not important.
Thick, milky with good bloom.	Transparent with low turbidity.
Non-isotropic.	Isotropic, for the most part.
Particle or micelle size of 5-100 μm.	Particle of micelle size of 0.01-0.1 μm.
Low tolerance for electrolytes, unless specific anionics are used.	High tolerance for electrolytes.
Limited stability (4-24 hrs).	Infinite stability.
Good to excellent biological efficacy.	Superior biological efficacy.
Low emulsifier level.	High emulsifier level.

From Narayanan, K.S. and Chaudhiri, R.K., Pesticide formulations and Application Systems, Vol. 12, Derissety, B.N., Chasin, D.G., and Berger, P.D., Eds., American Society for Testing and Materials, Philadelphia, 1993, 85. With permission

The Helmholtz free energy change (dF) for an interfacial system is given by:

$$dF = - SdT - PdV + \gamma_{ij}dA + \Sigma\mu_i d_{ni} - W_d dA \qquad (16)$$

where μ is the chemical potential of the ith component and W_d is the work of desorption (work required to desorb the component per unit area). For the formation of a micro-emulsion, there is a large increase in area of the interface (i.e. dA is very large) because of smaller droplet size. For a process at constant T,V, and composition ($dT = dV = dn_i = 0$), Equation 16 reduces to Equation 17

$$dF/dA = \gamma_{ij} - W_d \qquad (17)$$

For the process to be spontaneous, $dF/dA < 0$; and consequently $\gamma_{ij} \rightarrow 0$ and $W_d >> 0$. The term W_d includes the effects of change in entropy, changes in surface charge density, and molecular interaction between the components at the interface during the process accompanying the area increase to accomplish droplet size reduction. Under certain circumstances, γ_{ij} could assume negative values, at least on a transitory basis. Negative interfacial tension can occur by creating a surface pressure on the dispersed droplets, as a consequence of close packing of the hydrophobic groups from the emulsifiers on an oil droplet. The consequence of such an effect is the dispersed oil droplet becomes smaller. The driving force for the formation of a microemulsion comes from a large value of W_d.

Creation of a large W_d can often be achieved by using a combination of ionic surfactant (anionic or zwitter ionic/pseudo ionic) and an insoluble (water insoluble) fatty derivative or a cosolvent/coemulsifier, preferably an emulsifier that would form a complex with the first component. Thus, in general, a microemulsion consists of at least five components (water, oil, surfactant, coemulsifier, and cosolvent). For agricultural formulations, active ingredients form the additional components. Design and formation of a microemulsion is essentially a process of orientation of the molecules at the interface towards a state of $\gamma_{ij} \rightarrow 0$ and $W_d \rightarrow$ maximum positive value. Thus, once the ME is prepared, it is considered infinitely stable, since a true ME is a thermodynamically stable system.

During the molecular orientation process, there is little control over (ΔS), which can assume either positive or negative values at room temperature. ΔS is usually in the order of $+10^2$ kJ/mole and ΔF is in the order of -5 to -20 kJ/mole. Thus, the stabilizing energy in the formation of a microemulsion is not a large driving force and the success in forming a microemulsion is very much dependent on the chemical constitution of the micellar surface, with optimized composition of the ME droplet.[96,97]

The emulsifier system usually contains optimized mixed surfactants often obtained by experimental design and optimization techniques. Presence of cosolvents/hydrotropes is also advantageous. The amount of emulsifier is considerably large compared to the emulsion system. Strong adsorption of the coemulsifier/emulsifier system can induce formation of liquid crystals. The presence of liquid crystals can be detected by microscopic observation through a crossed polarizer. The anisotropic liquid crystalline phase will be radiant when observed through polarized light. The type of liquid crystal geometry can be derived from the interplaner distances detectable from low-angle X-ray-scattering measurements.[103]

1. Role of Liquid Crystals in Stabilization

The stabilizing factor of the liquid crystals is limited to protection against coalescence. The liquid crystals should be located at the interface with appropriate contact angles. The lamellar structure will reduce the Van der

Waals attractive forces. The liquid crystal network will reduce the mobility of the dispersed droplets. Small changes in the emulsifier concentration can produce large changes in the relative proportion of the oil, water and liquid crystalline phases; For example, see Figure 14.

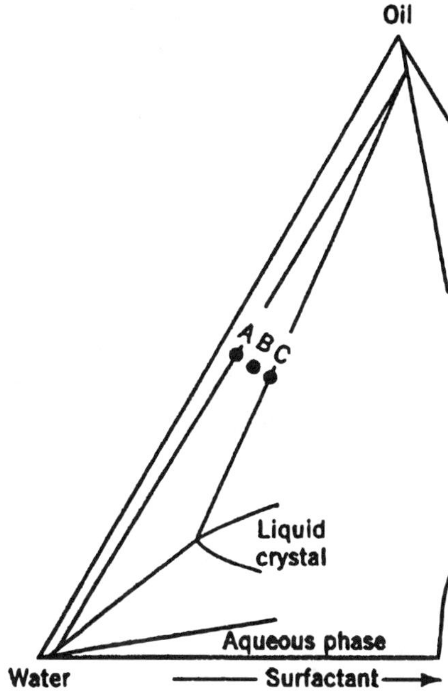

Figure 14 Increase of the emulsifier concentration in an emulsion with a liquid crystal leads to a drastic increase of the amount of liquid crystal. (From Friberg, S. E. and Jones, S., Kirk Othmer Encyclopedia of Chemical Technology, Vol. 9, Kroschwitz, J. I, Ed., John Wiley & Sons, New York, 1993. With permission.)

Consider the points A, B and C. At point A, the emulsifier level is 2% and the amounts of water and oil are each 49%, yielding a two-phase system. At point B, emulsifier concentration is raised to 4%, the oil is 47%, the aqueous phase is 29% and 24% of the total is liquid crystal phase. At point A, 2% emulsifier would provide a protective film of 0.07 μm to the emulsion droplet of 5 μm average size, assuming 100% adsorption. At point B in the presence of the liquid crystal (24%), the thickness of the protective layer is increased to 0.85 μm. The constituent of the thick film is 7% emulsifier, 75% water, and 18% oil. At point C the aqueous phase has completely disappeared and there is only oil and liquid crystals. The stabilizing phase is part of the emulsion itself.

2. Role of Hydrotropes/Cosolvents

Hydrotropes are organic compounds that increase the solubility of other organic substances in water. They have structures similar to surfactants in that they have hydrophilic and hydrophobic groups in the molecule, but the hydrophobic groups are shorter and branched. Some examples are sodium benzene/xylene sulfonate, 1-hydroxy-2-naphthoate, lower alkyl pyrrolidones, and isopentanol. For solubilization to occur, the micelle should be closer to spherical and lamellar structure necessary for liquid crystal formation should be avoided. In the lamellar or liquid crystalline matrix much solubilization cannot occur.

The shape of the micelles and distortion to various forms depend upon the area ratio of the hydrophobic group to the hydrophilic group given by Equation 18:

$$\text{Packing ratio} = P = V_{Hy}/ A_H \cdot l_{Hy} = A_{Hy}/ A_H \qquad (18)$$

where V_{Hy}, A_{Hy}, l_{Hy} are the volume, area, and length of the hydrophobic chains of the surfactant, and A_H is the cross-sectional area of the hydrophilic head group[63] V_{Hy} and l_{Hy} can be calculated from empirical Equations.[113] A_H is a measurable quantity (ex, Langmuir balance). The geometric shapes of the micelles and The relative values of the packing ratio are summarized below in Equations 19 to 22:

$$\text{spherical structures} \quad R_p \leq 1/3 \qquad (19)$$

$$\text{cylindrical structures} \quad 1/3 < R_p \leq 1/2 \qquad (20)$$

$$\text{bilayers or vesicles} \quad 1/2 < R_p \leq 1 \qquad (21)$$

$$\text{inverse micelles} \quad 1 \leq R_p \qquad (22)$$

As hydrotropes have large head groups and smaller hydrocarbon area, the packing ratio is small ($R_p \leq 1/3$) and they assume spheroidal rather than lamellar structure. The inhibition of a liquid crystalline phase increases the solubility of the surfactant in the aqueous phase and the capacity of its micellar solution to solubilize materials.

The transition of the micellar → lamellar → invert micellar configuration can occur during the process of dilution, change in temperature, or composition change by addition of suitable reagents. Figure 15 summarizes the changes.

These changes can be monitored by following viscosity or conductance, which will go through a maxima.

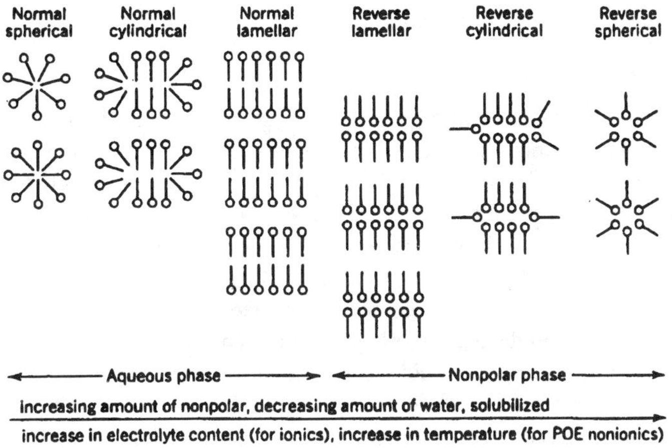

Figure 15 Effect of solubilization content and other molecular environmental factors on micellar structure. (From Matsumoto, S., Koh, Y., and Michiura, A., J. Disp. Sci. Tech., 6, 507, 1985. With permission.)

III. FORMULATION TOOLS - COMPONENTS AND METHODS

A. COMPONENTS

An agricultural formulation as an emulsifiable concentrate or microemulsifiable concentrate will contain the following typical components:

- Active ingredient(s)
- Solvent(s)
  - -Primary solvent
  - -Secondary solvent, cosolvent
- Emulsifiers
  - -Primary emulsifier (dispersing agent)
  - -Secondary emulsifier (secondary dispersing agent, wetting agent)
  - -Additional wetting agent
- Polymeric stabilizers (crystal inhibition)
- Preservatives (optional)
- Others

These concentrates should produce stable emulsions/microemulsions on dilution. Stability criteria depends on the use pattern.

B. ACTIVE INGREDIENTS

The general trend is to formulate active ingredients of wide biological spectra in a single formulation. The active ingredient should not chemically react in the medium. The formulator's choice of the active ingredient(s) is limited. The choice is dictated by the history of biological performance. Thus, the formulation matrix must be robust in being capable of formulating structurally different active ingredients.

C. SOLVENTS

Choice of the solvent is based on physical properties such as solubility of active ingredients, flash point, toxicity,[8] vapor pressure, compatibility with emulsifiers, solubility in water, thermal/chemical stability, and phytotoxicity. Most of the data are available from the suppliers. The solvent to be chosen must show high solubility for the active ingredients. On dilution of the concentrate, it is the solvated molecules of the active ingredients that are emulsified, and therefore higher solubility generally results in higher loading of the active ingredients.[78] Flash point of a formulation will generally be higher than the lowest flash point of the component. However all interactions of the components are to be considered in judging the flash point of a formulation. It is best to determine the flash point of a formulation experimentally.[14,66] Use of solvents with low vapor pressure is advantageous in preventing crystallization of solid active ingredients from the spray droplet after deposition on the target. Most solvents can evaporate as azeotropes with water. Solvents with low vapor pressure not only provide low VOC in the formulation, but may keep the active ingredients in the micellar form, when most of the water evaporates. See Figure 16 for relative evaporation rates of different solvent systems and Figure 17 for the relative advantage of using a nonevaporative solvent system in the formulation.[76]

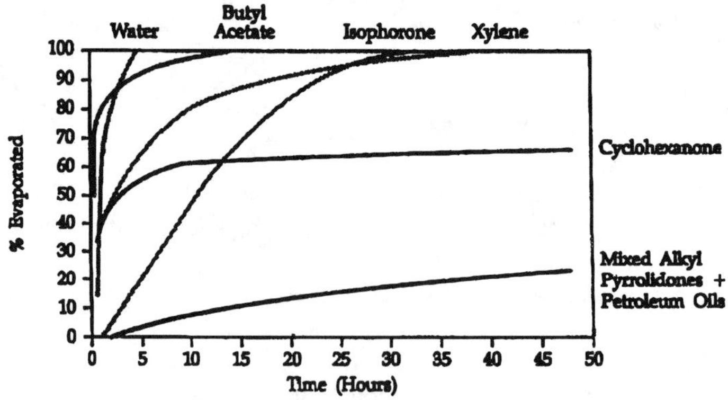

Figure 16 Relative evaporation rates of solvents.[76]

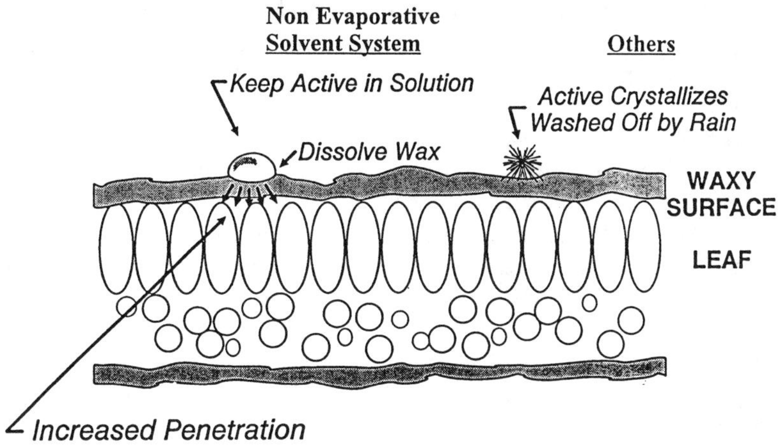

Figure 17 Pictorial representation of the advantage of using a nonvolatile solvent system , wherein the active ingredients are kept dissolved/emulsified as a gel after water evaporates. Use of evaporative solvents can produce crystals.[76]

Solubility characteristics of the solvents can be evaluated from the Hanson's fractional solubility parameters.[2,78] The use of solubility parameters to optimize the emulsifier system was attempted by Meusburger[62] in formulating an o/w emulsion for DDT. A computer program has been developed for optimization of surfactant systems for the given active ingredient and solvent system based on the solubility parameters from group contribution.[62] Phytotoxicity of a number of hydrocarbon and oxygenated solvents is published.[46] The general trend is that xylene range solvents show highest phytotoxicity.[47,55] Aliphatic hydrocarbons showed lower phytotoxicity to host crops (soybean, corn, wheat, and cotton) compared to aromatic hydrocarbons. Oxygenated solvents that have higher polarity are generally more phytotoxic. Surface tension of the solvent and the evaporation rate would also affect its phytotoxicity. As some of the solvents, particularly those that are surface active, can enhance the biological activity of the formulated pesticides, and consequently will require a lower dose of the active ingredients, phytotoxicity has to be evaluated with the formulation at appropriate dose level (or reduced spray volume rate).[53,78]

Solvent-based formulations are generally preferred for foliar-applied pesticides, as they tend to enhance uptake and translocation via penetration, cuticular solubilization and stomatal entry. The uptake is also influenced by the surfactants via cuticular diffusion[3,92,106] or stomatal infiltration.[26]

Some of the newer solvents that are relatively safe and approved by the U.S. EPA and/or approved in Europe are alkylpyrrolidones,[78] alkyl byphenyls,[120] and tetrahydrofuran derivatives.[23]

1. Mixed Solvents

Use of mixed solvents is an alternative approach for formulating EC's and microemulsion concentrates[77,79-85]

The solvent combination is such that one of the solvents is a highly polar solvent (with Hansen's fractional solubility parameters: polar component f_p is > 0.3 and dispersive component $f_d < 0.5$), the second solvent being hydrophobic (with $f_d > 0.6$ and $f_p < 0.25$) and surface active and optionally a third highly hydrophobic solvent with $f_d > 0.8$. Mixed solvent approach is useful to simulate several existing solvents (often toxic) by using a limited number of well-studied and environmentally safe solvents. Solvent mixtures can be designed in terms of their Hansen's fractional solubility parameters. Figure 18 summarizes the Hansen's fractional solubility parameters for some of the common solvents. Appropriate mixtures of solvents, for example, solvent 1, 2, and K can simulate many other solvents listed.[78]

The solvent combination is so chosen to provide excellent solubility for the active ingredients. The ECs prepared by using appropriate surfactants provide excellent stability on dilution, especially if one of the hydrophobic solvents is surface active.

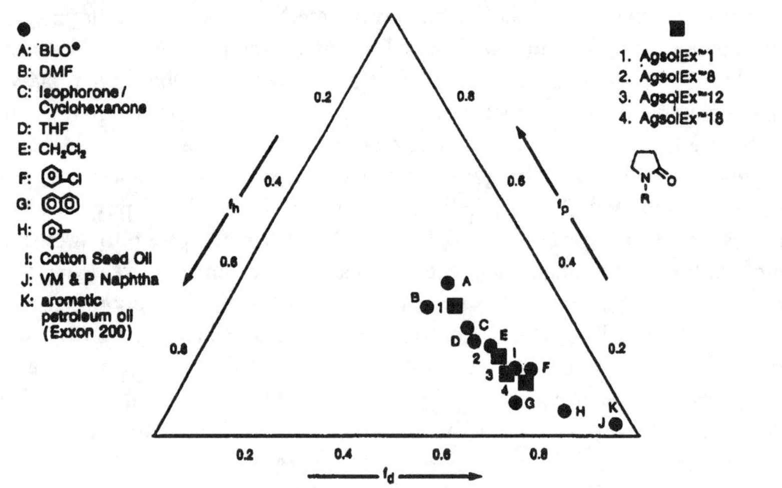

Figure 18 Hansen's solubility parameters for N-alkyl pyrrolidones and common solvents. (From Narayanan, K.S. and Chaudhiri, R.K., Pesticide formulations and Application Systems, Vol. 12, Derissety, B.N., Casin, D.G., and Berger, P.D., Eds., American Society for Testing and Materials, Philadelphia, 1993, 85. With permission)

D. EMULSIFIERS

Choice of the emulsifier in a formulation, particularly as applied to ECs and microemulsion concentrates or microemulsions, depends upon the properties of the emulsifiers. The following properties of an emulsifier will assist the formulator in making an appropriate selection in a particular formulation: the structure, polydispersity, HLB value, PIT, solubility in water and oil phase, Kraft point, surface tension, interfacial tension with model system (oil/water), dynamic surface tension, critical micelle concentration (CMC), synergy with other emulsifiers, wetting efficiency, capacity to form complexes with solvents and polymeric stabilizers. Generally, suppliers will provide some model formulations as guidelines.

The emulsifiers are one of the most important inert components in agricultural formulations. The major effect in formulating is to identify the most appropriate combinations of emulsifiers for a given system of active ingredients (a.i.s) and solvents (oil or water).

The criterion is to obtain the required stability of the concentrate, even on dilution with water that has a normally high salt content. Since the composition is very complex, the choice of emulsifiers is not obvious, and a background, knowledge and experience in the field of formulation will be very helpful. A screening process is essential before optimization of the emulsifier system. Another criterion for the end-use formulation is compatibility with other commercial pesticide formulations in the tank mix.[5,17]

1. Hypophilic-Lypophilic Balance (HLB) of Surfactants

The HLB value gives a measure of the relative simultaneous attraction of the emulsifier for water and oil (the two phases present in an emulsion). HLB values can be either calculated in terms of empirical group contributions given in Table 4 using Equation 23, or by experimental determination by comparing performance of the experimental emulsifier with standard mixtures of known HLB (e.g., sodium oleate HLB = 20, and oleic acid HLB = 1).

$$\text{HLB} = 7 + \Sigma \text{ hydrophilic group number} - \Sigma \text{ hydrophobic group number} \qquad (23)$$

Table 4
Davies' HLB Group Numbers

Groups	Group Numbers
Hydrophilic Groups	
$-OSO_3^-Na^+$	38.7
$-COO^-K^+$	21.1
$-COO^-Na^+$	19.1
N (tertiary amine)	9.4
Ester (sorbitan ring)	6.8
Ester (free)	2.4
-COOH	2.1
-OH (free)	1.9
-O-	1.3
-OH (sorbitan ring)	0.5
Lipophilic Groups	
-CH-	
$-CH_2-$	0.475
CH_3-	
=CH-	
Derived Groups	
$-CH_2CH_2O-$	0.33
$-CHCH_3CH_2O-$	-0.15

From Davis, J.T., Proc. Int. Congr. Surf Act., 2nd ed., 1, 1957, 426. With permission

Extensive bibliographies are available on HLB of several surfactants, their trade names, along with tabulated values.[7,60] Figure 19 shows the effect of HLB of the emulsifier and the type of resulting emulsions.

Table 5 summarizes the relationship between dispersibility of the emulsifier and HLB

Table 5
HLB and Dispersibility of Emulsifiers

Dispersion by observation	HLB
No dispersibility in water	1-4
Poor dispersion	3-6
Milky dispersion after vigorous agitation	6-8
Stable milky dispersion	8-10
Translucent to clear dispersion	10-13
Clear solution	>13

From Jones, S., Kirk Othmer Encyclopedia of Chemical Technology, Vol. 8, Guayson, M., Ed., John Wiley & Sons New York, 1979, 900. With permission.

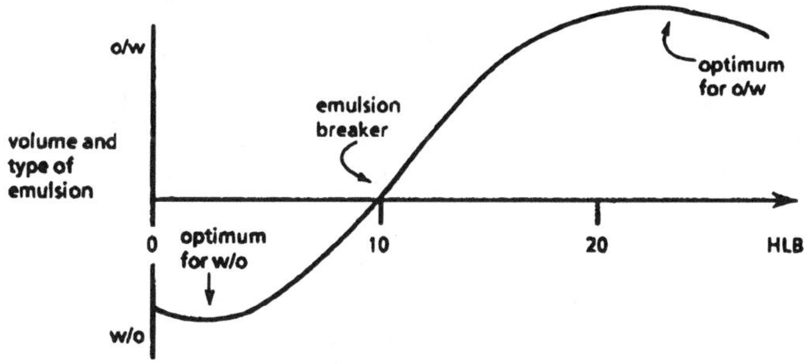

Figure 19 Variation of type and amount of residual emulsion with HLB number of emulsifier. (From Ross, S. and Morrison, I. D., Colloid Systems and Interactions, Wiley Interscience, New York, 1988. With permission.)
Table 6 summarizes the effect of HLB and the type of application.

Table 6
Summary of Applications at Different Ranges of HLB

HLB range	Application
3.5-6	W/O emulsifier
7-9	Wetting agent
8-18	O/W emulsifier
13-15	Detergent
15-18	Solubilizer

From Davis, J.T., Proc. Int. Congr. Surf. Act., 2nd ed., 1, 1957, 426. With permission

HLB values, which are helpful in screening emulsifiers as a first-step process to determine the type of emulsions, are not predictive of emulsion stability. Emulsion stability is a complex function of drop size, interfacial viscosity, interfacial film elasticity, magnitude of electrostatic and steric repulsion, internal phase volume, and the cross sectional area of the micellar components. Therefore, application of the HLB concept cannot solve the practical emulsion problems, but can be used as the first step in the screening process.

Choice of emulsifiers in terms of cohesive energies has been attempted for formulating ECs.[62] A formulator typically chooses the best solvent option for the active ingredients (taking into consideration cost, toxicity, environmental impact) and emulsifying agents that optimize the desired end result. Optimization is therefore done by manipulating surfactants, usually a mixture of nonionics and anionics. Triangular coordinate diagrams are useful in setting up mixture design for composition-based optimization.[4,64,111]

E. POLYMERS

Polymer properties are a complex function of several variables. The following information will be very useful in designing their application: backbone composition, molecular weight distribution, polydispersity (i.e., ratio of weight average and number average molecular weight), solubility in common solvents, θ solvent for the polymer, viscosity in solutions, rheology of polymer solution, glass transition temperature, thermogravimetric phase transition or decomposition temperature, interaction with surfactants.[33]

Use of polymers as stabilizers in emulsion/suspension/dispersion systems is gaining increasing importance. As most of the applications of polymers are in water dispersible granules (WDG), suspension concentrates, and flowables. Some of the trends will be reviewed here. Polymers can provide additional stability to the o/w interface by adsorption to the interface by increasing the steric repulsive energy. Generally block copolymers are used (e.g. EO/PO blocks).[118] Several block polymers are described in the literature[67,88] as candidates for steric stabilization. The polymer should have an anchoring group and stabilizing chains with the correct amount of tails/loops in the adsorbed phase.

Some alternating copolymers have received EPA exemption from tolerance for agricultural applications. These are methyl vinyl ether maleic anhydride/acid/half ester copolymers. These polymers are film forming and are mildly surface active.

F. POLYMERIC SURFACTANTS

1. Comb/Graft Polymers

Certain graft polymers or comb type polymers can replace the function of emulsifiers, with the polymers themselves having appropriate hydrophobic/hydrophilic grafts. Silicones can be considered as a special case of a graft copolymer. The silicone backbone being the hydrophobic part, the $(EO)_x$ being the hydrophilic part (Figure 20.)

Alkylated (graft) pyrrolidone copolymers have very low interfacial tension between oil/water and excellent adhesion to hydrophobic surfaces. This class of polymers has received EPA clearance from tolerance. The surface activity, and oil/ water solubility depend upon the alkyl chain length and the percent content of the alkyl group in the polymer.

Another example of a surface active polymer is $(EO)_x$ groups on both sides of a hydrophobic moiety such as bisphenol (Carbowax 20M)[103] as shown in Figure 21.

This type of structure is similar to the Gemini type of surfactants (see Section III. J. 6).

Another example of polymeric surfactants is polymethacrylic/acrylic acid, esterified with (EO)x. Polymethacrylic acid acts as the hydrophobic backbone and the (EO)x of suitable length offers the stabilizing moiety. Poly-12- hydroxy stearic acid (PHSA) esterified with polyalkylene glycols is an example of block polymeric surfactants. This class of surfactants has better oil solubility and can be tailor made for different applications.[10]

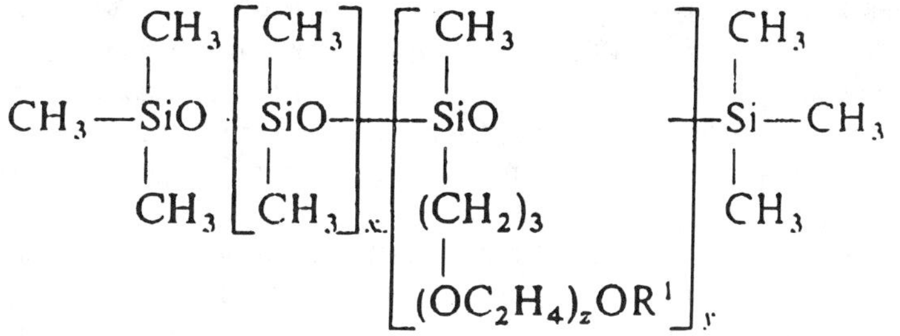

Figure 20 A typical silicone comb polymeric surfactant. (From Rosen, M.J., Surfactants and Interfacial Phenomena, 2nd ed., John Wiley & Sons, New York, 1989, chap. 8. With permission.)

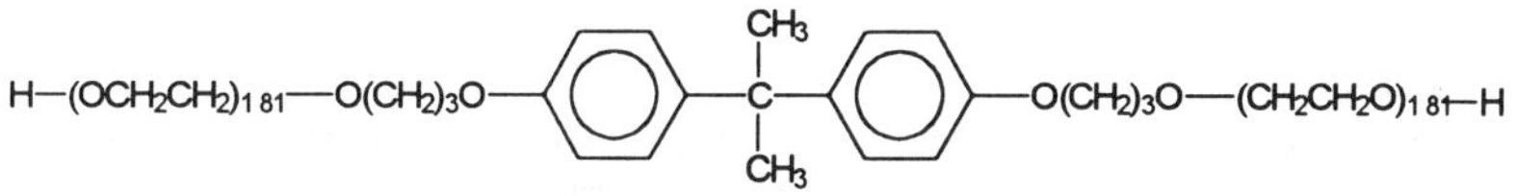

Figure 21 Structure of cabowax 20 M. (From Saroja, Surfactant and Emulsion Stability, PhD Thesis, Clarkson University, A Bell and Howell Company, Michigan, 1993, 48. With permission)

G. PRESERVATIVES

The use of preservatives is sometimes necessary in agricultural liquid formulations. Even though the concentrate contains a pesticide which is a toxicant (for example, herbicide), the matrix could act as a growth media, especially in aqueous systems containing nutrients for bacterial growth. Some of the oxygenated solvents/cosolvents and ethoxylated surfactants can act as a growth media for fungal/bacterial growth. Some of the common preservatives are parabens (alkyl esters of *p*-amino benzoic acid), or methylol derivatives of glycine, hydantoin, and others).

H. OTHER ADDITIVES

Other additives include bittering agents such as denatonium benzoate (DB), which has been used in denaturing alcohol. Use of bittering agents in pesticide formulations as a deterrent against ingestion has been recently recognized. A recent paper reviewed the compatibility and stability of DB in pesticide formulations.[89]

I. POLYMER-SURFACTANT INTERACTION

While choosing the polymer as a protective colloid, interaction between the polymer and surfactants is to be taken into account. This can be analyzed via the effect of polymer on the surface tension of the surfactant.[44,48] See Figure 22 for the effect of polymer on surface tension for SDS/PVP system.

Two effective CMCs are observed for the polymer/surfactant combinations. T_1 (the first transition concentration) is less dependent on the concentration of the polymer than T_2 (the second transition concentration). A second measure of interaction is the binding isotherm.[109] Figure 23 shows the binding isotherm for the SDS/PEO system.

Measurement of electrical conductance, electrophoresis, and viscosity can also provide information about activity of the polymer in binding and relative coverage.[33] Figure 24 summarizes the different types of interactions that can be observed with polymer/surfactant systems and their relative orientations.[34,35]

J. NEWER SURFACTANTS

Some of the promising newer surfactants particularly useful in agricultural formulations are listed here.

1. Silicones

These are polymeric 'comb' type surfactants which have a hydrophobic Si-based backbone with pendent hydrophilic chains based on EO/PO. The general structure is shown in Figure 20. They produce extremely low surface tension (~ 25 dynes/cm), interfacial tension, and contact angles ($< 20^o$). These are extremely good wetting agents and are used for enhanced spreading.[36,65] Figure 25 exemplifies the orientation effect of the relatively flat hydrophobe, with the silicone backbone on the surface with penetrating hydrophilic EO groups into the aqueous phase.

2. Surface Active Pyrrolidones

These (*N*-octyl and *N*-dodecyl, pyrrolidones) are recently developed, commercially available nonionic surface active solvents with excellent solvent power for most pesticides. They can form ion pairs with anionics and can form very stable micelles/invert micelles leading to the formation of microemulsions/solubilization of several pesticides.[53,77,78,79-85,100,101,126]

These are also excellent wetting agents. They can enhance diffusion of active ingredients across the cuticles.[105] These are approved for use in Europe and U.S. EPA classification is pending. These unique solvent/surfactants are also biodegradable.

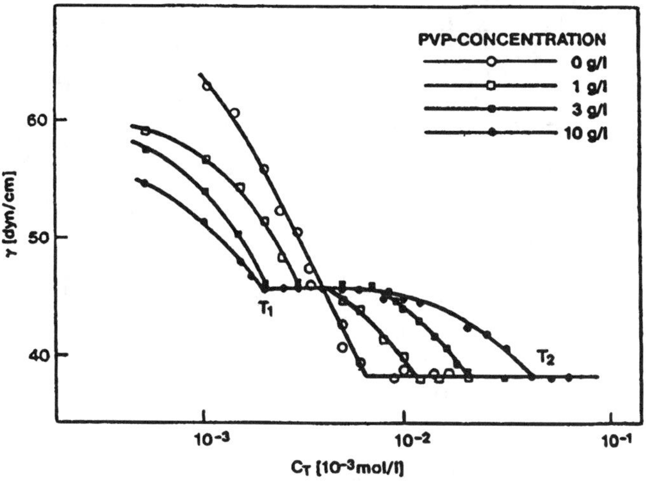

Figure 22 Surface tension (γ)/concentration plot of SDS in the presence of PVP at various concentrations. (From Lange, H., Kolloid Z. Z. Poly., 243, 101, 1971. With permission)

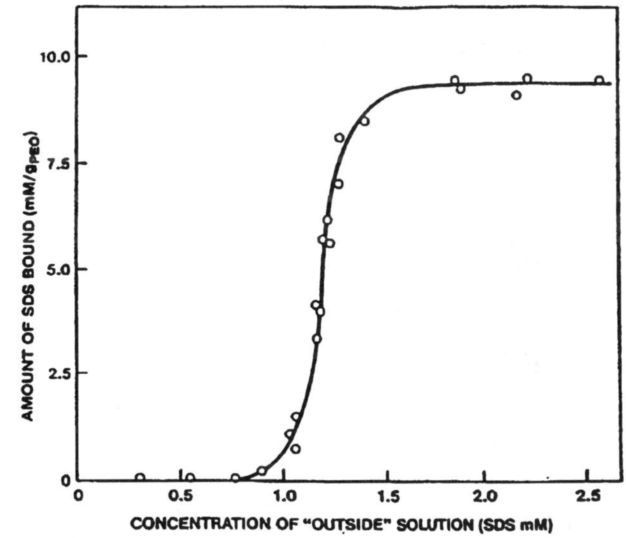

Figure 23 Binding isotherm of PEO-SDS system in 0.1 M Na Cl. (From shirahama, K., Colloid Poly m. Sci., 252, 978, 1979. With permission.)

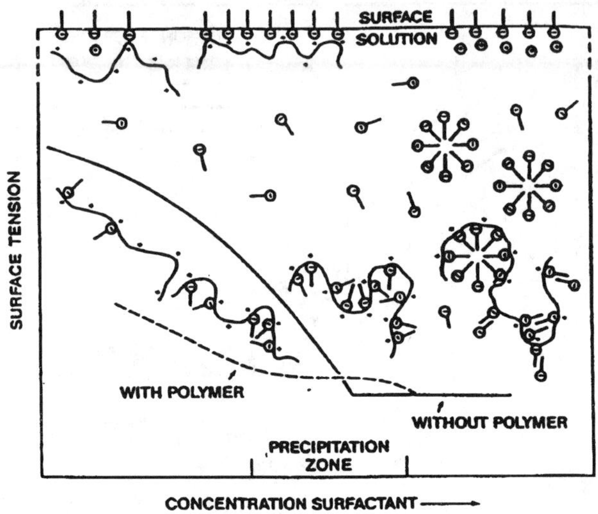

Figure 24 Conditions in the bulk and surface of solutions containing a polycation (fixed concentration) and anionic surfactant. Full line is the hypothetical surface tension-concentration curve of the surfactant alone; dotted line is that of mixture with polycation. Simple polymeric (gregen) cations are dipicted only in surface zone. (From Goddard, E. D., Colloids Surfaces, 19, 255, 1986. With permission.)

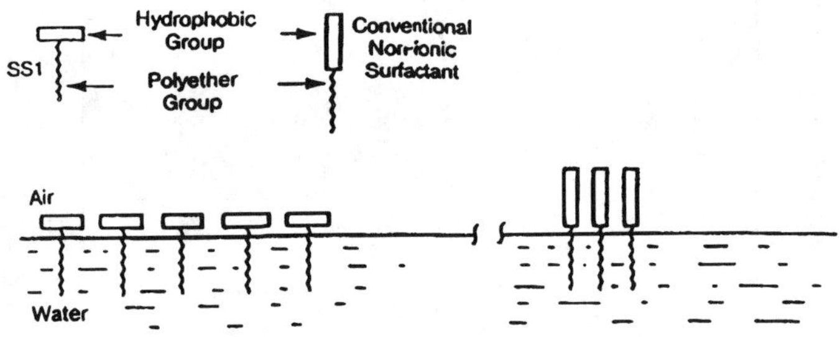

Figure 25 Relative orietation of a polyoxyethylene polysiloxane copolymeric surfactant (SS1) and a conventional surfactant. (From Goddard, E. D., Adjuvants and Agrochemicals, Foy, C. L., Ed., CRC Press, Boca Raton Fl, 1992, 373. With permission)

3. Alkyl Polyglycosides

Some of the recently developed biodegradable/food grade surfactants are alkyl polyglycosides. These are long chain acetals of polysaccharides.[1] Figure 26 shows the general structure.

The properties are similar to alcohol ethoxylates, but solubility is higher in water and in the presence of electrolytes. Unlike the alcohol ethoxylates, alkyl polyglycosides do not possess the inverse solubility temperature gradient, therefore they possess higher temperature tolerance.

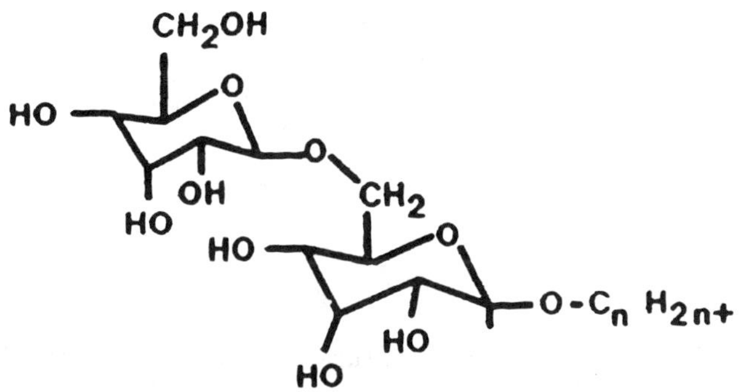

Figure 26 Example of alkyl($C_{10}H_{21}$) polyglycoside molecule. (From Aleksejczyk, R. A., Pesticide Formulations and Application Systems, Vol. 13, Devissety, B. N., Chasin, D. 6., and Berger, P. D., Ed., American Society for Testing Materials, Philadelphia, 1993, 22. With permission.)

4. Sucroglycerides

These are designed as food grade biodegradable mixed systems containing polyglycerides and transesterified sucrose fatty esters. The use of these low toxicity emulsifiers has been demonstrated.[25] Figure 27 shows the transesterification route from triglycerides for the preparation of sucroglycerides.

5. Styrylated Phenyl Ethoxylates and Derivatives

A modification of alkyl phenyl ethoxylates and phosphate esters of the above is to replace the hydrophobe (alkyl group) by styryl substitution. An example is shown in Figure 28.

6. Gemini Surfactants

These are surfactants that have two hydrophilic groups and two hydrophobic groups per molecule (Figure 29 and 30).

Gemini surfactants have very low CMC and are very efficient in reducing surface tension (i.e., low C_{20} value, which is the concentration required to reduce the surface tension by 20 dynes/cm). These surfactants are excellent hydrotropes and can solubilize hydrophobic water insoluble

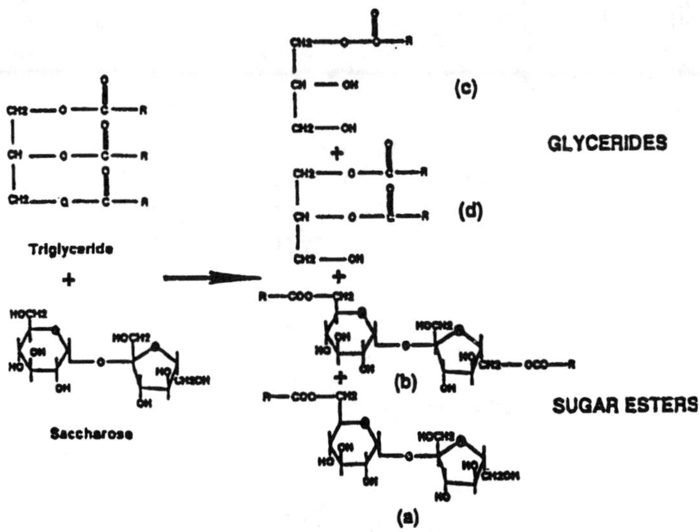

Figure 27 Examples of glyceride, and sugar ester surfactants. (From Fiard, J. F., Mercier, J. M., and Prevotat, M. L., (From Chasin, D. G., and Boger, P. D., Ed., American Society for Testing Materials, Pesticide Formulations and Applications Systems, Vol. 13, Devissety, B. N., Philadelphia, 1993, 33. With permission.)

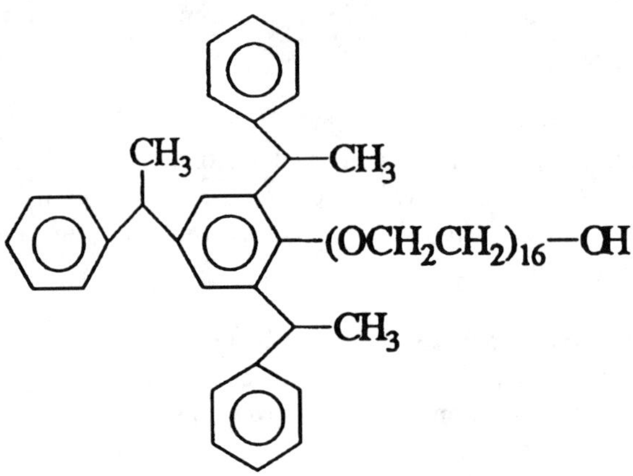

Ethoxylated tristyrylphenol(SOPROPHOR BSU™).

Figure 28 Example of ethoxylated tristyrylphenol surfactant. (From Derian, P. J., Guerin, G., and Fiard, J. F., (From Pesticide Formulations and Application Systems, Vol. 13, Devissety, B. N., Chasin, D. G., and Berger, P. D., Eds., American Society for Testing and Materials, Philadelphia, 1993, 73. With permission.)

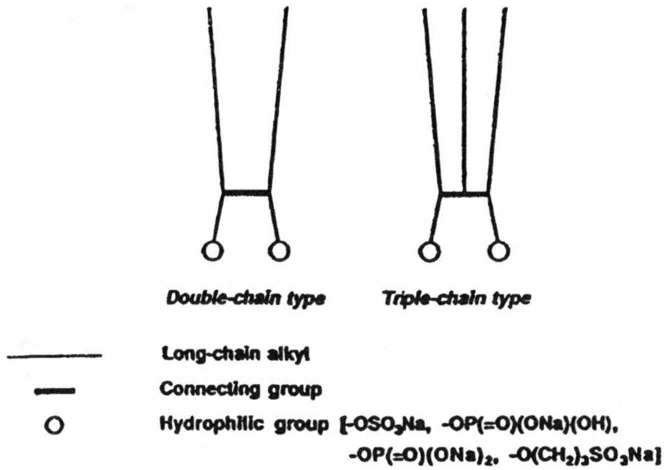

Figure 29 Molecular models of a noval series of Gemini surfactants with 2 or 3 hydrophobic chains connected by a short chain and 2 hydrophilic groups. (From Zhu, Y, Masuyana, A., Kobata, V., Nakatsuj, Y., Okahara, M., and Rosen, H., J. Colloid Interface Sci., 158, 40, 1993. With permission.)

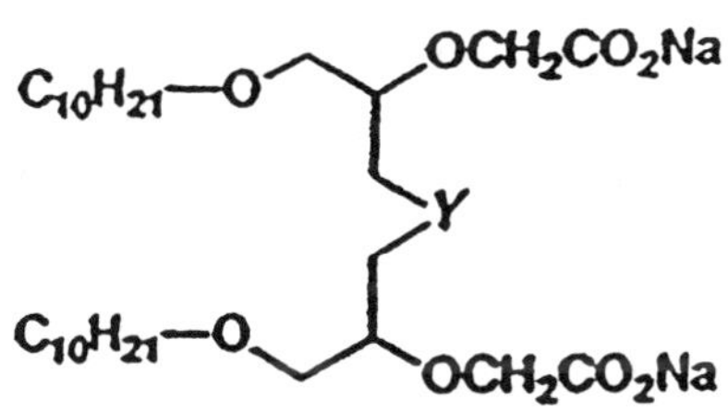

4

4	-Y- =	
a		-O-
b		$-OCH_2CH_2O-$
c		$-O(CH_2CH_2O)_2-$
d		$-O(CH_2CH_2O)_3-$
e		$-O(CH_2)_4O-$

Figure 30 Examples of double chain copmpounds (4), Gemini class of surfactants. (From Zhu, Y., Masuyama, A., Kobata, Y., Nakatsuji, Y., Okahara, M., and Rosen, M. J., J. Colloid Interface Sci., 158, 40, 1993. With permission.)

organics efficiently when present above their CMCs. As the CMCs are an order or two orders of magnitude lower than conventional surfactants, only a very small concentration of Gemini will be required.[61,95,99,125]

IV. RECENT LITERATURE EXAMPLES OF EMULSIONS AND MICROEMULSIONS

A. EMULSIFIABLE CONCENTRATES (ECs)

An EC is usually formulated by first preparing a concentrated stable solution of the pesticide(s) in a single solvent. The solution is optimized with respect to mixed surfactant systems to produce the desired properties on dilution. Some of the newer solvents, for example, *N*-octyl and *N*-dodecyl pyrrolidones are rare examples of surface active solvents. Use of these solvents in formulating ECs and microemulsion concentrates hase been demonstrated.[15,69-85]

The use of mixed solvents in the formulation of emulsifiable concentrates has been demonstrated[78] by optimizing the compositions in terms of the proportion of the solvents by the use of a single emulsifier for a number of active ingredients with widely different chemical structures. Figure 31 shows the optimum solvent compositions for a fixed weight of active ingredients and a fixed weight of the emulsifier systems.

Further optimization can be accomplished by using mixed surfactants system. The steps involved in formulating an EC follow the sequence: choice and optimization of solvent system, choice and optimization of emulsifier, fine tuning after stability evaluation and feedback from biological efficacy.

B. MICROEMULSION CONCENTRATES (MECs) AND MICROEMULSIONS (MEs)

The formation of MEC or ME is generally accomplished by following phase diagrams.[111] Figure 32 shows a schematic phase diagram identifying general phase regions for a microemulsion system.[95]

In agricultural formulations, if one is interested in preparing water-based MEs, particular attention is paid to the area in the phase diagram represented by the lower left-hand portion (Figure 32). As a large number of compositions are to be prepared and their phase behavior to be studied, use of robotics has been applied to formulate microemulsions successfully,[21] especially while investigating the entire system.

A few published examples of microemulsions for pesticides will be reviewed. Use of mixed *N*-alkylpyrrolidones and optimized surfactants (Igepal CO630 and EO/PO/EO blocks, Pegol L31) has been successfully demonstrated for the formation of stable mixed pyrethroids and synergist (D allethrin, permethrin, tetramethrin, and piperonyl butoxide) as microemulsions in water.[77] Figure 33 shows the optimized zone for the surfactant, solvent portion of the microemulsion composition.

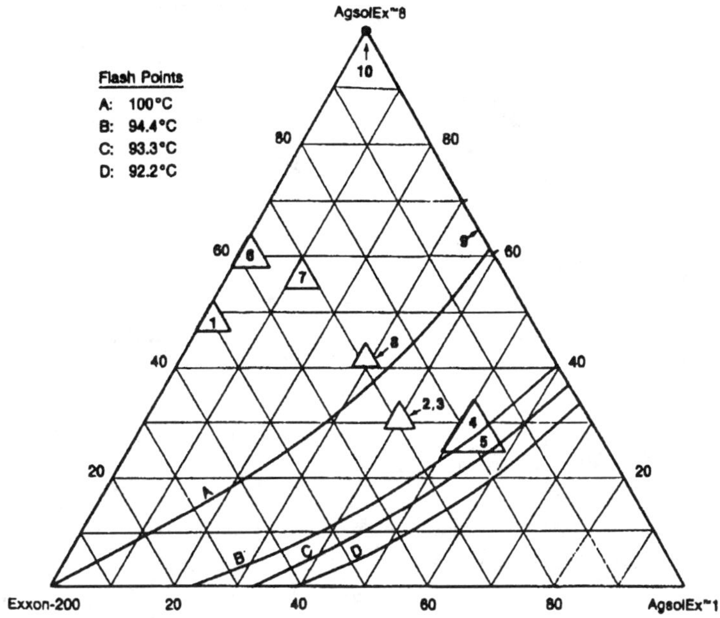

1.	Dichlofluanid: 10% Gafac RE-610: 5% Silwat: 5%	6.	Prodiamine: 10% Gafac RE-610: 10%
2.	Atrazine: 5% Gafac RE-610: 5%	7.	Prodiamine: 15% Gafac RE-610: 10%
3.	Atrazine: 5% Dual: 20% Gafac RE-610: 15%	8.	Triforine: 12% Gafac RE-610: 8%
4.	Metolachlor: 40% Gafac RE-610: 15%	9.	Carbaryl: 10% Gafac RE-610: 10%
5.	Pendimethalin: 30% Gafac RE-610: 18%	10.	Thidiazuron: 15% Gafac RE-610: 10%

Balance is made to 100% with the appropriate solvent compositions.

Figure 31 ECs formulated via optimized solvent systems. (From Narayanan, K. S. and Chaudhuri, R. K., Emulsifiable concentrate formulations for multiple active ingredients using N-alkylpyrrolidones, in Pesticide Formulations and Application Systems, Vol. 11, Bode, L. E. and Chasin, D. G., Eds. ASTM STP 1112, American Society for Testing and Materials, Philadelphia, 1992, 73. With permission.)

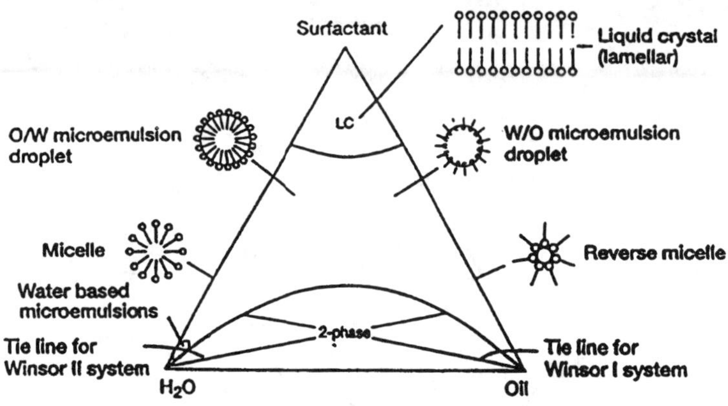

Figure 32 Schematic triangular phase diagram. (From Robinson, B.H., Chem. Br., 26, 342, 1990. With permission.)

An alternate system containing *N*-octylpyrrolidone and SDS is also a very useful system for preparing microemulsions. For example, carbaryl can be microemulsified in the above system, and can be stabilized by buffering using hydrophobic acid such as Gafac RE610 (nonyl phenyl ethoxylated phosphate esters). *N*-alkylpyrrolidones are the key ingredients in the microemulsion systems.[69-75] A correlation was attempted between the structural properties of several solvents [in terms of molar volume, aromaticity, chain linearity, presence of polar groups] and selection of surfactant types and adjustment of HLB in predicting microemulsion phase behavior.[37]

Figure 34 shows the particle size distribution, which is consistent with a w/o/w working model for the microemulsion system.

Use of tristyryl phenol based surfactants for formation of a stable water based pyrethroid microemulsion has been demonstrated. The optimized surfactant system consisted of tristyryl phenyl ethoxylate, phosphate ester of the above, neutralized with TEA and cyclohexanol/isobutanol and others as the cosolvents.[21,43]

Permethrin and cypermethrin have been microemulsified in water by the use of polyalkylphenol ethoxylate, (tristyrylphenol ethoxylate) dodecyl benzene sulfonate and isobutanol/octanol as cosolvent. The flash point of the above system was reported as > 100°C, which suggests a closely packed micellar system encapsulating the isobutanol component.[91]

Use of fatty acid methyl esters [C_8-C_{12}] as solvents, vegetable oil ethoxylate, EO/PO blocks as nonionic surfactants, and Ca dodecyl benzene sulfonate as an anionic surfactant has been demonstrated for formulation of MECs for trifluralin, dicofol, and hydroprene.[110]

The literature contains evidence for improved efficacy of pesticides formulated as microemulsions and their relative advantages.[49,58,90,94,117]

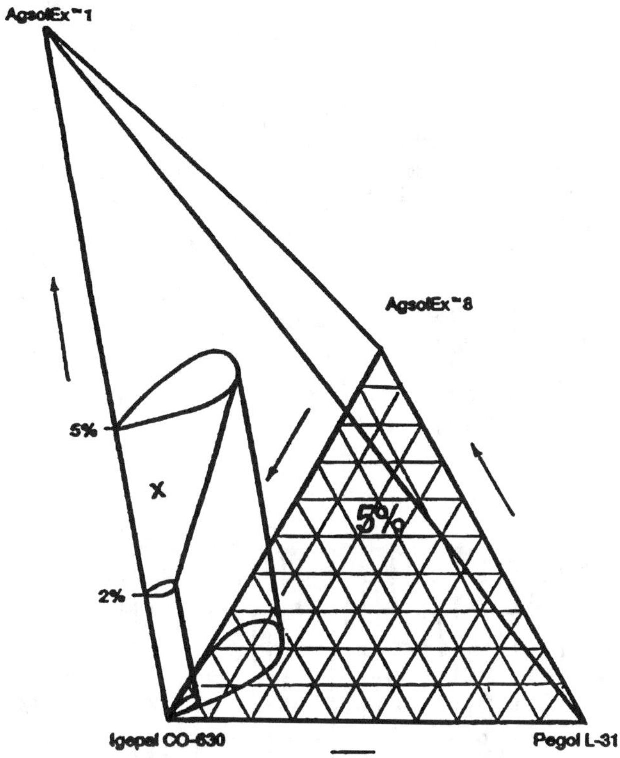

Composition:	
Emulsion:	D-Althethrin = 0.05%; Permethrin = 0.15% Tetramethrin = 0.2%; Piperonyl Butoxide = 1% AgsolEx 1 = 2 to 5%; Surfactant System as Shown; 5% qs to 100% with Water Optional: Film Forming Agent 0 to 1% (Agrimer 30 or Agrimer VA6)
Substrate:	Mixed Pyrethroids and Synergists in the above ratio: 12.4 to 16.7% Total Surfactant: 43.8 to 59.5% AgsolEx 1: 43.8 to 23.8%

Figure 33 Optimized surfactant-cosolvent system for a water based microemulsion for mixed pyrethroids. Compositions of the surfactant-cosolvent that produced stable microemulsions are given by the cone X, cosolvent 2-5%; total surfactant 5% in the microemulsion. Projection on the plane of the triangle gives the surfactant compositions (5%). (From Narayanan, K. S. and Chaudhuri, R. K., Pesticide Formulations and Application Systems, Vol. 12, Devissety, B. N., Chasin, D. G., and Berger, P. D., Eds., American Society for Testing and Materials, Philadelphia, 1993, 85. With permission.)

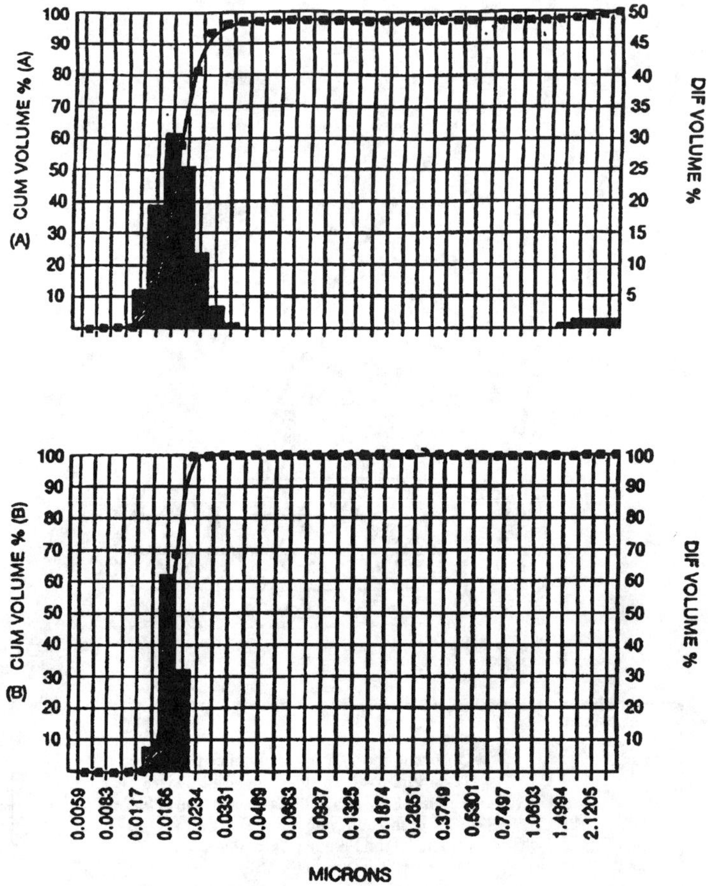

Figure 34 Particle size distribution of water-based micoemulsion. (From Narayanan, K. S. and Chaudhuri, R. K., Pesticide Formulations and Application Systems, Vol. 12, Devissety, B. N., Chasin, D. G., and Berger, P. D., Eds., American Society for Testing and Materials, Philadelphia, 1993, 85. With permission.)

However, MECs are more expensive, as much higher amounts of emulsifiers are required compared to EC formulations. The loading is typically 15 to 20% in microemulsions.

Table 7 shows the commonly used ingredients in EC/MEC formulations. Those marked with an asterisk are relatively new inerts.

V. OPTIMIZATION

Agricultural formulations (EC, MEC) contain high loading of the active ingredients, solvent, cosolvents, mixed surfactants, and polymers in the concentrates. The mode of use is invariably in aqueous medium (usually water) containing a high level of electrolytes (1000 ppm hardness as $CaCO_3$ equivalent). Often the dilution is performed in the presence of fertilizers.

The objective of an acceptable formulation is to realize the highest loading of active ingredients, and use the least effective quantities of inert ingredients described above (solvents, cosolvents, emulsifiers, polymers). In addition to the storage stability of the concentrate and its characterization, the formulation should be compatible with other commercial formulations and should show acceptable stability of suspension in water to maintain a uniform concentration of the active ingredients during its use.

Standard methods for evaluating stability and compatibility are described elsewhere and documented in the literature.[1,4,5,11,14,17,21,43,76-78,91,110,111] Recently, new techniques have been developed to evaluate emulsifiable concentrates and microemulsion concentrates. These test methods are based on measurements of contact angle, dynamic surface tension, and particle size distribution.[114] It is generally the practice to monitor the physical stability of the formulation on dilution. However, it is preferred to relate the physical stability (i.e., absence or minimum of separation of the active ingredients with time) to distribution of the active ingredient in the diluted system via analytical determination of the active species. As fast analytical methods (HPLC, GC) are available, the species-specific method should be incorporated in the testing protocol.

As pesticide formulations are complex systems and specific computer programs are not readily available, it is often customary to use partial optimization limited to three components, using triangular coordinates and mixture compositions, wherein the total weight fraction is unity. The points can be plotted based on performance and optimized areas of composition or contours can be generated.[78,111] In cases where a large number of formulations are generated, use of robotics for preparation of formulations and screening the same has been reported.[11,21]

Most of the computer aided designs are suited for analysis of process variables or biological results, wherein factorial designs are used as a model. Use of such models are not suited for agricultural formulations,[124] as attempts to apply factorial designs often lead to impractical mixtures. A mixture design often reduced to three broad components such as set of three emulsifiers for a fixed active ingredient/solvent composition;[77,111] or a set of three solvents for a fixed weight of active ingredients and emulsifiers;[78] or (1), solution of the active ingredient in a solvent: (2), water; and/or (3) cosolvent and emulsifiers[91] as the third component. The total weight fraction of the three broad components equal to unity is often useful from a practical standpoint. Use of computer program in agricultural formulations development has been demonstrated in a few cases.[4,62,64] The general approach is to use the three-component mixture design, and fix the data as contour plots.

Table 7
Ingredients Commonly Used in EC/MEC Formulations

Ingredients	**Examples**
Solvents:	Xylenes, vegetable oils, aromatic petroleum oils, aliphatic hydrocarbons, esters, alcohols, ketones (isophorone, cyclohexanone), methyl esters of C_8 -C_{18} fatty acids. DMF and others; N-methyl pyrrolidone *, N- octyl pyrrolidone *, N-dodecyl pyrrolidone *, γ butyrolactone, alkyl biphenys *, tetrahydrofurfuryl alcohols and ethers *.
Surfactants,	Ethoxylated fatty alcohols (C_8 - C_{18} alcohols), nonionics:ethoxylated alkyl phenols (C_8, C_9, C_{12} - C_{16} alkyl), ethoxylated dialkyl phenols (C_8, C_9, C_{12} - C_{16} alkyl), ethoxylated castor oil, ethoxylated alkyl amines, ethoxylated tristyryl phenols *, [sucrose esters + glycerylesters *], sucrose ethers (C_8 - C_{18}
Surfactants, anionics:	alkyl) *.alkyl (C_8 - C_{18}, usually C_{12}) sulfates, alkyl (C_8 - C_{18}, usually C_{12}) benzene/toluene/xylene sulfonates, alkylphenyl ethoxylated phosphates/ sulphates (C_8, C_9, C_{12}- C_{16} alkyl),dialkylphenyl ethoxylated phosphates/sulphates (C_8, C_9, C_{12} - C_{16} alkyl), ethoxylated tristyryl phosphates/sulphates *, as sodium, calcium, amine, or alkanol amine salts.
Block polymers:	$(EO)_x$ $(PO)_y$ $(EO)_z$ or $(PO)_x$ $(EO)_y$ $(PO)_z$ where x,y, and z vary from 0 to 100, and x + y + z = 10 to-100 Polymeric surfactants, (comb/graft): Silicones with alkyl silicone backbone and $(EO)_x/(PO)_y$ pendent groups, with x/y varying from 10 to 100 and molecular weight in the range 500 to 5000, [poly dimethyl siloxane poly alkyleneoxide copolymer], ethoxylated/propoxylated alkyl phenyl blocks*,alkylated poly vinyl pyrrolidones (C_4-C_{20} alkyl) and % pyrrolidone in the range: 10 to 90 *.

A generalized polynomial for fitting the data is shown in Equation 24:

$$\phi = \sum^{n} \beta_i X_i$$

$$+ \sum^{n} \beta_{ij} X_i X_j$$

$$+ \sum^{n} \beta_{ijk} X_i X_j X_k$$

$$+ \sum^{n} \beta_{iijk} X_i X_j (X_i - X_j) \quad (24)$$

$$\Sigma X_i = \text{constant} \quad (25)$$

$$\phi = \Pi \phi_i$$

β's are the coefficients of interactions, X_i is the weight fraction of the ith component in a n-component system; ϕ is the composite response function and is the product of several individual weighted responses, ϕ_i (such as stability, cost,. biological performance, and others)

The generalized cubic Equation 24 reduces to special cases as follows:

$$\text{for linear Equation, } \beta_{ij} = \beta_{ijk} = \beta_{iijk} = 0 \quad (26)$$

$$\text{for quadratic fit, } \beta_{ijk} = \beta_{iijk} = 0 \quad (27)$$

$$\text{for special cubic fit, } \beta_{iijk} = 0 \quad (28)$$

for a simplified case of three components, where n = 3, assuming a special cubic model Equation 24 reduces to Equation 29

$$\phi = \beta_1 X_1 + \beta_2 X_2 + \beta_3 X_3 + \beta_{12} X_1 X_2 + \beta_{13} X_1 X_3\ \beta_{23} X_2 X_3 + \beta_{123} X_1 X_2 X_3 \quad (29)$$

$$X_1 + X_2 + X_3 = \text{constant} \quad (30)$$

For most purposes as shown for a reduced three-component systems, the simplified Equation 29 would be sufficient.[5,64,78] The response function can be plotted in a triangular diagram as contours. Computer-aided design and analysis of results for formulations will be used increasingly as programs and software become available.

VI. FUTURE TRENDS

Formulation science has become an advanced technology. Use of computer-aided experimental designs for formulations and their evaluations is a common practice today. Solvent-based liquid concentrates will continue to be a major formulation type because of inherent advantage the solvents offer for biological efficacy and convenience of use of such formulations. The environmentally and toxicologically safe solvents (not included in EPA's Lists 1 and 2 will eventually replace the other solvents included in Lists 1 and 2. Formulations of the future will be designed for increased biological activity and will use lower levels of active ingredients. Such formulations will contain multiple active ingredients for broad biological activity spectra. The well-designed adjuvant systems will be included in the concentrate as part of formulation package, so that the enduser receives full instruction without having to use other additives. The advantages of adjuvant incorporation in the concentrate will include: increased spreader sticker activity, penetration, translocation where needed, reduced leaching, reduced drift, and biodegradability of the inert system. A review on "issues on inert ingredients" was recently published.[124]

A direct conversion of the above system would be the design of an instantly microemulsifiable solid/gel that would disperse into water instantly to produce emulsion/dispersion with infinite or long-term stability.

Finally, the industry would generate several inert carrier systems with multifunctional advantages, as described earlier, as inert matrices designed for several classes of active ingredients or species-specific formulations. The design of inert systems will parallel the design of the active ingredients.

VII. CONCLUSIONS

We have discussed the elements of emulsions, microemulsions and factors leading to their formation, stability, and the components that constitute them. We have reviewed the availability of the tools, newer solvents, surfactants, polymers, and methods of formulation and their evaluation. Formulations of the future will incorporate optimized components to offer additional benefits of adjuvants as needed. Formulation is an advanced technology, which is evolving constantly. There is increasing interdisciplinary interaction in creating a successful formulation. There is a greater acceptance by the industry for the new generation of inert ingredients.

ACKNOWLEDGMENTS

I wish to thank International Specialty Products, Inc., for permission to publish this work. I thank Dr. Robert, Ianniello Director, Pharmaceuticals, Agricultural, and Beverage Technology, at International Specialty Products for critically reviewing the manuscript. I also thank Ms. Melanie C. Sze, Supervisor, Technical Library, for her assistance in compilation of bibliography section; Ms. Suzanne Currie, and Ms. Lisa Ruit, Administrative Coordinators, for their able and prompt secretarial work.

REFERENCES

1. **Aleksejczyk, R. A.,** Alkyl polyglycosides: versatile, biodegradable surfactants for the agricultural industry, in *Pesticide Formulations and Application Systems,* Vol. 13, Devisetty, B. N., Chasin, D. G., and Berger, P. D., Eds., ASTM STP 1146, American Society for Testing and Materials, Philadelphia, 1993, 22.
2. **Barton, A. F. M.,** Ed., *CRC Handbook of Solubility Parameters and Other Cohesion Parameters*, CRC Press Inc., Boca Raton, FL, 1983.
3. **Bauer, H. and Schonherr, J.,** Determination of mobilities of organic compounds in plant cuticles and correlation with molar volumes, *Pestic. Sci.,* 35, 1, 1992.
4. **Becher, D. Z.,** Formulation test methods and statistical experimental design, in *Pesticide Formulations and Application Systems,* Vol. 14, Hall, F. R., Berger, P. D., and Collins, H. M. Eds., ASTM STP 1234, American Society for Testing and Materials, Philadelphia, 1995.
5. **Becher, D. Z.,** Pesticide compatibility, in *Pesticide Formulations and Application Systems,* Vol. 11, Bode, L. E. and Chasin, D. G., Eds., ASTM STP 1112, American Society for Testing and Materials, Philadelphia, 1992, 121.
6. **Becher, P.**, *Emulsions: Theory and Practice*, Reinhold, New York, 1965.
7. **Becher, P.,** Hydrophile-lipophile balance: an updated bibliography, in *Encyclopedia of Emulsion Technology, Applications*, Vol. 2, Becher, P., Ed., Marcel Dekker, New York, 1985.
8. **Becher, P.,** Philosophy of formulation, in *Pesticide Formulations and Application Systems,* Vol. 8, Hovde, D. A. and Beestman, G. B., Eds., ASTM STP 980, American Society for Testing and Materials, Philadelphia, 1989, 54.
9. **Boehn, J. T. C. and Lyklema, J.,** The adsorption behavior of polyelectrolytes at liquid-liquid interfaces and the properties of polyelectrolyte-stabilized emulsions, in *Theory and Practice of Emulsions Technology,* Smith, A. L., Ed., Academic Press, New York, 1976, 23.
10. **Bognolo, G., Chasin, D. G., Hoorne, D., and Rogiers, L. M.,** Polymeric surfactants and their application in agrochemical formulations, in *Proc. 11th ASTM Symposium on Pesticide Formulations and Application Systems,* San Antonio, TX, Nov., 1990.

11. **Bolts, M. F.,** Developing formulations with robotics, in *Pesticide Formulations and Application Systems,* Vol. 13, Devisetty, B. N., Chasin, D. G., and Berger, P. D., Eds., ASTM STP 1146, American Society for Testing and Materials, Philadelphia, 1993, 188.
12. **Boyd, J., Parkinson, C., and Serman, P.**, Factors affecting emulsion stability, and the HLB (hydrophilic-lipophilic balance) concept, *J. Colloid Interface Sci.*, 41, 359, 1972.
13. **Brewer, M. M. and Robb, I. D.**, Interactions between macromolecules and detergents, *Chem. Ind.*, 530, 1972.
14. **Catanach. J. S. and Hampton, S. W.,** Solvent and surfactant influence on flash points of pesticide formulations, in *Pesticide Formulations and Application Systems,* Vol. 11, Bode, L. E. and Chasin, D. G., Eds., ASTM STP 1112, American Society for Testing and Materials, Philadelphia, 1992, 149.
15. **Chaudhuri, R. K., Narayanan, K. S., and Dahanayake, M.,** Delivery System for Agricultural chemicals, U.S. Patent 5,160,528, 1992.
16. **Clayfield, E. J. and Jumb, E. C.,** An entropic repulsion theory for random copolymer dispersant action, *Macromolecules,* 1, 133, 1968.
17. **Collins, H. M. and Munie, L. A.,** Improvements in dry flowable tank mix compatibility, in *Pesticide Formulations and Application Systems,* Vol. 11, Bode, L. E. and Chasin, D. G., Eds. ASTM STP 1112, American Society of Testing and Materials, Philadelphia, 1992, 134.
18. **Davis, J. T.,** A quantitative theory of emulsion type. 1. Physical chemistry of the emulsifying agent, in *Proc. Int. Congr. Surf. Act.*, 2nd ed., 1, 1957, 426.
19. **Davis, J. T. and Rideal, E. K.,** *Interfacial Phenomena,* 2nd ed., Academic Press, New York, 1963, chap. 8.
20. **Davis, J. T. and Rideal, E. K.**, Disperse systems and adhesion, *Interfacial Phenomena*, 2nd ed., Academic Press, New York, 1963, chap. 8.
21. **Derian, P. J., Guerin, G., and Fiard, J. F.,** Microemulsions of pyrethroids: phase diagrams and effectiveness of tristyrylphenol based surfactants in, *Pesticide Formulations and Application Systems,* Vol. 13, Devisetty, B. N., Chasin, D. G., and Berger, P. D., Eds., ASTM STP 1146, American Society for Testing and Materials, Philadelphia, 1993, 73.
22. **Deryagin, B. V. and Landau, L. D.,** Theory of stability of strongly charged lyophobic sols and of adhesion of strongly charged particles in solutions of electrolytes, *Acta Physicochim. U.S.S.R.* 14, 633, 1941.
23. **Doyal, K. J., McKillip, W. J., Shin, C. C., and Richard, D. A.,** Comparison of the cyclic ester-alcohol tetrahydrofurfuryl alcohol to other known solvents, in *Adjuvants and Agrichemicals,* Foy, C. L., Ed., CRC Press, Boca Raton, FL, 1992, 225.

24. **Federal Register**, See List 1 and 2, *40.CFR 154.6*, dated April 22, 1987.

25. **Fiard, J. F., Mercier, J. M., and Prevotat, M. L.**, Sucroglycerides: new biodegradable surfactants for plant protection formulations, *Pesticide Formulations and Application Systems*, Vol. 13, Devisetty, B. N., Chasin, D. G., and Berger, P. D., Eds., ASTM STP 1146, American Society for Testing and Materials, Philadelphia, 1993, 33.

26. **Field, R. J. and Bishop, N. J.**, Promotion of stomatal infiltration of glyphosate by an organosilicone surfactant reduces the critical rainfall period, *Pestic. Sci.*, 24, 55, 1988.

27. **Florence, A. T., Elworthy, P. H., and Rogers, J. A.**, Stabilization of oil-in-water emulsions by nonionic detergents. VI. The effect of a long chain alcohol on stability, *J. Colloid Interface Sci.*, 35, 34, 1971.

28. **Freed, Y. H. and Witt, J. M.**, Physicochemical principles in formulating pesticides relating to biological activity, in *Pesticidal Formulation Research, Advances in chemistry Series 86*, American Chemical Society, Washington, D.C., 1969, 70.

29. **Freshwater, D. C., Scarlett, B., and Groves, M. J.**, Particle size of pharmaceutical emulsions, *Am. Cosmet. Perfum.*, 81, 43,1966.

30. **Friberg, S. E. and Jones, S.**, Emulsions, in *Kirk Othmer Encyclopedia of Chemical Technology*, Vol. 9, Kroschwitz, J. I., Ed., John Wiley and Sons, New York, 1993, 393.

31. **Friberg, S. E., Mandell, L., and Frontell, K.**, Mesomorphous phases in systems of water-nonionic emulsifier-hydrocarbon, *Acta Chem. Scand.*, 23, 1055, 1969.

32. **Friberg, S. J. Mandell, L., and Larsson, M.**, Mesomorphous phases, a factor of importance for properties of emulsions, *J. Colloid. Interface Sci.*, 29, 155, 1969.

33. **Goddard, E. D.**, Polymer/surfactant interaction, *J. Soc. Cosmet. Chem.*, 41, 23, 1990.

34. **Goddard, E. D.**, Polymer-surfactant interaction. I. Uncharged water-soluble polymers and charged surfactants, *Colloids Surfaces*, 19, 301, 1986.

35. **Goddard, E. D.**, Polymer-surfactant interaction. II. Polymer and surfactant of opposite charge, *Colloids Surfaces*, 19, 255, 1986.

36. **Goddard, E. D. and Padmanabhan, K. P. A.**, A mechanistic study of the wetting, spreading, and solution properties of organosilicone surfactants, in *Adjuvants and Agrichemicals*, Foy, C. L., Ed., CRC Press, Boca Raton, FL, 1992, 373.

37. **Graff, J. L., Bock, J., and Robbins, M. L.**, Effects of solvent phase behavior, in *Pesticide Formulations, Innovations and Developments*, Cross, B. and Scher, H. B., Eds., ACS Series 371, American Chemical Society, Washington, D.C., 1988, chap. 15.

38. **Graham, D. E. and Phillips, M. C.,** Conformation of proteins at interfaces and their role in stabilizing emulsions, in _Theory and Practice of Emulsions Technology,_ Smith, A. L. Ed., Academic Press, New York, 1976, 75.
39. **Groves, M. J., Kaye, B. H., and Scarlett, B.**, Size analysis of sub-sieve powders using a centrifugal photosede mentometer, _Br. Chem. Eng._, 9, 742, 1964.
40. **Hartley, D.,** Ed., _The Agricultural Handbook,_ 2nd ed., Royal Society of Chemistry, London, 1987.
41. **Herb, C. A., Bergess, E. J., Chang, K., Morrison, I. D., and Grabowski, E. F.**, Using quasi-elastic light scattering to study particle size distributions in submicrometer emulsion systems, in _Particle Size Distribution,_ Provder, P., Ed., ACS Symposium Series 332, American Chemical society, Washington, D.C., 1987, 89.
42. **International Union of Pure and Applied Chemistry,** _Manual of Colloid and Surface Science,_ Butterworths, London, 1972.
43. **Isabelle, V., Bonneau, G., Mailhe, P., and Perrin, M. A.,** Tristyrylphenol surfactants in agricultural formulations: properties and challenges in applications, _Pesticide Formulations and Application Systems,_ Vol. 14, Hall, F. R., Berger, P. D., and Collins, H. M., Eds., ASTM STP 1234, American Society for Testing and Materials, Philadelphia, 1995.
44. **Jones, M. N.,** Dye solubilization by a polymer-surfactant complex, _J. Colloid Interface Sci._, 26, 532, 1968.
45. **Jones, S.**, Emulsions, in _Kirk Othmer Encyclopedia of Chemical Technology,_ Vol. 8, Garyson, M., Ed., John Wiley and Sons, New York, 1979, 900.
46. **Krenek, M. R. and King, D. N.,** The relative phytotoxicity of selected hydrocarbon and oxygenated solvents and oils, in _Pesticide Formulations and Application Systems,_ Vol. 6, Vander Hoven, D. I. B. and Spicer, L. D., Eds., STM STP 943, American Society for Testing and Materials, Philadelphia, 1987, 3.
47. **Krenek, M. R. and Rohde, W. H.,** An overview-solvents for agricultural chemicals, in _Pesticide Formulations and Application Systems_, Vol. 8, Houde, D. A. and Beestman, G. B., Eds., ASTM STP 980, American Society for Testing and Materials, Philadelphia, 1988, 113.
48. **Lange, H.,** Interaction between sodium alkyl sulfates and polyvinyl pyrrolidone in aqueous solutions, _Kolloid Z. Z. Polym._, 243, 101, 1971.

49. **Lankford, W. T. and Davis, L. G.,** The effect of cosurfactant HLB on the toxicity of Cypermethrin microemulsion formulation to Blalella Germanica, in *Proc. 7th Int. Congr. Pesticide Chemistry*, Vol II, Freshe, H., Kesseler, Schmidtz, E., and Conway, S., Eds., Hamburg, Germany, 1990, 42.
50. **Leeds Northrup,** *Instruction Manuals* 179521 and 179523, 1990.
51. **Levius, H. P. and Drommond, F. G.,** Elevated temperature effects on emulsion stability, *J. Pharm. Pharmacol.*, 5, 743, 1953.
52. **Lissant, K. J., Ed.,** *Emulsions and Emulsion Technology*, Marcel Dekker, New York, 1974.
53. **Login, R. B., Chaudhuri, R. K., Haldar, R. K., Hashem, M. M., Helioff, M. W., and Tracy, D. J.,** Surface Active Lactams, *U.S.* Patent, 5,093,031, 1992.
54. **Madani, K. and Friberg, S. E.,** Van der Waals interactions in three phase emulsions, *Prog. Colloid Polym. Sci.*, 65, 164, 1978.
55. **Manthey, F. A. and Nalewaja, J. D.,** Relative wax solubility and phytotoxicity of oil to green foxtail [*Setaria viridis* (L.) Beauv.], in *Adjuvants and Agrichemicals,* Foy, C. L., Ed., CRC Press, Boca Raton, FL, 1992, 463.
56. **Matsumoto, S., Kita, Y., and Yonezawa, D.,** An attempt at preparing water-in-oil-in-water multiple-phase emulsions, *J. Colloid Interface Sci.*, **57**, 353, 1976.
57. **Matsumoto, S., Koh, Y., and Michiura, A.,** Preparation of water/olive oil/water emulsions in an edible form on the basis of phase inversion technique, *J. Disp. Sci. Tech.*, 6, 507, 1985.
58. **Matsunaga, T., Tanka, Y., Abe, Y., and Tsujji, K.,** Microemulsion and delivery system of pyrethroids for a better household environmental concern, in *Proc. 7th Int. Congr. Pesticide Chemistry*, Vol II, Freshe, H., Kesseler, Schmidtz, E., and Conway, S., Eds., Hamburg, Germany, 1990, 63.
59. **McCarthy, J. F.,** Regulatory issues and adjuvants, in *Adjuvants for Agrichemicals*, Foy, C. L., Ed., CRC Press, Boca Raton, FL, 1992, 245.
60. Anen., McCutcheon's Emulsifiers and Detergents, Vol. 1, MC Publishing Co., Glen Rock, NJ, 1993, 229.
61. **Menger, F. M. and Littau, C. A.,** Gemini surfactants: a new class of self-assembling molecules, *J. Am. Chem. Soc.*, 115, 10083, 1993.
62. **Meusburger, K. E.,** Computerized optimization of emulsifiers for pesticide emulsifiable concentrates, in *Advances in Pesticide Technology*, ACS Symp. Ser. 254, Scher, H. B., Ed., American Chemical Society, Washington, D.C., 1983, chap. 9.
63. **Mitchel, D. J. and Ninham, B. W.,** Micelles, vesicles and microemulsions, *J. Chem. Soc.* Faraday Trans., II, 77, 601, 1981.

64. **Mookerjee, P. K.**, Computer-assisted correlation analysis in development of pesticide formulation, in _Advances in Pesticide Formulation Technology_, Scher, H. B., Ed., ACS Symp. Ser. 254, American Chemical Society, Washington, D.C., 1983, 105.
65. **Murphy, D. S., Policello, G. A., Goddard, E. D., and Stevens, P. J. G.,** Physical properties of silicone surfactants for agricultural applications, in _Pesticide Formulations and Application Systems,_ Vol. 13, Devisetty, B. N., Chasin, D. G., and Berger, P. D., Eds., ASTM STP 1146, American Society for Testing and Materials, Philadelphia, 1993, 45.
66. **Myers, D.,** _Surfactant Science and Technology_, VCH Publishers, New York, 1988.
67. **Namnath, J. S.,** Characterizing the fire safety of emulsive liquid pesticide formulations by using ASTM D56 tag closed cup flash point method, in _Pesticide Formulations and Application Systems,_ Vol. 12, Devisetty, B. N. Chasin, D. G., and Berger, P. D., Eds., ASTM STP 1146, American Society for Testing and Materials, Philadelphia, 1993, 180.
68. **Napper, D. E.,** _Polymeric Stabilization of Colloidal Dispersions,_ Academic Press, London, 1983.
69. **Narayanan, K. S.,** Polymers for instant dispersions for the herbicide metolachlor and other chloroacetanilides, in _Pesticide Formulations and Application Systems,_ Vol. 14, Hall, F. R., Berger, P. D., and Collins, H. M., Eds., ASTM STP 1234, American Society for Testing and Materials, Philadelphia, 1995.
70. **Narayanan, K. S.,** Inert Matrix Composition Microemulsifiable Concentrate and Aqueous Microemulsion, U.S. Patent 5,389,297, 1995.
71. **Narayanan, K. S.,** Water-Based Microemulsion Formulation of Carbamate Ester, U.S. Patent 5,338,762, 1994.
72. **Narayanan, K. S.,** Water-based Microemulsion Formulations, U.S. Patent 5,317,042, 1994.
73. **Narayanan, K. S.,** Stable, Clear, Efficacious Aqueous Microemulsion Compositions Containing a High Loading of a Water-Insoluble, Agriculturally Active Chemical, U.S. Patent 5,300,529, 1994.
74. **Narayanan, K. S.,** Method of Stabilizing Aqueous Microemulsions Using Surface Active Hydrophobic Acid as a Buffering Agent, U.S. Patent 5,298,529, 1994.
75. **Narayanan, K. S.,** Cold Stabilization of Aqueous Microemulsions of a Water-Insoluble Agriculturally Active Compound, U.S. Patent 5,266,590, 1993.

76. **Narayanan, K. S.,** N-alkylpyrrolidones for emulsifiable concentrates, microemulsions, and superior adjuvant formulations in _13th Int. Symp. Pesticide Formulations and Application Systems,_ Japan Society of Pesticide Science, Japan, Nov., 1993,1.

77. **Narayanan, K. S. and Chaudhuri, R. K.,** N-alkylpyrrolidone requirement for stable water based microemulsions, in _Pesticide Formulations and Application Systems,_ Vol. 12, Devisetty, B. N., Chasin, D. G., and Berger, P. D., Eds., ASTM STP 1146, American Society for Testing and Materials, Philadelphia, 1993, 85.

78. **Narayanan, K. S. and Chaudhuri, R. K.,** Emulsifiable concentrate formulations for multiple active ingredients using N-alkylpyrrolidones, in _Pesticide Formulations and Application Systems,_ Vol. 11, Bode, L. E. and Chasin, D. G., Eds. ASTM STP 1112, American Society for Testing and Materials, Philadelphia, 1992, 73.

79. **Narayanan, K. S. and Chaudhuri, R. K.**, Delivery System for Agricultural Chemicals, U.S. Patent 5,283,229, 1994.

80. **Narayanan, K. S. and Chaudhuri, R. K.**, Delivery System for Agricultural Chemicals, U.S.Patent 5,354,726, 1994.

81. **Narayanan, K. S., Chaudhuri, R. K., and Anderson, L. R.,** Alkoxyalkyl Lactams as Solvents for Macro and Microemulsions, U.S. Patent 5,385,948, 1995.

82. **Narayanan, K. S., Chaudhuri, R. K., and Dahanayake, M.,** Delivery System for Agricultural Chemicals," U.S. Patent 5,250,499, 1993.

83. **Narayanan, K. S., Chaudhuri, R. K., and Dahanayake, M.,** Delivery System for Agricultural Chemicals," U.S. Patent 5,176,736, 1993.

84. **Narayanan, K. S., Chaudhuri, R. K., and Dahanayake, M.,** Delivery System for Agricultural Chemicals," U.S. Patent 5,166,666, 1992.

85. **Narayanan, K. S., Chaudhuri, R. K., and Dahanayake, M.,** Delivery System for Agricultural Chemicals, U.S. Patent 5,071,463, 1991.

86. **Osipow, I.**, Emulsions, in _Surface Chemistry_, Reinhold, New York, 1962, chap. 11.

87. **Ottewill, R. H. and Rastogi, M. C.,** Stability of hydrophobic sols in the presence of surface-active agents. II. Stability of silver iodide sols in the presence of cationic surface-active agents, _Trans. Faraday Soc._, 56, 866, 1960.

88. **Piirma, I.,** Polymeric surfactants and colloidal stability, in _Polymeric Surfactants,_ Surfactant Series Vol. 42, Schick, M. J. and Fowkes, F. M., Eds., Marcel Dekker, New York, 1992, chap. 1.

89. **Payne, J. S. and Tracy, M. J.,** Denatonium benzoate (DB) as bittering agent in formulation, *Pesticide Formulations and Application Systems*, Vol. 14, Hall, F. R., Berger, P. D., and Collins, H. M., Eds., ASTM STP 1234, American Society for Testing and Materials, Philadelphia, 1995, 40.
90. **Prince, L. M.**, *Microemulsions-Theory and Practice*, Academic Press, New York, 1977.
91. **Rebmann, V. and Fiquet, L.**, Pyrethroid microemulsions, *Pesticide Formulations and Application Systems*, Vol. 13, Berger, P. D., Devisetty, B. N. and Hall, F. R., Eds., ASTM STP 1183, American Society for Testing and Materials, Philadelphia, 1993, 30.
92. **Riederer, M. and Schonherr, J.,** Effect of surfactants on water permeability of isolated plant cuticles and on composition of their cuticular waxes, *Pestic. Sci.*, 29, 85, 1990.
93. **Ries, H. E., Jr.**, Monomolecular films, *Sci. Am.*, 204, 152, 1961.
94. **Rife, H. E.**, Insecticidal Pyrethroid Compositions Having Increased Efficacy, U. S. Patent 3,954,977, 1976.
95. **Robinson, B. H.,** Microemulsions, properties and novel chemistry, *Chem. Br.* 26, 342, 1990.
96. **Rosano, H. L., Cavallo, J. L., and Lyons, J. B.,** Mechanism of formation of six microemulsion systems, in Surfactant Series 24, Rosano, H. L. and Clausee, M., Eds., Marcel Dekker, New York, 1987, 259.
97. **Rosano, H. L. and Nixon, A. L.,** Interfacial stoichiometry of a microemulsion system, *J. Phy. Chem.*, 93, 4536, 1989.
98. **Rosen, M. J.**, Emulsification, in *Surfactants and Interfacial Phenomena*, 2nd ed., John Wiley and Sons, New York, 1989, chap. 8.
99. **Rosen, M. J.,** Geminis: a new generation of surfactants, *Chem. Tech.*, 30, 1993.
100. **Rosen., M. J. and Murphy, D. S.,** Synergism in binary mixtures of surfactants, *J. Colloid Interfacial Sci.*, 129, 468, 1989.
101. **Rosen, M. J., Zhu, Z. H., Gu, B., and Murphy, D. S.,** Relationship of structure to properties of surfactants. 14. Some N-alkyl pyrrolidones at various interfaces, *Langmuir*, 4, 1273, 1988.
102. **Ross, S. and Morrison, I. D.**, *Colloidal Systems and Interactions*, Wiley Interscience, New York, 1988.
103. **Saroja, S.,** Surfactant Structure and Emulsion Stability, PhD Thesis, Clarkson University, A Bell and Howell Company, Michigan, 1993, 48.
104. **Sato, T.,** Stability of dispersions, *J. Coatings Tech.*, 65, 113, 1993.
105. **Schonherr, J.,** Technische. Universitat, Munchen, Germany, Private communication, 1993.

106. **Schonherr, J. and Bauer, H.,** Analysis of effects of surfactants on permeability of plant cuticles, in *Adjuvants and Agrichemicals,* Foy, C. L., Ed., CRC Press, Boca Raton, FL, 1992, 17.
107. **Sherman, P.**, *Emulsion Science*, Academic Press, New York, 1968.
108. **Shinoda, K. and Arai, H.,** The correlation between phase inversion temperature in emulsion and cloud point in solution of nonionic emulsifier, *J. Phys. Chem.*, 68, 3485, 1964.
109. **Shirahama, K.,** The binding equilibrium of sodium alkyl sulfates to poly(ethylene oxide) in 0.1 m sodium chloride solution at 30^{0}C, *Colloid Polym. Sci.*, 252, 978, 1974.
110. **Skelton, P. R.,** Pesticide microemulsion concentrate formulation utilizing fatty acid methyl esters as solvent alternatives, *Pesticide Formulations and Application Systems,* Vol. 13, Berger, P. D., Divesetty, B. N., and Hall, F. R., Eds., ASTM STP 1183, American Society for Testing and Materials, Philadelphia, 1993, 114.
111. **Skelton, P. R., Munk, B. H., and Collins, H. M.,** Formulations of pesticide microemulsions, in *Pesticide Formulations and Application Systems,* Vol. 8, Hovde, D. A. and Beestman, G. B., Eds., ASTM STP 980, American Society for Testing and Materials, Philadelphia, 1992, 36.
112. **Smith, A. I.**, Ed., *Theory and Practice of Emulsion Technology*, Academic Press, New York, 1976.
113. **Tanford, C., J.**, Micelle shape and size, *Phys. Chem.*, 76, 3020, 1972.
114. **Tann, R. S., Memula, S. G., and Friloux, K. M.,** The use of novel techniques in the evaluation and development of liquid pesticide formulations, in *Proc. 14th ASTM Symposium on Pesticide Formulations and Applications Systems*, Fort Worth, TX, 1993.
115. **Thomson, W. T.**, *Agricultural Chemicals*, Vol. I, II, III, and IV, Thomson Publications, Fresno CA, 1989.
116. **Tinsworth, E. F.**, Regulation of pesticides and inert ingredients in pesticide products, in *Adjuvants for Agrichemicals*, Foy, C. L. Ed., CRC Press, Boca Raton, FL, 1992, 237.
117. **Urlon, J. T.**, Method of Producing Microcolloidal Aqueous Emulsions of Unsaturated Organic Insecticidal Compounds, U. S. Patent 3,954,967, 1976.
118. **Utz, C. G., Drewno, G. W., and Hollis, P. R**., Using nonionic surfactants in aqueous formulations, in *Pesticide Formulations and Application Systems,* Vol. 13, Devisetty, B. N., Chasin, D. G., and Berger, P. D., Eds., ASTM STP 1146, American Society for Testing and Materials, Philadelphia, 1993, 133.
119. **Vervey, E. J. W. and Overbeek, J. T. G.,** *Theory of the Stability of Lyophobic Colloids*, Elsevier, Amsterdam, 1949.
120. **Vincent, R. M.,** BVA Oils, personal communications, 1992.

121. **Void, R. D. and Groot, R. C.,** The effect of electrolytes on the ultracentrifugal stability of emulsions, *J. Colloid Interface Sci.*, 19, 384, 1964.

122. **Void, R. D. and Mittal, K.,** The effect of lauryl alcohol on the stability of oil-in-water emulsions, *J. Colloid Interface Sci.*, 38, 451, 1972.

123. **Walstra, P. and Becher, P.**, Eds., *Encyclopedia of Emulsion Technology*, Marcel Dekker, 1983, 74.

124. **Welch, C. B., Leifer, K. B., and Levine, T. E.,** New issues on pesticide formulations and application systems, in *Proc. 14th ASTM Symposium on Pesticide Formulations and Applications Systems*, Fort Worth, 1993.

125. **Zhu, Y., Masuyama, A., Kobata, Y., Nakatsuji, Y., Okahara, M., and Rosen, M. J.,** Double chain surfactants with two carboxylate groups and their relation to similar double chain compounds, *J. Colloid Interface Sci.*, 158, 40, 1993.

126. **Zhu. Z. H., Yang, D., and Rosen., M. J.,** Some synergistic properties of N-alkyl-2-pyrrolidones. A new class of surfactants, *J. Am. Oil Chem. Soc.*, 66, 998, 1989.

Chapter 9

Suspoemulsion Technology and Trends

Joseph R. Winkle

CONTENTS

ABSTRACT

Suspoemulsions or suspension emulsions are formed via the addition of an emulsified phase containing one or more active ingredients with a bulk phase (typically water) that also contains one or more active materials as suspended solids. Perhaps the greatest challenge in the preparation of suspoemulsions is the physical stabilization of the system. This stability can be imparted through careful choice of surfactants, dispersants, rheological builders and stabilizers, and process control parameters. Besides handling and inventorying advantages, suspoemulsions of biologically complementary active ingredients offer a unique formulation alternative for fitting into integrated pest management (IPM) and resistance strategies.

1-8493-7678-5/96/$0.00+$.50

I. INTRODUCTION

As we approach the turn of the century, issues facing agrichemical manufacturers and distributors are rapidly focusing more on exposure of products to humans, animals, ecosystems, and the environment; stewardship programs dealing with packaging, containers and delivery systems; toxicological significance of inerts utilized in formulations, especially solvents; and responsible care programs. From the biological perspective, resistance management and cross contamination potentials have received the recent spotlight. Somewhat overshadowed, but certainly not lost in all the regulatory and environmental pressures, is the basic premise of supplying to the enduser a chemically and physically stable presentation of a biologically active system that will meet his particular needs. New chemistries have evolved in the past decade that are many times more active (low g ha^{-1} rates) than their predecessors, but they often do not alone yield the broader spectrum of control that is desired. As a consequence, combination products that are consistent with the aforementioned issues are in increasing demand.

Combination products offer several potential advantages, the primary ones being broad spectrum of biological control for given targets, a "built-in" resistance management tool, elimination of tank mixing, and reductions in the number of containers and total inventory. Many combinations can be formulated in conventional systems such as suspension concentrates (SC), emulsifiable concentrates (EC), soluble liquids (SL), wettable powders (WP), and water-dispersible granules (WG). When the properties of the active ingredients to be combined are not compatible due to physical or chemical constraints, a more elaborate and oftentimes complex system must be employed. Suspension emulsions, or suspoemulsions (SE), are a recent example of a more intricate liquid-based combination product system. A suspoemulsion consists of a discontinuous oil phase (liquid active ingredient(s) or dissolved active ingredient(s) in a solvent) that is uniformly dispersed in a continuous aqueous phase containing finely dispersed particles of another active ingredient(s). In some cases, a third pesticidal phase can be introduced if the active is soluble in water; this scenario may also be subject to more complications, such as electrolyte stability, when compared to traditional systems.

This paper will review the current status of suspoemulsions in the agrichemical industry as derived from the open literature. Methods of preparation, technical issues, technology advancements, and other advantages and disadvantages will be discussed. Opinions on the future utilization of these combination systems will also be given.

II. UTILITY, COMPOSITION, AND PREPARATION

Before defining the components that typically comprise a suspoemulsion, it is worthwhile to look at other potential reasons to consider this type of delivery system.

A. SUSPOEMULSION ADVANTAGES

Due to the fact that water is the continuous phase, suspoemulsions have reduced skin and eye toxicity, greater flash points, better compatibility with high density polyethylene (HDPE) containers, and lower solvent content when compared to traditional emulsifiable concentrates.

As previously mentioned, suspoemulsions allow for mixtures of active ingredients of widely differing properties (biological, chemical, physical). The preferred formulation of liquid and lower melting solid active ingredients is as a liquid. Thus these particular materials, in those cases where the other active ingredient(s) is a high-melting, insoluble solid, may be best suited as suspoemulsions rather than dry combination products. Many times, benefits in the spray tank are realized as well since suspoemulsions may provide superior mixing and suspension characteristics compared to tank-mixed products that notoriously exhibit poor compatibility.

B. SUSPOEMULSION DISADVANTAGES

The major disadvantages of suspoemulsions are (1) the time required to develop stable systems may be quite long due to the complexity involved; (2) manufacturing of suspoemulsions can be quite costly with only narrow windows of operation; and (3) hydrolytic stability of the active ingredient(s) can be a major problem for many would-be suspoemulsion concepts and must be investigated at the early stages of development.

C. REQUIREMENTS FOR STABLE SYSTEMS

1. Solubility

Ideally, the aqueous-dispersed solid active ingredient should be insoluble in the discontinuous oil phase and be of very low solubility in water. The emulsified liquid or solubilized active ingredient, conversely, should be practically insoluble in the aqueous phase. All active ingredients must also be chemically compatible. In selecting the preferred solvent, if needed for the emulsion phase, care should be taken in choosing one having high solvency throughout the probable temperature storage regime, a high flash point, and specific gravity that will favorably aid in physical stability. The reality of many suspoemulsions is that the solid, dispersed phase has some solubility in the oil phase, and water solubilities are high enough (>500 ppm) to allow for Ostwald ripening. Compatibility issues can thus arise, and each system will consequently have its own unique challenges.

2. Emulsifiers/Dispersants

Perhaps the most critical selection in composing the formulation ingredients is the selection of compatible surfactants and dispersants for the system. In most cases, materials for emulsifying the oil phase will be different from the polymers typically used to disperse the solids. Serious problems can be encountered if desorption of the dispersant occurs and the emulsifier exchanges for the dispersant. Coalescence and/or flocculation will undoubtedly result. Phase transfer of the suspended particle into the oil also is a possibility if the dispersant is not tightly adsorbed and the oil phase wets the particle surface. This can lead to coalescence of the droplets as well as a potential for crystal growth if the active has partial solubility in the oil phase. Thus, a tight anchoring of carefully selected emulsifiers/dispersants for respective phases is essential for stability.

Alternatively, one could select a single surfactant that would function as an emulsifier for the oil phase as well a dispersant for the solids; this would alleviate the negative coating effects. From a practical standpoint though, there are not that many cases where one surfactant will suffice.

3. Other Ingredients

As required, materials such as biocides, anti-foaming agents, freezing point depressants, structure builders such as modified clays and silicates, and viscosity modifiers such as naturally occurring gums can be added to the system in the appropriate phase.

4. Preparation of Suspoemulsions

On a laboratory scale, the most typical method of preparation is to form stable suspension concentrates and emulsion phases separately, and then combine with mixing. This is also the most controlled method of preparation since each phase is defined very closely. Stability in each separate phase, however, does not guarantee stability in the combined state. In another method, the air-milled solid to be dispersed is slowly added to the aqueous emulsion with high shear mixing. Conversely, the oil phase can be added to the aqueous dispersed phase and emulsified in situ. Lastly, milling of the solid to be dispersed in the presence of the emulsion phase or in the oil phase alone has been described.

III. REVIEW OF LITERATURE

In reviewing the publications on suspoemulsions that have been issued in the last decade or so, several good summary papers already exist.[6,7,9,10,12] As would be expected, no ubiquitous teachings or findings arise as each system is uniquely based on the chemistries of distinct oil and dispersed phases. Important literature relating to emulsion stabilization comes closest to having universal application and will be discussed at the end of this section. The following discussions present summaries of recent patents and papers, although they oftentimes deal with unique resolution of specific problems. In European Patent 01430998B1, a stable suspoemulsion containing alachlor (2-Chloro-2',6'-diethyl-N-(methoxymethyl)-acetanilide) and atrazine (6-chloro-N^2-ethyl-N^4-isopropyl-1,3,5-triazine-2,4-diamine) is described that represents an improvement over the originally launched commercial formulation sold under the tradename LASSO® and Atrazine.[8] The original suspoemulsion suffered from solid atrazine settling out over time, with difficulty encountered in resuspending the particles. This was of particular concern when the formulation was held in bulk storage tanks. As a consequence, the composition was reevaluated. As noted in Table 1 several changes were employed in the novel suspoemulsions described in the patent (compositions A and B) as compared to the prior art formulation (Composition C). The use of hydrophilic clays to build rheological properties, the addition of a solvent with lower specific gravity, and the addition of a small amount of inorganic oxide to stabilize the emulsion phase all contributed to a very striking change in the suspension properties of the suspoemulsion. After 9 months of storage in a warehouse, compositions A and B were virtually free of any settled atrazine and also were easily resuspended after five inversions; the original composition C exhibited significant quantities of flocculated atrazine that were hardpacked.

Researchers at American Cyanamid have explored ways to mitigate the strong flocculation of imidazolinone particles that occurs when combined with acetanilides as a suspoemulsion.[2] Suspoemulsions were prepared in a three-step process. The first part consisted of the aqueous phase formation where the solid imidazolinone was added to an aqueous phase containing the dispersant and antifoam agent. Bead milling or similar comminution processes were required if the technical product had not previously been airmilled. The second phase consisted of forming the oil phase by adding solvent to the molten acetanilide, followed by addition of surfactant and emulsifiers. The third step involved the slow addition of the oil phase to the stirred aqueous phase. Typical amounts of surfactants and dispersant as found in most emulsions and suspension concentrates were not adequate when the two phases were combined.

As a consequence, much higher than normal levels of emulsifier and dispersant were examined. Increasing amounts of the sodium alkylarylsulfonate dispersant, Morwet® 3028, were tried as shown in Table 2; all systems flocculated. Various other dispersants such as modified lignosulfonates, sulfosuccinates, and ethoxylated carboxylates were also tested with no successful alleviation of the flocculation. The nature and level of surfactants were then evaluated.

Table 1
Suspoemulsion Composition of Alachlor and Atrazine

Component	A	B	C
Alachlor (94.3% A, B; 95.3% C)	29.48	29.41	28.96
Atrazine (97%)	17.20	17.15	17.11
Chlorobenzene	16.95	16.91	16.70
C9 Aromatics	11.28	11.25	---
Flomo® LHF	3.69	3.68	5.00
MgO	0.10	0.10	---
Antifoam	0.02	0.02	0.02
Bentone® EW	0.39	---	---
Veegum® T	---	0.50	---
Ethylene glycol	1.50	1.50	1.50
Methyl violet	0.005	0.005	0.005
Water	19.385	19.475	30.70
	100.00	100.00	100.00

Flomo® LHF is a registered product of Henkel Corp.
Bentone® EW is a registered product of RHEOX Inc.
Veegum® T is a registered product of R.T.Vanderbilt.

Surprisingly, nonylphenolethoxylates at high levels had a dramatic impact on the flocculation behavior of the air-milled imidazolinones tested. Compositions were constructed using the base formulation as shown below:

Ingredient	**By weight (%)**
Alachlor	37.46
Aromatic® 200	20.54
Toximul® 8320	1.00

Results of varying the amount of the surfactant FloMo® 9N are demonstrated in Table 3. Compositions using 6% by weight and greater of the surfactant gave substantially reduced flocculation and passable results on wet sieve analysis. Although not clearly commented on in the patent application, the probable reason for success of this system relates to stabilization of the oil phase and prevention of coating of the imidazolinone particles with the oil. The insensitivity of dispersant type and level suggests that there was no sequestering of the dispersant by the emulsifiers at normal weight percent. Information relating to other emulsifiers within the same family (e.g., higher and lower nonylphenol ethoxylates) would have been helpful in trying to determine the relationship of the surfactant with imazaquin. [(R8)-2-(4-isopropyl-4-methyl-5-oxo-2-imidazolin-2yl)quinoline-3-carboxylic acid] It is not evident from the data presented what effect viscosity may also have played in the stabilization.

Table 2
Suspoemulsions Containing an Acetanilide and an Imidazolinone at Varying Dispersant Rates

Ingredient	**A1**	**A2**	**A3**	**A4**
Alachlor	36	36	36	36
Aromatic oil	26	26	26	26
Nonionic Surfactant (Flomo® 9N)	2.5	2.5	2.5	2.5
Butyl polyalkylene oxide block copolymer	1.5	1.5	1.5	1.5
Water	30.35	29.75	26.75	22.75
Imazaquin (technical)	2.23	2.23	2.23	2.23
Antifoam	0.02	0.02	0.02	0.02
Dispersant (Morwet® 3028)	0.04	1.00	4.0	8.0
	100.00	100.00	100.00	100.00

Note: Flomo® 9N is a registered product of Witco Corp.

Ingredient	By Weight (%)
Flomo® 9N	2.00 - 8.00 (Varying in 1% increments)
Water	34.38 - 28.38
Imazaquin	2.38
Morwet® 3028	0.10
Antifoam	0.04

Aromatic® 200 (Exxon) is a mixture of high boiling point aromatic petroleum distillates.

Toximul® 8320 (Stepan Chemical Co.) is a butyl polyalkylene oxide block copolymer.

FloMo® 9N (Henkel) is a poloxyethylene nonylphenol ether.

Morwet® 3028 (Witco) is a sodium alkyl aryl sulfonate.

The positive effects of rheological builders in suspoemulsion formulations was also noted by Wigger and Guckel.[18] In their studies, they examined three suspoemulsions from a physical property standpoint as well as from milling characteristics. The solid dispersed phase was phenyl pyridazinone in all systems while the oil phase consisted of three different liquids - chloroacetylaniline, thiolcarbamate, and paraffin oil. A bead mill was used to process the phenyl pyridazinone to three different median diameter sizes of 0.7, 1.3, and 1.8 microns, and the liquids were directly emulsified into these suspensions using high shear mixing. Viscosities, particle size distributions, interfacial tensions and energy dissipation densities were measured for all systems. Both specific phase volume ratios and the addition of oil-soluble emulsifiers allowed for further comminution of the suspoemulsion relative to standard cases. Separation of phases occurred in two of the systems as a result of creaming and sedimentation. Addition of small amounts of xanthan gum reduced this effect.

In another example, latex dispersions were used to stabilize various oil phases.[7,11] Particle size distributions of the oil phase in the latex dispersion after high-shear mixing gradually changed from a broad, bimodal distribution to a narrow single pattern that approximates the original latex distribution.

Stability of the emulsion phase over a variety of temperature regimes with time was superior. The order of addition of the oil phase and latex dispersion generally did not matter. Two suspoemulsions (1) fluoxypyr{[(4-amino-3,5-dichloro-6-fluoro-2-pyridinyl)oxy]acetic acid}, ioxynil (4-hydroxy-3,5-diiodobenzonitrile), and isoproturon [3-*p*-cumenyl-1,1-di-)methylurea (I);3-(4-isopropylphenyl)-1,1-dimethylurea]; and (2) fenpropimorph {(±)-*cis*-4-[3-(4-*tert*-butylphenyl)-2-methylpropyl]-2,6-di-)methylmorpholine} and lindane (gamma isomer of 1,2,3,4,5,6-

hexachlorocyclo-)hexane) were prepared using a polystyrene latex as the oil phase stabilizer with similar outstanding stability. The authors also commented that added benefit might be imparted to seed coating treatments if the latex employed was of the type that formed film.

A. RELATED EMULSION TECHNOLOGIES

Besides traditional means of oil phase stabilization through proper selection of emulsifiers, novel methods have been employed recently that have directly benefited concentrated emulsions and suspoemulsions on perhaps a more universal scope. The use of polymeric surfactants (graft copolymers)[5] has provided stability of emulsions by anchoring the oil phase via hydrophobic moieties on the polymer backbone and by electrostatic or steric repulsion of neighboring droplets through the water-soluble, hydrophilic parts of the polymer. Recently, Wessling et al.[15] have disclosed novel graft copolymers comprised of a reactive polymeric surfactant base polymer and a nonionic, hydrophobic grafted composition that enhanced the stability of various emulsified phases as compared to prior systems that tended to coalesce. Other patents have also been recently issued on the use of structured latex particles in imparting stability to emulsions.[16,17]

Table 3
Effect of Surfactant Level on Imidazolinone Flocculation

Flomo ® 9N in Formulations %	Dispersability	Appearance	Concentrate Microscopy[a] (micrometers)	Wet[2] Sieve Test[b]
2	Good	Flocs	100 +	Fail
3	Good	Flocs	50-100	Fail
4	Good	Good	30-50	Fail
5	Good	Good	>25	Fail
6	Good	Good	10	Pass
7	Good	Good	<5	Pass
8	Good	Very viscous	None	Pass

[1]Imazaquin particle size.

[2]Wet sieve test: 7g of concentrate diluted in 93ml water, 30 mins standing, 10 inversions, pour through nest of 100-, 200-, and 325- mesh sieves.

Investigators at Rhone-Poulenc have utilized titanium dioxide to stabilize oil-in-water macroemulsions in which multiple active ingredients were dissolved in either the oil phase alone or, in some cases, in both oil and water phases.[3] They also prepared novel suspoemulsions by milling solid active ingredients directly in the stabilized TiO_2 emulsions. Other means of stabilizing oil phases have employed the use of polyvinyl alcohol;[4,19] the use of dialkylphthalate esters as solvents to stabilize what previously had been a molten addition of oil to the aqueous phase;[1] and the use of a water-immiscible solvent with high solubility for suspended hydrophobic particles used at low levels that precluded complete solubilization of the active.[14] Although there is a fair amount of literature on emulsion stabilization, there is little information on applications toward stabilizing suspoemulsion systems.

B. RELATED SOLID PHASE TECHNOLOGIES

While the stability of the emulsion phase is certainly critical and has been the benefactor of many recent technological advances, stability of the dispersed phase with materials that are not readily desorbed from the particle surface is of equal importance. Tadros has recently reported in U.S. Patent 5139773 the use of polymeric "comb" surfactants that strongly adsorb onto the surface of various dispersed solids present in example suspoemulsions.[13]

The polymeric surfactant used is a graft copolymer of polymethylmethacrylate-methacrylic acid with methoxy polyethylene oxide methacrylate. The polyethylene oxide chains act as the "teeth" in the polymeric "comb" backbone. As shown in Table 4, a variety of chemistries were successfully investigated in these studies, suggesting the versatility of this class of dispersing agents for adsorbing onto solid surfaces of varying hydrophobicity and electron density.

Table 4
Suspoemulsions Prepared Using Polymeric "Comb" Surfactants

Suspension pesticide	Emulsion pesticide	Additional Suspension pesticide
Flutriafol	Propiconazole	
Flutriafol	Imazalil	
Diclobutrazol	Imazalil	
Carbendazim	Prochloraz	
Propyzamide	Fluazifop-P-butyl	
Diclobutrazol	Prochloraz	
Chlorothalonil	Tridemorph	Hexaconazole
Chlorothalonil	Fenpropidin	Hexaconazole

IV. FUTURE TRENDS

The primary technical challenges in preparing stable suspoemulsions are specific to the proposed mixing partners, but advances in emulsion stabilization and in particle-particle repulsion will have broad applicability in helping to overcome physical phenomena such as coalescence, Ostwald ripening, and flocculation. The need still exists, however, to better understand pesticide phase movement in suspoemulsions and how that relates to the mobility/desorption behavior of the surfactants and dispersants employed. This basic understanding directly relates to processing aspects of suspoemulsions as well. The critical control and assurance of quality products that are necessary in suspoemulsion preparation require an in-depth knowledge of the system. Because this is a relatively new area, little information from patents, journal articles, and commercial products is available to help formulation scientists and engineers draw upon a collective data base to further the science. Advances are more typically achieved within the framework of a specific system.

As mentioned earlier in this chapter, combination products have been increasing globally and have been accepted for what they offer in a broader spectrum of control and in tools to manage resistance effectively. Suspoemulsions have been prevalent in Europe, particularly as a formulation option for combination fungicides. Many of the sterol biosynthesis inhibitors are either liquids or low melting solids and do not easily lend themselves as dry combination partners; consequently, a liquid system such as suspoemulsions is the primary choice. The global trend to use safer solvents and less solvent is also in concert with suspoemulsion usage.

Although product lines in North America are not replete with suspoemulsion examples, the drive toward more environmentally and toxicologically friendly liquid systems is already prevalent. Suspoemulsions offer an excellent alternative to many oil-based suspensions, especially if driven that way by regulatory bodies and the marketplace.

REFERENCES

1. **Albrecht, K.** and **Frisch, G.**, European Patent 0117999, 1986.
2. **Dexter, R.W.** and **Elsik, C. M.**, South African Patent Application No. 906980, 1990.
3. **Dookhith, M.** and **Linares, H.**, U.S. Patent 5206021, 1993.
4. **Fuyama, H.** and **Tsuji, K.**, U.K. Patent GB 2025770A, 1980.
5. *Anon.*, ICI Surfactants Report 20-44E, 1993.
6. **Mulqueen, P. J.**, **Banks, G.**, and **Press, D. L.**, *Pesticide Science* and *Biotechnology*, Blackwell Scientific, 1987, 273.
7. **Mulqueen, P. J.**, **Paterson, E. S.**, and **Smith, G. W.**, *Pesticide Science,* 29, 451, 1990.
8. **Prill, E. J.**, European Patent 0143099, 1989.
9. **Rogiers, L. M.**, ICI Specialty Chemicals Report RP34/89E, 1989.
10. **Seaman, D.,** *Pesticide Science*, 29, 437, 1990.

11. **Smith, G.W.**, **Mulqueen, P.J.**, **Paterson, E.S.**, and **Cuffe, J.**, International Patent W089/03176, 1989.
12. **Tadros, T. F.,** *Pesticide Science*, 26, 51, 1989.
13. **Tadros, T.F.**, U.S. Patent 5139773, 1992.
14. **Wada, T.**, **Ogura, Y.**, and **Kume, R.**, European Patent 0253682, 1988.
15. **Wessling, R. A.**, **Pickelman, D. M.**, and **Wujek, D. G.**, European Patent 0357149, 1990.
16. **Wessling, R.A., Pickelman, D.M.**, and **Wujek, D.G.**, U.S. Patent 5089259, 1992.
17. **Wessling, R.A.**, **Pickelman, D.M.**, and **Wujek, D.G.**, U.S. Patent 5188824, 1993.
18. **Wigger, A.** and **Guckel, W.**, *Pesticide Science*, 25, 401, 1989.
19. **Wirth W., Niessen, H. J.**, **Goosens, J. W. S.**, and **Schulze, N.**, U.S. Patent 4824663, 1989.

Chapter 10

Current Packaging Trends and Related Technologies

Steven Gleich

CONTENTS

1-8493-7678-5/96/$0.00+$.50

ABSTRACT

Global packaging trends are driven by several forces in dynamic equilibrium. These include societal focus on the environment, government regulation, and customer and market needs. Overlaying all of this are, of course, geographic and cultural preferences and differences.

A useful metaphor for packaging is as a "bridge" between the formulated active ingredient and its ultimate application to the target . This metaphor requires that we view both the formulation-packaging and packaging-application relationships.

Different packaging technologies relate specifically to the most important trends and are more or less successful in responding to the above driving forces. Packaging source reduction technologies include recycling, bulk systems, returnable-refillables and ways to avoid use of select packaging components. The evolution of biodisappearing materials, including water-soluble films and biodegradable plastics, provide other possible approaches to source reduction. New packaging-application designs can result in such desirable features as closed systems and controlled rate of pesticide release. New formulation technologies lead to and require new packaging systems. The important relationship between pesticide use rate and packaging requirements leads to the realization that the total amount of packaging requirements are better viewed as a function of acreage treated than per pound of active ingredient used.

The way the future unravels depends on both the relative emphasis given to the market and the public's perception. We can help shape that future with the technologies we choose to emphasize as well as the degree to which we can help public perception and reality coincide.

I. INTRODUCTION

This chapter will discuss U.S. as well as global trends in the packaging of pest control agents which are applied to agricultural crops. The trends are influenced by perceptions and issues having to do with the two broader subjects of: (1) packaging and (2) chemical and biological pesticides. Trends, of course, occur from the interaction of many forces. In fact, we lawmakers, businessmen, academicians, farm consultants, dealers, and farmers of the world community create these very forces. How we think and behave makes a great deal of difference towards which trends strengthen and which decline. Our work in packaging of pest control agents aims at the same goals as those of agriculture as well as society at large. These include:

- Production of safe and abundant food.
- Protection of our natural resources, human and animal health and safety.
- Economic sustainability of our farm enterprises, Agricultural businesses, industries and nations.

How well we balance our attention toward these *interwoven goals* significantly influences our ability to attain any of them.

We will, therefore, describe current trends, the values which drive them and the technologies that these values encourage. In addition, the concept of "packaging" is broadened beyond the traditional view and related technologies are suggested which greatly impact management of packaging. These technologies illustrate alternative approaches toward attainment of the above goals. Where appropriate, perspectives are offered on the effectiveness of different approaches.

II. THE FORCES WHICH DRIVE PACKAGING TRENDS

Major forces which drive trends in packaging of pest control agents include public perception, government regulation, customer needs and values, and relative cost versus benefit of alternative solutions.

A. PUBLIC PERCEPTION

Public perception is heavily influenced by availability and accuracy of information as well as long held beliefs. A characteristic of these beliefs is that they change more slowly than the actual facts on which they're based. For instance, packaging in the modern world is often perceived as reducing, rather than adding value. This comes about, in part, because the package must be discarded once the item it carries is consumed. Invisible is the immense value packaging brings to contain, protect, preserve, and distribute perishable goods. In fact, where adequate packaging does not exist, up to 40% of the food raised never survives to be consumed by human beings.[29] This same quality of packaging is necessary for protection, storage, and supply of crop protection agents. Another public belief is that the hazard of chemical pest control agents present is high relative to the other hazards we face in modern society.[25,43] Relative to automobile travel and cigarette smoking, however, chemical pest control agents (CPCs) pose significantly lower risks. In the U.S., we arguably have the world's safest and most abundant supply of food, yet we paradoxically have the greatest sense of fear and most intense regulatory processes regarding food quality and safety. In fact, the high quality, availability, and low cost of food is due in part to CPCs, which often provide health benefits that outweigh the risks resulting from their use. Another belief held by the public is that we are throwing more and more packaging away each year. In fact, on a per capita basis, packaging waste to landfills has been decreasing.[29,33]

B. GOVERNMENT REGULATION

Legislative trends in the U.S. and Europe are very similar, though the focus for implementation is different due to differences in the economic and political environments. In general, existing regulations and proposed legislation encourages, in order of priority:[27]

- Reduction of packaging weight and volume
- Reuse of packaging
- Recycling of packaging materials

In all cases, land filling and incineration are intended to be significantly reduced as disposal means.

With respect to rigid CPC containers, both the U.S.[22,46] and Europe[39] are setting standards for rinsing them to preset levels of active ingredient removal, with the U.S. currently proposing a removal percentage of 99.9999% (1 ppm) versus a level of 99.99% in Europe. Because of this degree of stringency, the U.S. also aims to encourage the development and use of CPC formulations that facilitate the removal of CPC residues from containers to the proposed level. The value of this effort is questionable, since protection of the environment depends mainly on the farmer actually rinsing the container.

In addition to the above, the proposed U.S. regulations aim to encourage the use of bulk and mini-bulk facilities which significantly reduce the need for nonrefillable containers. Europeans are less attracted to the returnable/refillable option, because of their smaller average farm size, but compliance with regulations may nevertheless drive them in that direction. All the "refillable" options include concerns about the procedures for insuring safe handling and refilling of used containers.

The systemic concept of "life-cycle analysis" (or eco-balance) is being used in the attempt to create an objective understanding of the overall environmental impact different types of materials or products may have through all the stages in its life cycle.[44] Such an analysis is rather detailed. For plastic packaging it includes such steps as extraction of raw materials, natural gas processing, organic chemicals production, plastics polymerization, plastics molding and forming, package filling by user, product purchase by consumer, and disposal. This approach aims to quantify what today is a rather subjective set of judgments regarding the relative impact of different packaging options. Even the best of these analyses require that some difficult judgments be made; e.g., how much carbon monoxide equals a gallon of dirty water.[13]

C. CUSTOMER VALUES

In order for a product system to be competitive, it must, first and foremost, provide benefits to customers. In addition, the system must comply with government regulation. Although these two requirements may seem contradictory, it is possible to meet both. In fact, it is necessary. From our own market research, customers share three related universal goals: (1) increased profit, (2) reduced risk, and (3) more effective use of time. These goals can be achieved in a variety of ways, one of which is packaging. Some important benefits which packaging systems can bring include:

- Increased safety to users
- Greater handling convenience
- Greater measuring convenience
- Easier compliance with regulation; reduced or more easily managed waste
- Reduced cost
- Better integration into the overall farm system

As we proceed toward a discussion of technologies, we must keep in mind the direction being set through government regulation as well as the specific needs of customers in a wide variety of unique global markets. Although their goals may be similar, the solutions which are likely to work on a 5000 acre U.S. corn and soybean enterprise may be quite different from those which meet the needs on a 100-acre French grain farm.

III. AN EXPANDED VIEW OF PACKAGING

A useful metaphor for packaging is "a bridge" between the formulated active ingredient and the application method. Together, the formulation, package and application method constitute a "*delivery system*" which preserves the efficacy of an active ingredient and translate it to the target pest. The packaging portion of that system therefore must integrate with the formulation it protects as well as integrate with the application method it supplies. For many of today's systems, application simply involves the addition of the formulation into a mix tank with water, followed by spraying of the mixture onto the crop. As the next generation of active ingredients, formulations, and application methods are introduced, packaging will have to be developed in conjunction with those other related technologies. This point of view will be made clear as the discussion proceeds.

IV. PACKAGING TECHNOLOGY TRENDS

A. BULK, MINI-BULK, AND SMALL VOLUME RETURNABLES

The forces which drive changes in packaging systems have led to the increased use of existing technologies such as bulk and mini-bulk systems as well as the further refinement of small volume returnables (SVRs). Opportunities exist with this technology to design completely closed systems. Challenges which arise include the infrastructure to manage multiple containers of varying sizes, the assurance of no tampering and the potential cross contamination possible in refilling of nondedicated containers.[12] The integrity of bulk storage facilities is another issue in itself. Although the use of SVRs and mini-bulk tanks are growing in acceptance in the U.S. with its large farm operations, there are significant doubts about the practicality of this solution in Europe with its greater diversity in number of products and smaller average farm size.[6,26,36] The container management program in Canada is also encouraging a move away from non-refillables. Sales of pesticides in refillable containers more than doubled in 1992 to 60,000 refillables in use.[7] Since these systems are intended for extended use, compatibility of container material and the formulation it holds is critical.

B. RECYCLING

Significant efforts in the U.S. and Europe have led to increases in the recyclable content of plastic as well as paperboard containers, with the challenge of providing recycle material properly free of contaminants and an infrastructure which can economically sustain itself. In 1992 the Agricultural Container Research Council was formed in the U.S. with support from government, environmental groups, and industry. The organization has been highly instrumental in setting up the infrastructure for plastic container rinsing and recycling. Container collection systems are now a reality in 38 states. Recycling has grown from 1.5 MM lb yr^{-1} in 1992 to 3.5 MM lb yr^{-1} in 1993 and is projected to reach 5.5 MM lb yr^{-1} in 1994.[2] This is still only a little over 25% of the 20 MM lb yr^{-1} of plastic being produced, but shows continued growth. One significant challenge, of course, is that the cost of recycled plastic is several times that of virgin plastic, making the recycling business hard to justify on purely economic grounds. Improvements in recycling technology continue to be made; examples include new bottle and film reclaim systems,[16,37] and automated separation and sort systems.[45] Chemical recycling, which is the breakdown of polymers into their building blocks followed by repolimarization, is a new technology with some promise to improve the economics of recycling and to improve the quality of plastics which result.[40]

C. WATER SOLUBLE PACKAGING AND BIODEGRADABLES [32,35,42]

Although water soluble films (WSFs) are often included as part of the emerging class of biodegradable materials, they should really be treated separately, since WSFs have been commercially viable for many years and constitute the special class of materials which completely dissolve in water. Other biodegradable materials, covered later in this section, possess a wide range of properties, yet still remain to be accepted in the marketplace.

1. Water Soluble Packaging

Although water soluble bags (WSBs) have existed commercially for 20 years, it is only recently that there has been a significant improvement in the quality of films which dissolve under a wide range of water temperatures and have the integrity to remain unbroken during shipping and storage. More and more products are being packaged in WSBs with the benefit to customers of safe handling of product and uncontaminated packaging waste left behind. This technology, along with the bulk and mini-bulk technologies, provides another means for eliminating the use of nonreturnable containers. WSBs also provide an opportunity to combine normally incompatible products in the same package through the use of a "twin-pack" design.

Another positive feature of WSBs is their ability to mitigate the negative features of existing dusty or higher toxicity formulations. Another quality considered positive by some customers is the unit dose convenience of the water soluble package. Others see this as a limitation especially when they wish sub-unit doses to be available.[35] This is especially important for small farms.

The agrochemical industry seems to favor packaging dry formulations and gels in WSBs as evidenced by their proliferation. Ciba-Geigy has even introduced its Tilt[10] agricultural fungicide propionazole {(±)-=1-[2-(2,4-dichlorophenyl)-4-propyl-1,3-dioxolan-2-ylmethyl]-1H-1,2,4-triazole}as a gel in a water soluble pouch.[8,38]

In addition to the water soluble film, Air Products, Inc. developed a polyvinyl alcohol (PVOH) resin which can be blow molded into water soluble bottles. The material is compatible with a wide range of dry products as well as hydrophobic organic liquid formulations. Container sizes as large as 2.5 gallons have been produced. Empty bottles can be rinsed and completely dissolved in water.[42,35] Cost for the resin will be approximately $1.50 per lb, which is considerably more expensive than virgin polyethylene, but may prove competitive if costly container decontamination regulations are imposed.

2. Other Biodegradables

A new class of materials is available which have the unique property of being completely biodegradable, unlike the first starch-polyethylene blends which left shredded PE behind and gave the public the impression that biodegradability was not practicable. Not only does the current class of materials completely degrade, many are also based on renewable resources, unlike PVOH which uses petroleum based feedstocks. This review will not cover all the biodegradable plastics available on the market. Rather, two examples are given of materials which represent a wide range of properties.

a. Polylactic Acid (PLA)

Currently, ECOCHEM (DuPont-ConAgra joint venture) and Cargill, produce PLA.[32,34] Raw material "lactic acid" is produced by bacterial fermentation of sugars derived from corn, potatoes, grains or milk. The resulting plastic has a high molecular weight and initially resists water. It is compostable and susceptible to hydrolysis through which it loses its strength and plasticity. High quality film as well as injection molded parts can be produced from PLA. Projected selling prices are in the range of $1-3 per lb.

b. Polyhydroxybutyratevalerate (PHBV, Biopol[47])

Zeneca has developed and is producing a unique polymer produced during fermentation by common soil bacteria.[5,32,34,41] The bacteria feed on glucose and produce polyhydroxybutyrate (PHB). Addition of simple organic acids enable the copolymer to be produced which improves the ductility of the material. As with PLA, "Biopol" is totally compostable. It's degradation is activated by microorganisms and is much less susceptible to hydrolysis than other degradable polyesters. PHBV can be formed into films, injection molded and blow molded. Zeneca claims that within 10 years they will be able to harvest PHBV directly from crops.[41] The current price is very high at $8 to 10 per lb, with projections down to $4 for higher volumes.

One of the great challenges (and strengths) of some of these materials (PVOH and PLA) is their sensitivity to moisture. They, themselves must be protected as they protect their contents in order for any useful shelf life to be achieved. The attractiveness, however, of a packaging material derived from the soil and returning to that same soil, is great enough to make this class of materials worth pursuing and supporting.

Other materials are, of course, being developed.[34] Among them Flexel's cellulose film, Novon's starch-based polymer, Union Carbide's polycaprolactone, and Dow's ethylene-carbon monoxide photodegradable copolymer.

D. CONTAINERS COMPATIBLE WITH CLOSED SYSTEMS

Increased attention to user safety has led to totally enclosed systems where the user never comes into direct contact with the product. Packaging, in this case, must be designed to integrate with the application equipment and, in most cases, the empty package returned to the manufacturer or dealer for refilling. In all cases, the farmer must be equipped with necessary unloading and/or application devices. Four examples are discussed here; two liquid and two solid dispensing systems.

1. Closed Liquid Dispensing Systems

These have already been discussed in the section on returnable/refillables. Small volume returnables which link directly into sprayer induction systems are being used in Europe. Examples are Schering's SVR and Ciba-Geigy's LinkPak[11] (formerly Ciba-Link) systems.[26] "LinkPak" incorporates a special dry break coupling that simply slots into a female valve incorporated into the lid of the induction hopper. Beer-keg-style SVRs can include metering systems which are either included in the container design or are separate units kept on the farm.[26] Capacities of these units range from the small 10-liter "LinkPak" through intermediate sizes of 15, 30 and 60 gallons to the larger 250-gallon closed systems for custom applicators .[27]

2. Closed, Solids Dispensing Systems

American Cyanamid introduced their so called Lock 'N Load[4] system in 1992. It was designed to eliminate applicator exposure[14] to their soil insecticide, Counter[4] {terbufos (S-tert-butylthiomethyl O,O-diethyl phosphorodithioate)} through the use of a closed-system, returnable packaging concept. The system was developed in conjunction with John Deere. Empty containers were returned to Cyanamid for refilling.

DuPont, Zeneca and Ingersoll-Dresser Pump company are currently in the final stages of developing the next generation of closed, solids dispensing system under the trade name SmartBox.[21] The "SmartBox" incorporates a returnable/refillable container with a built in metering device. It has an on-the-go rate adjustment, end row shut-off and ground speed compensating flow control. The device can be easily retrofitted to all major brands of planters. The system is currently being developed for soil-applied insecticides for corn.[20]

E. CONTROLLED APPLICATION

One stated goal for the world's developed nations is the reduction in the total amount of pesticides being applied. All other things being equal, reduction of the application rate can be achieved in several ways. One of these requires the ability to control the application rate based on specific needs for a particular part of the field. Several of the above closed-system technologies also incorporate controlled metering technologies. The DuPont-Zeneca-Ingersoll Dresser "SmartBox" is an example of a system which might be directly linked to a satellite tracking system and database to control application of material to the soil based on actual rather than averaged needs.

F. INTERACTION OF ACTIVE INGREDIENT, FORMULATION, AND PACKAGE

Packaging technology must be developed in conjunction with the development of new formulations and active ingredients to produce the front end of a complete delivery system. The more integrated these activities can become, the more effective the results.

1. Active Ingredient

a. Chemicals

Of course, everything starts with the active ingredient. Although one might not think of it this way, the invention of low-use-rate chemistry, notably the sulfonylurea herbicides, has, among other things, profoundly reduced the amount of packaging generated per unit area of crop treated. For instance, some older liquid herbicides, such as the triazines, use about 1 lb of packaging for every acre of corn treated. Sulfonylureas (SU), a much more recent class of herbicides introduced by DuPont throughout the 1980s, use, on the other hand, as little as 0.01 lb of packaging per acre treated; a hundredth the amount. The difference is almost entirely due to the lower use rate of the SU chemistry.

b. Biologicals

The desire of the developed countries to reduce use of chemicals as well as such issues as the development of resistant populations of insects and protection of beneficial species, have led to increased interest in biological means of control. Although one can think of the biologicals as just another class of active ingredients, many of them are fragile to formulate, package and store. Current product systems tend to utilize existing formulation technologies. However, one can envision advances being made both in formulation technology and packaging to enhance the shelf life and extend the efficacy of these actives.

2. Evolution of Formulation Technology

Trends in formulation technology have a very definite effect on the packaging technologies which are finding favor as well as creating needs and opening up opportunities for new materials and designs not yet in existence.

Formulations appear to be moving away from volatile organic liquids and toward granular forms or aqueous based liquids. Gels, suspensions of microcapsules, and the like are also being developed and tested in the market.[23,30] *Gel* formulation technology is evolving and offers a wide range of forms with varied properties somewhere between solids and liquids. Opportunities for new packaging designs seem fertile in this area alone. Gels have also been shown compatible with WSF. *Water soluble film* has also found increasing favor because of its compatibility with dry formulations and its capacity to neutralize the dustiness of wettable powder formulations, rendering them safe. *Effervescent tablets* have been developed for a number of applications including disease control in pomme fruit and vineyards (TOPAS[9]) and in cereals (EXPRESS[19]). Protection of effervescent tablets requires careful packaging design. The EXPRESS system consists of a unique combination of built-in cushioning and water vapor protection to keep the tablets ready for use during an extended period of storage. Some formulation development is going on with the goal of improving the rinsing characteristics of the containers in which they are kept.[31] Considering the direction the EPA is taking with its proposed container regulations, we can expect work in this area to intensify.

3. Packaging for Mixtures

The number of products in which two active ingredients are combined is increasing. This is done for a variety of reasons including market segmentation, enhanced efficacy, broadened application window, and management of resistant pest species. Formulations can be created in those cases for which the active ingredients are both physically and chemically compatible. Occasionally, however, where the actives are not compatible or blendable, unique packaging designs can accomplish the goal. One example is Dow Elanco's COMBO[18] in which the water soluble liquid, picloram (4-Amino-3,5,6-trichloropyridine carboxylic acid)is contained in a plastic bottle, and DuPont's insoluble metsulfuron-methyl (methyl-2-[[[[(4-methoxy-6-methyl-1,3,5-triazine-2-yl)amino]-carbonyl] amino] sulfonyl]-benzoate) contained in a cleverly designed DuPont's cap. When the product is ready for use a seal on the cap is broken and the solid is combined with the liquid for addition to a spray device.[17]

Combinations of solid formulations are enabled with water soluble films where each formulation is isolated in a separate water soluble bag. The strength of the formulations and the amounts in each bag are in the desired proportions so the attached packs can be added to the spray tank simultaneously.

G. PACKAGING DESIGNS AND MATERIALS [30]

Although there is a decided shift away from nonreturnable containers, they are important to many farmers, especially those with smaller operations in the U.S. and in Europe and in developing countries around the world. Small nonreturnable containers will probably therefore be around for a long time to come. This leaves open opportunities to continue to find ways to reduce their weight and increase their rinsing effectiveness and barrier properties. Molds have been designed with reduced height and squarer shapes. They require less material, yet provide adequate integrity to pass required shipping tests. Counter to this, of course, is the importance the U.S. and European regulatory bodies attach to ensuring the strength and durability of containers. The question, "What constitutes adequate strength and durability", will be important to define how much lightweighting of primary containers is possible.

Alternate designs, such as the flexible pouch, which is showing up all over the world, may be applicable for certain agricultural applications, assuming adequate secondary protection can be provided.

The use of water soluble film as an internal bag liner for some products with higher use rates may be valuable in obtaining a "clean" outer bag or box which can be disposed of as household trash.

Another place to look for weight reduction is in the area of secondary packaging. Although considerable corrugated material is currently recycled, the opportunity exists for elimination of secondary packaging altogether through the creation of returnable/recyclable shippers. Just as with the primary returnable/refillables, the issues of useful life and the management of many containers have to be resolved.

Lightweighting is an example of a technology which addresses both environmental and business needs; less material is put into the environment, fewer resources are used, and shipping costs diminish for lower-weight products.

V. CHALLENGES FOR THE FUTURE

Clearly, there will be no "one size fits all" approach for packaging solutions going into the future, since the global marketplace consists of a wide variety of markets with widely varying needs. Manufacturers of crop protection products, for the most part, deal globally and so must respond to customers everywhere, as well as to different regulatory approaches. Regulatory approaches in developed countries will create the experience base from which developing countries will take the lead. The trends that will influence packaging of crop protection products include:

- Reduction in the use of chemical pest control agents
- Reduction in the amounts of chemical pest control agents which are allowed into the environment
- Source reduction in the amount of packaging
- Decrease in the number and amount of nonrefillable containers
- Increase in the number and volume of returnable/refillable containers and systems
- Increased assurances for the safe shipment, storage, and handling of pest control products

In addition, there must also be attention to the economic sustainability of the farm community and of the agricultural industry which provides the products and services to them. All the goals can be met if we encourage and support an approach in which government and industry resources are used in a way that creates balanced overall benefits. We should not encourage approaches which generate little environmental improvements for society at great expense to industry. The trends toward imbalance are seen most strongly in the U.S., Canada, Europe, and Australia, but can be expected to take hold even more rapidly in Latin America and developing countries throughout the world.

Germany, for instance, has taken the lead in setting and attempting to carry out a law which bans all packaging that cannot be recycled or reused. This had left them with a mountain of allegedly recyclable trash for which there was no market. One solution attempted was the illegal export of waste to France where it was incinerated.[15] Since this experience, Germany has softened its stance on plastics incineration.[24]

Recent estimates by the Agricultural Container Research Council suggest that the container rinsing requirements proposed in the EPA Container Regulations for Pesticides, currently under review, will cost industry $100 MM.[1] The impact of successful implementation for society will most certainly not achieve the desired results of ensuring that CPC containers are rinsed free of their contents, because there is nothing to ensure that farmers will rinse them. The regulation merely requires that industry demonstrate they can be rinsed to remove 99.9999% of their original contents.[22]

In another study the Agricultural Container Research Council (ACRC) has shown that properly rinsed pesticide containers along with recycling are best used as an alternate fuel for incineration rather than the costly practice of landfilling.[3] Incineration of 20 MM pounds of properly rinsed high density polyethylene containers is equivalent to saving 3,700,000 gallons of oil.

An eco-balance analysis conducted by the University of Victoria in British Columbia compared the environmental impact of polystyrene cups versus paper and found [stet] paper to consume significantly more natural resources than the plastic.[28]

The main point of these examples is that things are not always what they may seem or what the public currently believes to be true. A more objective approach is critical if we are to plot a course in which the interests of the public, the environment, the farmer, and industry are truly met.

REFERENCES

1. Agricultural Container Research Council, Feedback to EPA RE: Proposed Container Regulations; May 1994, Annapolis, Maryland.
2. Agricultural Container Research Council, Recycling Plastic Pesticide Containers, ACRC-9302, Annapolis, Maryland.
3. Agricultural Container Research Council, White Coal: Empty HDPE Pesticide Containers As An Energy Source, January 1994, Annapolis, Maryland.
4. American Cyanamid, Inc., Trademark , Princeton, New Jersey.
5. Biodegradable Bottle Gets Major Launch, News Perspective, *Packaging Digest*, May 1993, 2.
6. **Blackbeard, J.**, Agrochemicals may come in refillable containers, *Arable Farming*, January 1993, 24.
7. Anon., Canadian Container Management Program on Target, *AGROW*, Issue 196, 11/19/93.
8. Anon., Ciba Stresses Role of New Formulations, *AGROW*, Issue 192, 9/17/93.
9. Ciba-Geigy, Inc., Registered Trademark, Basel, Switzerland.
10. Ciba-Geigy, Inc., Registered Trademark, Basel, Switzerland.
11. Ciba-Geigy, Inc., Trademark, Basel, Switzerland.
12. Cross Contamination Threatens U.S. Mini-Bulk Usage, *AGROW*, 190, August 20, 1993, 16.
13. **Crossen, C.** How 'tactical research' muddied diaper debate, *Wall Street Journal*, B1, 5/17/94.
14. Cyanamid Cuts Worker Exposure, *AGROW*, Issue 146, p37, 11/01/91.
15. **Demetrakakes, P.**, European packaging laws: A Pandora's box, *Packaging*, February 1993, 51.
16. **Dinger, P.**, Bottle reclaim systems, *Modern Plastics*, mid November 1993.

17. Dow Elanco, Inc., promotional literature, 1993, Indianapolis, Indiana.
18. Dow Elanco, Inc., Registered Trademark, Indianapolis, Indiana.
19. E.I. duPont de Nemours and Company, Registered Trademark, Wilmington, Delaware.
20. E.I. duPont de Nemours and Company, AG Products Internal Reports, 1994, Wilmington, Delaware.
21. E.I. duPont de Nemours and Company, Trademark,Wilmington, DE.
22. Federal Register, Standards for Pesticide Containers and Containment, EPA-Proposed Rules, Vol. 56, No. 29, February 11, 1994.
23. **Frei, B.** and **Nixon, P.**, Novel formulation and packaging concepts-customer need or marketing tool, *Brighton Crop Prot Conf- Pests and Diseases-1992*, 4B-2, 1992, 321.
24. Germans Soften Stance on Plastics Incineration, ECN Environmental News, *European Chemical News*, July 12, 1993, 21.
25. **Graham, J. D.**, "Regulation: a risky business", *Wall Street Journal* Editorial, 5/18/94.
26. **Henly, S.**, The will to refill, *Crops*, Vol. 11, (11), June 12, 1993.
27. **Holmberg, M.**, Packaging priorities, *Successful Farming*, March 1993, 52.
28. **Horking**, Relative Merits of Polystyrene Foam and Paper in Hot Drink Cups; Implications for Packaging", *Environmental Management*, 13(6), November 1991.
29. Institute of Packaging Professionals, Fundamentals of Packaging Technology Course, 1993, Herndon, Virginia.
30. **Keck, B.**, Searching for new barriers; chemical packaging in the U.S., *European Packaging*, 1, April 1992, Vol. 3.
31. **Little, D.**, Griffin comes of age, *Farm Chemicals*, October 1993, 14.
32. **McCarthy-Bates, L.**, Biodegradables blossom into field of dreams for packagers, *Plastics World*, March 1993, 22.
33. Modern Plastics Encyclopedia '94, Special Issue and Buyer's Guide for *Modern Plastic Magazine*, mid November, 42, 1993.
34. **Moore, J. W.**, Degradable plastics, *Modern Plastics*, Mid-December 1992, 58.
35. **Ogando, J.**, Biodegradable polymers crop up all over again, *Plastics Technology*, August, 1992, 60.
36 Opinions Split on European Pack Recycling, *AGROW*, Issue 197, 12/03/93.
37. **Pollack, C.**, Film reclaim systems, *Modern Plastics*, mid November 1993.
38. Pouch/ Clamshell/Corrugated Trio Solves HazMat Shipping Dilemma, *Packaging Technology and Engineering* "Greenwatch" November 1993, 16.
39. Proposal for a Council Directive on Packaging and Packaging Waste, *Official Journal of the European Communities*, 92/C263/01, COM(92) 278 final - SYN 436.
40. **Randall, J. C.**, Chemical Recycling, *Modern Plastics*, mid November 1993.

41. **Roberts, M.**, Zeneca doubles capacity of new polymer, *Chemical Week*, October 20, 1993.
42. **Schmuck, D.**, Containers can count, *Farm Chemicals*, March 1992, 14.
43. **Stossel, J.**, Are we scaring ourselves to death, *20/20 Documentary, ABC News*, 5/94.
44. Tellus Institute, Tellus Packaging Study; Assessing the impacts of production and disposal of packaging and public policy measures to alter its mix , May 1992, Boston, Massachusetts.
45. **Tomascek, T.**, Automated separation and sort, *Modern Plastics*, mid November 1993.
46. U.S. Environmental Protection Agency, Container Study- Report to Congress, EPA 540/09-91-116, May 1992 PB-91-110411.
47. Zeneca, Inc., Registered Tradename, Wilmington, Delaware.

Chapter 11

The Market for Agricultural Pesticide Inert Ingredients and Adjuvants

Edward G. Hochberg

CONTENTS

I. INTRODUCTION

This discussion is divided into two parts, additives (inerts) and adjuvants. Additives, or so-called inerts, are used by the primary formulators of agricultural pesticides. Adjuvants are defined as formulated compounds used to enhance the effectiveness of the active pesticide. These adjuvants are also referred to as tank mix additives as they are mixed with the formulated pesticides at the point of field crop application. Their function is to improve the efficacy of the active pesticide ingredient by optimizing its delivery or placement onto the target plant or insect, by increasing retention time, penetration, and wetting, and by plant uptake or translocation.

II. PESTICIDE ADDITIVES (INERTS)

In 1992 approximately 1.0 billion lb, 100% active basis, of chemical additives or so-called inerts, with a market value of $179 million (U.S.) at the wholesale level, were consumed by U.S. formulators of agricultural pesticides. These companies produced some 3 billion lb of formulated pesticides requiring approximately 1.1 billion lb of active ingredients.

The functions of these various inerts are to assist in getting the active ingredients or toxicants into aqueous or solvent solutions, to get them more easily onto the target plant or insect, and to aid in their pesticidal effectiveness.

An important part of our investigation was to determine the poundage

1-8493-7678-5/96/$0.00+$.50

consumption of the formulated pesticides by physical forms.

Table 1 shows the estimated 1992 U.S. formulated pesticide poundage consumption by physical forms. Also shown are projected total changes in their consumption levels by the end of 1997.

By the end of 1997, total agricultural pesticide poundage consumption is projected to drop by 6.7% to 2.8 billion lb. This poundage decline is predicted on the assumption that over the coming years more potent and effective active ingredients will be utilized, requiring less formulated product. These will be applied in grams/acre rather than pounds/acre.

Table 1
U.S. Shipments of Formulated Agricultural Pesticides by Physical Form 1992

Physical form	MM lb	%	1992-1997 Changes
Emulsifiable concentrates (EC)	370	12.3	-15%
Water soluble solutions	900	30.0	-20%
Granulars	700	23.3	-14%
Wettable powders	100	3.3	-0-
Liquid flowables	170	5.7	-10%
Water dispersible granules	300	10.0	+33%
Others including:	460	15.4	+17%
Emulsion-in-water (EW)			
Water soluble granules (WSG)			
Suspensions-in-water			
Tablets			
Microencapsulation			
Biotech materials			
Total 1992	3,000	100.0%	-6.7%
Total 1997	2,800		

Source: Hochberg and Company, Inc.

A. INERT (ADDITIVE) CONSUMPTION

As mentioned, the U.S. pesticide formulators consumed an estimated 1.0 billion lb of inerts valued at $179 million in 1992. Approximately 42% of this expenditure was for various surface active chemicals functioning as surfactants, emulsifiers, dispersants, solubilizers, antifoams/defoamers, and compatibilizers. Other important inerts included solvents, carrier/diluents, antifreeze, clay deactivators, preservatives, and thickeners/suspending agents.

Table 2 lists the various inerts consumed in 1992 in U.S. dollars and gives their total poundage. Also given are the projected dollar volumes for these inerts by the end of 1997. An average annual inflation rate of 3% was used between

1992 and 1997. By the end of 1997 the total consumption of inerts is projected to decline to 900 million lb, 100% basis, having a current market value of almost $197 million. Also given in Table 2 are the expected total percent poundage changes anticipated between 1992 and 1997 for the various inerts.

Table 2
Additive (Inert) U.S. Consumption In Pesticide Formulations
1992-1997

Additive (inert)	**1992 ($Million)**	**1997 ($Million)**	**Total poundage change**
Surface active agents; surfactants, disperants, emulsifiers, solubilizers, antifoams, compatibilizers	75.0	88.0	3%
Solvents	32.0	33.0	-15%
Carrier/diluents	55.0	56.5	-10%
Antifreeze	8.0	9.0	0
Clay deactivators	4.5	4.5	-10%
Preservatives	2.0	2.0	0
Thickeners/suspending agents	2.5	3.5	+20%
Total	$ 179.0	$ 196.5	-10%
Inflation rate: 3%			
Total pounds (millions)	1,000	900	-10%

Source: Hochberg and Company, Inc.

The $75 million of surfactant chemicals included:

Nonionics

- Tallow amine ethoxylate (for glyphosate)
- Nonyl and octyl phenol ethoxylates
- Alcohol ethoxylates
- EO/PO block copolymers
- Castor oil ethoxylates

Anionics

- Calcium dodecyl benzene sulfonate
- Sodium naphthalene sulfonate
- Dioctyle sulfosuccinate

Some assumptions used to derive our data were:

100% of the domestic shipment of active ingredients were formulated in the U.S.

50% of the exported active ingredients were formulated in the U.S.
The EPA has been promulgating four lists of inerts. These lists segment those inerts which are

1. Of toxicological concern and require more testing.
2. Potentially toxic.
3. Of unknown toxicity.
4. Of minimal toxic concern.

The future use of the inert chemicals on Lists 1 and 2 will be dependent on the outcome of further test results and EPA action.

III. ADJUVANTS

Adjuvants or tank mix additives contain such constituent materials as surfactants, wetting agents, dispersants, emulsifiers, foaming agents, antifoams, pH buffers, polymeric adhesion promoters, film formers, antifreeze agents, alcohols, mineral oils, and modified vegetable oils.

Proper use of an adjuvant can provide a more uniform application of the active ingredient upon a specific target, reduce water surface tension, improve compound stability, and allow for better mixing together of organic solvent and aqueous systems or chemically diverse ingredients.

The nomenclature used to identify the adjuvants for our investigation was based upon their functions as we saw it. Effort is being made by the ASTM E-35 committee to bring some order to the adjuvant categories. For purposes of our investigation, agricultural adjuvants have been segmented into 11 categories (Table 3).

Table 3
Agricultural Pesticide Adjuvant Categories

Adjuvant category	Function
Spreader/activator/penetrants	Wet, penetrate
Spreader/stickers	Increase retention time
Drift/mist control agents	Greater deposition on target
Defoamers/antifoams	Suppress forms
Compatibility agents	Make stable solutions
Paraffinic oil concentrates/spray oils	Increase retention time and wetting
Vegetable oil concentrates	Increase retention time and wetting
Foam markers	Identify target areas
Buffering agents	Stabilize pH
Super wetters	Increases wetting and take up
Fertilizer absorption enhancers	Enhance performance of nitrogen-containing fertilizers

Source: Hochberg and Co., Inc.

In 1992, an estimated 330 million lb of adjuvants were consumed having a formulator/distributor market value of U.S. $196 million (Table 4). They required $117 million of constituent chemicals and materials. By the end of 1997, adjuvant consumption is expected to increase to $285 million. Poundage consumption will increase by a total of 21%. This increase is based on our assumption that adjuvants will have increased use with more water based pesticides and for the better application of the highly effective active ingredients that are applied in grams per acre rather than pounds per acre.

Table 4
U.S. Consumption of Adjuvants and Constituents
1992-1997
(Millions)

Item	1992	1997	Total change
Total adjuvant consumption ($) (formulator/distributor level)	$ 196	$ 285	-
Total adjuvant consumption (lb)	330	400	21%
Total constituent materials (4)k	$ 117	$ 169	-
Average annual inflation rate: 3%			

Source: Hochberg and Company, Inc.

In 1992, there were at least 80 companies supplying one or more formulated adjuvants. They either did their own formulating or repackaged and sold tolled materials. The four leading suppliers, Helena, Loveland, Terra, and Wilbur Ellis controlled 32% of the total adjuvant market in 1992. Table 5 lists the leading eight adjuvant suppliers along with their market shares.

Table 5
Leading Eight Adjuvant Suppliers
1992

Supplier	Market share
Helena/Marubeni	13%
Loveland/Conagra (United Agri)	8%
Terra International	6%
Wilbur Ellis	5%
Agsco/American Cyanamid	3%
Cenex/Land-O-Lakes	2%
Drexel	2%
Amway	2%
Leading eight subtotal	41%
Others	59%
Total	100%

Source: Hochberg and Company, Inc.

Table 6
U.S. Consumption of Agricultural Inerts and Adjuvants
1992-1997
(Millions)

Material	1992 lbs	1992 $	1997 lbs	1997 $	Total poundage change
Inerts (Additives)	1,000	$179	900	$197	10%
Adjuvants	330	$196	400	$285	+21%
Total	1,330	$375	1,300	$482	-2%
Total formulated pesticides	3,000		2,800		

Average annual inflation rate: 3%

Source: Hochberg and Company, Inc.

IV. SUMMARY

The total U.S. consumption of adjuvants and inerts, in 1992, was 1.33 billion lb having a market value of $375 million (Table 6). By the end of 1997 this total poundage will decrease (by 2%) to 1.3 billion lb having an inflated market value of $482 million. The average annual inflation rate between 1992 and 1997 is projected at 3%. Total consumption of formulated pesticides is projected to decrease from 3.0 billion lb in 1992 to 2.8 billion lb by the end of 1997.

Chapter 12

The Reduction of Pesticide Drift

Franklin R. Hall and Robert D. Fox

CONTENTS

I. INTRODUCTION

Pesticides are applied in agricultural systems for the purpose of protecting plants from injury by weeds, insects, disease, etc., which today still destroy almost 33% of all food crops. Applications are considered effective if they achieve the desired biological result and economic if there are a crop yield and a quality response above and beyond the cost of chemicals and their application. Yet the use of pesticides has also resulted in significant costs to public health and the environment. In general, the amount of pesticides released into the environment has risen about 1900% in the 50-year period between 1930 and 1980.[18] The improved efficacy of the more recent pesticides has allowed rates to be reduced in some instances from kilograms in the 1970s and 1980s to a few grams per hectare in the 90's. However, the

1-8493-7678-5/96/$0.00+$.50

efficiency with which these reduced amounts of agrochemicals are delivered to an array of target microstructures remains suspect.

Currently, approximately 3 million tons of pesticides were used in 1992 and this amount continues to increase yearly with the bulk (60-70%) of the use as herbicides.[28] Off-target losses have been estimated at ca. 50-70% for various airblast and aerial applications of insecticides to forests due to evaporation and drift. With the broad-spectrum nature of most pesticides, this waste of scarce resources and its potential impact on the environment continues to be of concern.

A. DEFINITION

Atomization of liquids for the purpose of applying a mist onto a target results in loss of a portion of the spray cloud in two ways: spray drift, consisting of airborne movement of liquid particles immediately after hydraulic or rotary atomization, and vapor drift which is associated with volatilization. This movement of active ingredient (AI), usually away from the intended foliar target, continues to be an important topic for agricultural researchers worldwide. Displacement of pesticides out of the intended target area is not only wasteful, but represents a loss in efficacy and leads to increased costs to the user and the environment (hence, society as a whole).

Pesticide drift is affected by several major factors, including chemical/physical properties of the solution, the equipment (nozzle type, number, pressure, and spray volume), the application technique, weather, and operator care and expertise. This paper deals principally with the characterization of spray drift as particulate rather than vapor drift, and the current and possible future strategies of mitigating that type of spray drift.

II. ESTIMATES OF SPRAY DRIFT

A. THE SPRAY CLOUD

In an attempt to assess the drift potential of sprays, numerous researchers have published droplet distribution data. Two most commonly used terms to describe such distributions are the Volume Median Diameter ($DV_{0.5}$) and the Number Median Diameter ($DN_{0.5}$), the diameters below which 50% of the total volume and number of drops of liquid sprayed occur in drops smaller than the $DV_{0.5}$ and $DN_{0.5}$, respectively. The interpretation of standard droplet distributions in terms of frequency distributions, cumulative or otherwise, is not especially intuitive. An alternative way to express droplet spectra data is given in Fig. 1[16]. The numbers across the main body of the graph are the percentage by volume, in the various size classes, for water, and

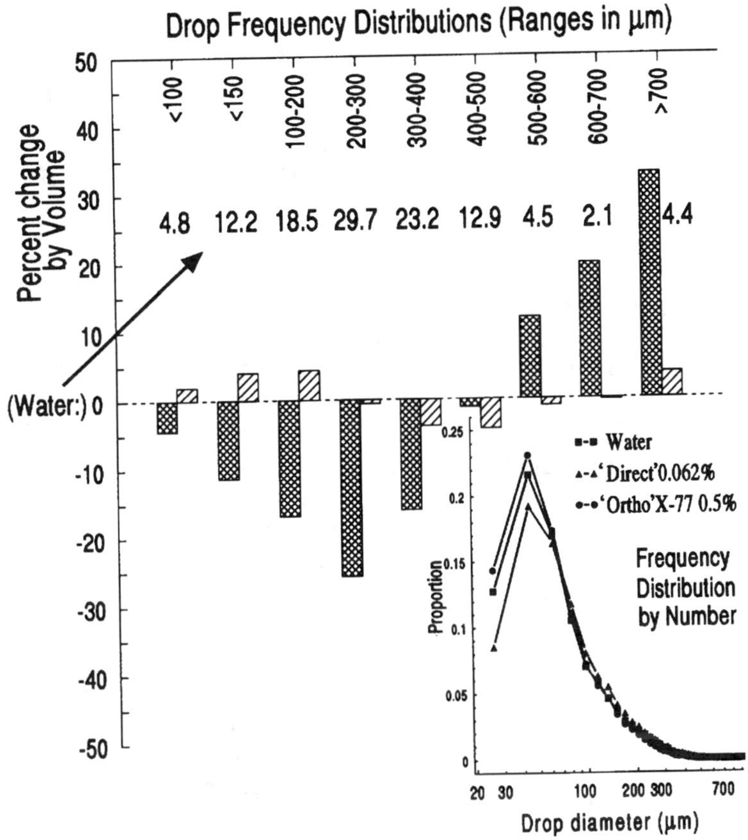

Figure 1. Droplet frequency spectra (proportion by number and volume for water, Ortho X-77, and Direct).[16] (From Hall, F., Chapple, A., Downer, R., Kirchner, L., and Thacker, J., Pestic. Sci., 38, 123, 1993. With permission.)

form the baseline against which all comparisons are made. For example, with Direct, a drift-control polymeric adjuvant, there is about 4.5% less volume in droplets that are <100 μm diameter compared to water where ca. 21% of the total volume sprayed is in drops within this size range. The information presented in this way is easier to understand with respect to the possible biological consequences of changes in atomization. Similar, but more limited information about adjuvant effects on a partitioned assessment (e.g., % volume, <100 μm) is presented in Table 1, where precise changes in volume or number of drops below a specified threshold can be elucidated. Other work by Chapple et al.[7] demonstrated that adjuvants can severely affect the spray pattern beneath a boom. Figure 2 shows such an effect for the overlap between two flat-fan nozzles using water compared to water plus two polymer-based drift control agents. Where evenness of deposit is desired, Dorr and Pannell[8] demonstrated that disruption of spray patternation may lead to inefficient placement of actives with economic consequences, even though drift may be reduced.

Polyacrylamide polymers such as Nalcoltrol™, and Nalcotrol II™ are popular drift reducing additives, however, some of these products have several characteristics not conducive to ease of use or reliable efficacy. For example, the polymers may disperse poorly, hydrate slowly, may be water quality sensitive, and degrade under shear stress (Figure 3).[6] The recent development of some new, dry, polymers for drift management, such as AgRHô® DR-2000, has addressed many of the shortcomings of the polyacrylamides.[9] For example, both AgRHô DR-2000 and Nalcotrol II™ reduced off-target drift compared to tap water alone by 71% and 73%, respectively, for unsheared mixtures at 4 m/s wind speed (Figure 4). In addition, sheared and unsheared AgRHô DR-2000 provided drift control equivalent to unsheared Nalcotrol II. When sheared, Nalcotrol reduced drift by 48% compared to water, while AgRHô DR-2000 maintained "unsheared" levels of drift control (71%).

Evaporation (particularly of small droplet sizes) of spray solutions is a key factor in how long particles remain in the air. Thus, studies of materials/formulation modifications to reduce the rate of evaporation have become even more important and critical for effective modeling experiments.[17] Figure 5 shows the potential change in rates measured under controlled temperature (T) and relative humidity (RH) conditions, but without the dynamics of air turbulence. Current efforts by LPCAT scientists are now completing an upgraded computer modulated evaporation method which also includes air turbulence as a function of droplet evaporation potentials.

Table 1. Percentage by Volume and Number of Droplets < 100 μm and Percentage by Volume > 300 μm, Calculated from PDPA Data for 13 Spray Formulations[a]

Formulation[21]	% By volume < 100 μm	% By number < 100 μm	% By volume > 300 μm
Water	2.5d	37.7de	54.0d
'Ortho' 0.0625%	2.9d	39.2d	54.1d
'Ortho' 0.5%	3.5c	35.1e	47.2b
'Plex'	6.5a	49.0a	38.8a
'NuFilm' 17	1.3e	18.0h	59.lef
'Windfall'	3.4c	42.6bc	55.1d
'Nalcotrol'	2.7d	38.6d	61.0fg
'Invade'	1.4e	23.2g	55.5d
'Penetrator'	1.7e	27.2f	58.3e
'Complex'	1.5e	24.9fg	61.6g
'NuFilm' P	1.7e	27.6f	60.9fg
'Windfall' + 'Plex'	6.0b	45.4b	40.3a
'Windfall' + 'Invade'	3.9c	40.lcd	50.5c

[a] Means followed by a different letter were significantly different at the 5% level. Student-Newman-Keuls means separation test. (From Hall, F., Chapple, A., Downer, R., Kirchner, L., and Thacker, J., Pestic. Sci., 38, 123, 1993. With permission.)

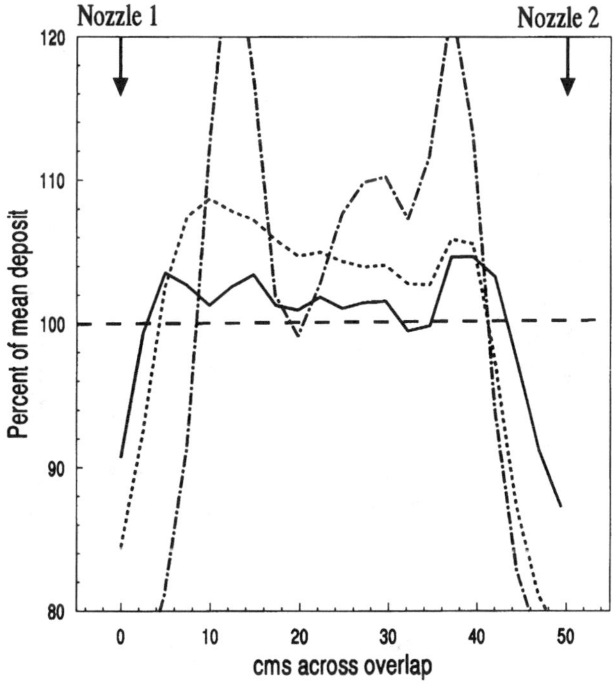

Figure 2. Swath deposit pattern across the overlap for two hydraulic nozzles (-) water, (---) windfall, and (----) Nalcotrol. (From Hall, F., Chapple, A., Downer, R., Kirchner, L., and Thacker, J., Pestic. Sci., 38, 123, 1993. With permission.)

B. DELIVERY SYSTEMS

The drift of spray out of the intended swath has been a concern ever since a farmer sprayed a grain field with 2,4-D which was upwind of his wife's tomato patch. Scientists have been measuring drift from boom, aerial, and airblast sprayers for the past 40 years or so. Most of the factors affecting drift have been determined. Current research is aimed at developing equipment and practices that will minimize drift while still providing adequate pest control. There have been hundreds of scientific and popular articles written on drift measurements and control. Also good summaries for ground and aerial delivery are provided.[10,14,19,22,37]

Modelling and wind tunnel validation efforts[29,30] showed the effect of wind velocity and release heights on droplet displacement (Figures 6 and 7). With a wind speed of 8 km/h, 200-μm droplets are moved about 0.15 m; whereas 100-μm droplets are transported about 3 m (Figure 6). Droplet displacement may not be a large problem when herbicide is applied to a section of wheat, but it is a concern when insecticide is applied by a single nozzle over a row of cabbages that are 20 cm wide, with 76 cm between rows.

Figure 7 shows that raising the nozzle discharge height increases droplet displacement. When the release height of a 100-μm droplet is increased from 51 to 102 cm, droplet displacement increased from 0.6 to 1.7 m.

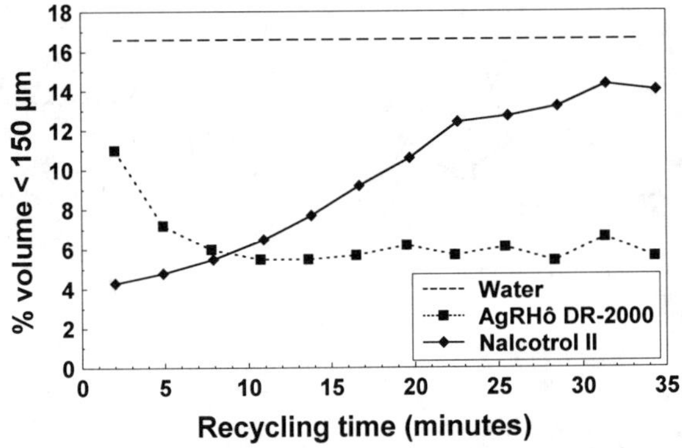

Figure 3. Effect of pump shear stress on the proportion of spray volume in driftable droplets.[9]

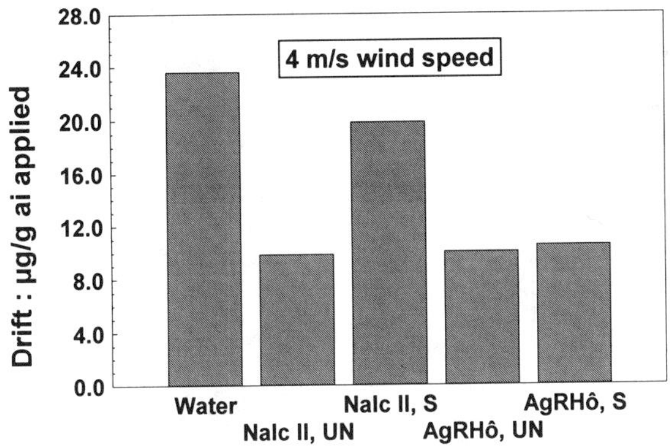

Figure 4. Effect of pump shear stress (S = sheared, UN = unsheared) on drift reduction by adjuvants at 4 m/s wind speed.[9]

Under some conditions, such as a stationary nozzle, the entire spray pattern may respond as a unit, and individual droplets within the pattern may be insulated from the effects of external forces such as air currents, etc.

Bode et al.[4] used statistical methods to identify the most important factors (from five climatic, four spraying systems and six combinations) affecting measured spray drift deposit and expected ground deposit beyond 2.4 m from the edge of the field, assuming a field 402 m wide was sprayed (Table 2). Significant factors affecting these measures of downwind deposit were boom height, wind, liquid spray volume and, if a 402 m wide strip were sprayed, the amount of Nalcotrol being the most significant. The F-values shown are a measure of the relative significance of each factor. Nozzle type was not included as a factor in the statistical analysis because not enough trials were made.

Gilbert and Bell[13] describe a system of collectors they have developed to measure airborne spray and ground deposits downwind of a spray site. Figure 8 is a plot of expected airborne spray drift for several spraying systems for a range of wind speeds. These results are based on a series of experiments they conducted over several years, but

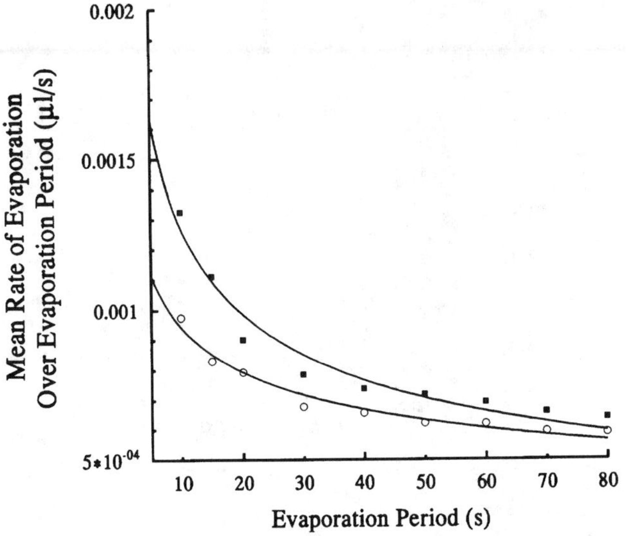

Figure 5. Static evaporation rates of (○) 38-F and (■) water at 35°C and 30% RH. (From Hall, F., Kirchner, L., and Downer, R., Pestic. Sci., 40, 17, 1993. With permission.)

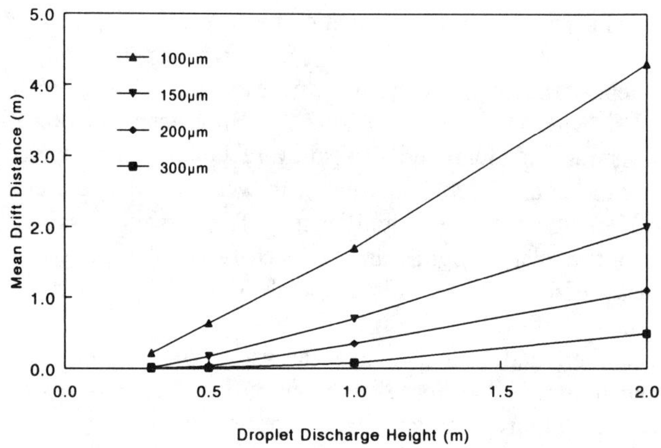

Figure 6. Effect of wind velocity on mean drift distance of droplets directed downward at 45 mph from height of 20 in.[29,30]

Table 2. Factors Affecting Downwind Drift (Nozzle Type Not Included).

Factors considered in analysis	Order of importance in explaining spray drift deposit	F value	Order of importance in explaining drift beyond 2.4 m (if 402 m were sprayed)	F value
Wind speed	Boom height	9.92	Concentration of thickener	12.85
Relative humidity	(Boom height) (Wind)	9.65	Wind speed	6.11
Temperature	Spray application volume	7.48	Boom height	4.89
Atmospheric pressure	Wind	7.10	Concentration of thickener/Wind	4.55
Richardson No. (stability)	RH	4.12	(Boom height) (Wind)	3.99
Boom height (43 cm and 58 cm)	(Wind) (Richardson No.)	3.90	RH	3.90
Spray application volume	(Wind) (Nozzle Press.)	2.80	Temperature/RH	3.26
Nozzle pressure	Atmospheric pressure	2.62	Atmospheric pressure	2.60
Concentration of thickener	Concentration of thickener	1.50	(Temp) (Wind)/ (RH)	2.38
6 other combinations of these 9 factors	Nozzle pressure	0.28	Temperature	0.51

Adapted from Bode, L. R., Butler, B. J., and Goering, C. F., Trans ASAE, 19(2), 213, 1976. With permission.

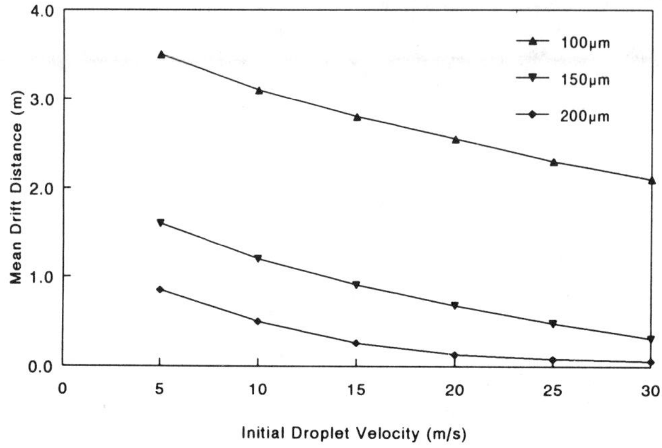

Figure 7. Effect of discharge height on mean drift distance of droplets directed at 45 mph from height of 30 in.[29,30]

drift quantities seem to be somewhat lower than have been reported by other researchers.

Payne et al.[26] tried to estimate the required width of buffer zones around lakes when spraying permethrin with a knapsack mistblower (about 50 μm $DV_{0.5}$). They report that the worst case for drift deposits is when spraying in light but not calm conditions, with a stable boundary layer, low relative humidity, and high air temperature.

Fox et al.[11,12] found that ground deposits from airblast drift experiments at 61 m downwind were about 1/250 deposits near the tree row. Beyond 61 m downwind, airborne spray deposits were about 30 times greater than ground deposits based on deposit per unit area of the collector. Riley and Wiesner[31] measured off-target spray losses resulting from applying pesticide on 6-m-tall trees with an air-assisted sprayer. They found that both number and size of droplets decreased rapidly with increased distance downwind, and they developed regression equations that can be used to predict worst case deposits from multiple row applications. Salyani and Cromwell[33] compared ground and airborne spray deposits resulting from spraying an orange grove with an orchard air sprayer, a fixed wing aircraft, and a helicopter. Averaged over all downwind locations, greatest airborne spray deposit and the least ground deposit was produced by the orchard sprayer applying the low liquid volume; there were not significant differences in drift from aerial and ground spraying.

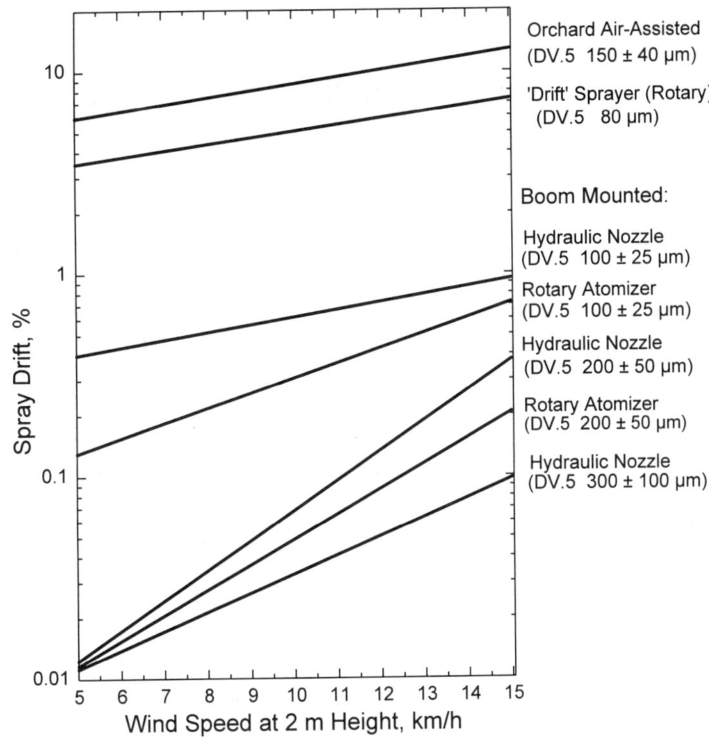

Figure 8. Airborne spray drift at 26 ft downwind as % of applied spray vs. wind speed.[13]

The data in literature, which is robust and complete, can be summarized as follows.

Ground Sprayers: Total drift from ground sprayers is estimated to be about 5% of applied spray. About half of this drift falls on the ground in the first 9.1 m downwind, and ground deposits fall off to about 1/10000 of the applied spray by 91 m. Airborne spray flux at 9.1 m downwind and at 1.5 m height is about 2 to 10 times ground deposits at the same downwind distance (on an area basis). Reducing the percentage of spray in the "driftable" size range (less than about 100 μm diameter) by changing nozzle type and lowering nozzle pressure can reduce drift. Lowering boom height also reduces drift. The use of adjuvants, etc., may also reduce drift; for example, adding a drift control additive can reduce ground deposits in the first 9.1 m by 50-75% of deposits using water alone.

Meteorological Factors: Temperature and relative humidity affect the evaporation rate of droplets. These factors will probably not have much effect on deposit in the first 9.1 m, but may affect long range dispersion of droplets. Wind speed is an important factor; increased wind speed will increase swath displacement and the total amount of spray carried out of the swath. However, increased wind speed may reduce ground deposits immediately downwind of the displaced swath. Atmospheric stability affects the vertical mixing of the spray cloud into the air and can affect the distance spray droplets drift. Very stable conditions (vertical mixing suppressed) is often associated with low wind velocities, so it is difficult to separate the effect of wind speed from stability. Most researchers show increased ground deposit at downwind distances less than 9.1 m when stability increased. With calm winds and very stable conditions, many experts advise against spraying. Under these conditions, small droplets can "hang" in the air and then be carried by slow moving air currents >50 to 80 m and the entire cloud deposited in high concentrations. This cloud may move in any direction so any sensitive area close to the sprayed field may be affected.

Aerial and Airblast Spraying: Aerial application results in about 5 to 10 times more ground deposit within 9.1 m and 5 to 10 times more airborne spray flux than ground boom spraying. Airblast sprayers usually produce less ground deposit and airborne spray flux than aircraft spraying. However, under certain conditions, such as using low volumes with small spray droplets, airblast sprayers may produce more drift than aerial applications.

In summary, there have been some excellent studies on drift in the literature, leading to increased efforts in prediction of spray drift. There are now available at least two commercial programs that predict spray dispersion and deposition when spraying forests with aircraft, the Forest Service Cramer-Barry-Grim model (FSCBG) and a model by Picot, et al.[27] Walklate[35] developed a random walk atmospheric diffusion model to predict spray movement from an axial flow fan orchard sprayer. Drift simulations agreed well with measured drift over stubble and orchard canopies. Recent studies being developed by the Spray Drift Task Force (SDTF) should fill in the gaps in airblast drift literature.

III. DRIFT MITIGATION

A. MECHANICAL

Wolf et al.[36] investigated both nozzle sizes and shields in field trials designed to determine the effectiveness of shields on the extent of off-site movement of pesticides. They found that increasing the application rate to 100 ℓ/ha by using Spraying Systems #8002 tips reduced drift of the unshielded sprayer by 65%. Decreasing application rate to 15 ℓ/ha by using 800017 tips increased drift by 29% despite the use of a shield. Off-target drift increased with increasing wind speeds for all sprayers, but the increase was less for shielded sprayers and coarser sprays. The decreased droplet size of spray from 110° tips increased drift when the boom height was the same as for 80° tips. High wind speeds, lower carrier volumes and finer sprays, 110° tips, and solid shields tended to decrease on-swath deposit uniformity, whereas a perforated shield or cones did not affect deposit uniformity (Figure 9).

Wolf et al.[36] concluded that shields, given their ability to reduce drift, can make an important contribution to efforts leading to reduced pesticide inputs into the environment. However, they do not eliminate drift, and drift losses are still a function of the coarseness of the spray. Shields, or any drift reduction device, must still be used responsibly and not under increased wind conditions. They concluded, "It will be equally important to stress the limitations of shields as to applaud their effectiveness. Ultimately, improved pesticide application still depends on the knowledge, care and judgement of the sprayer operator."

Others found that covered booms did not permit applicators to check the nozzles during spraying.[23] The use of spray adjuvants to increase average spray particle size[33], or a reduction in boom pressure or boom height (or change of nozzle from 80° to 110° without boom adjustments can result in uneven spray deposit patterns and may change biological effects.[9,20] Dorr and Pannell[8] showed the economic significance of spray pattern inefficiencies.

Attempts have been made to reduce drift via controlled droplet applicator (CDA) application (e.g., a more narrow spray droplet distribution, electrostatics, air assistance, and numerous novel nozzle types, as well as sprayers with sensors designed to detect the presence or absence of trees/plants and determine canopy height and width. Alternatively, drift control agents - adjuvants which shift the droplet spectra to larger sizes - may be used. However, these may actually adversely affect the biological effect since it has been repeatedly demonstrated that small droplets (< 200 μm diameter) are not only more effectively captured by foliage,[34] but they are also, dose for dose, more biologically active than large (> 300 μm in diameter) droplets.[1] A simple method of limiting the environmental impact of drift may be the use of buffer zones, i.e., untreated areas of land surrounding the orchard or field which serve to 'absorb' the drift. A buffer zone of about 15 m

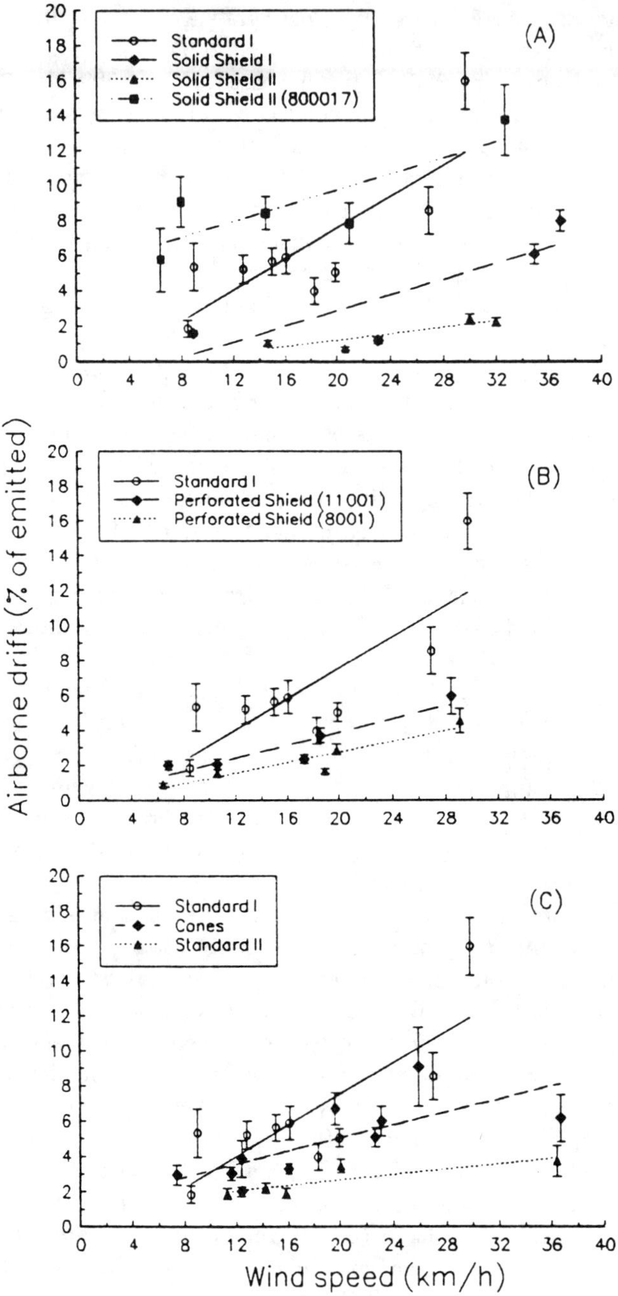

Figure 9. Off-target airborne drift amounts for various sprayers vs. unshielded 8001 tips (A) solid shields, (B) perforated shields, and (C) cones.[36] (From Wolf, T. M., Grover, R., Wallace, K., Sewchuk, S. R., and Maybank, J., Can. J. Plant Sci., 73, 1261, 1993. With permission.)

would appear to be sufficient for ground-based applications to tree crops. However, it should also be noted that tree height, leaf area index (including planting density), spray volumes, atomizing system, meteorology, user expertise, and formulation chemistry, all play a role in the ultimate exposure of areas surrounding an orchard.[15] Additionally, the capture efficiency of pesticide spray drift by border vegetation (row of trees, etc.) has been shown by Hall[15]. Depending upon tree size and planting density, the second row away from the treated row had from 99 to 100% reduction of applied pesticides. Leaving the outside four to six rows unsprayed, however, presents an economic dilemma for growers. Ultimately, the off-target fate of agrochemicals must be related to the potential toxicological impacts on non-target organisms and the surrounding environment. In this 1991 study, movement of toxicant up to 30 m from the outside rows of an orchard, if projected onto a ten to fifteen seasonal application schedule, could clearly result in highly significant toxic effects to non-target organisms.

Drift per se, may be an inevitable consequence of all liquid pesticide applications. However, reducing drift will not contribute to reductions in the environmental impact of pesticides - the ultimate goal - if it occurs at the expense of increased soil and/or ground water contamination. With many Integrated Pest Management (IPM) strategies, spraying the outside rows (and reducing middle block area treatments) is an effective technique to reduce pest migration into the orchard while at the same time reducing total AI toxicant load in the orchard.

B. USER STRATEGIES

The effectiveness with which spray drift is reduced is no better than the weakest component in the delivery protocol. Weather conditions have a major impact on the amount of drift, with the major factor being that of wind speed. However, wind direction, temperature, relative humidity, and atmospheric stability are also influencing factors as shown in the various drift models. The lack of standardized on-site wind monitoring devices is a detriment to improved drift mitigation practices on the farm.

In addition to weather, the equipment and how it is utilized (education, user expertise, etc.) is probably a most influential parameter. Common sense, clearly, must remain the primary factor in reducing off-target movement. If the air is very stable, use of low volumes (resulting in a small average pesticide size) will allow suspended particles to remain in the air - thus, it is better to spray when there is air movement, but below 24 km/h. However, recent aerial application studies by Payne[25] indicated that increased on-target deposition (needles) and reduced off-site deposits were achieved by using higher wind speeds. Thus, the complexities of these interactions between application and environmental variables (including target macrostructures) make it difficult for precise predictions. Growers can use larger spray volumes (large nozzles can reduce potential off-target contamination) but, depending upon the target requirements and the AI utilized, there may be a significant trade-off in transfer efficiency. Bode and Wolf[5] provided a summary of recommended procedures for reducing off-target movement of pesticides for ground boom type applicators (Table 3). These site-by-site and daily changes in conditions require specific decision making by the users every time they initiate the spraying process.

From a Tree Crops Workshop in 1993, Hall (unpublished, 1993) provided a series of recommendations for drift mitigation appropriate for airblast sprayers (Table 4). Again, it can be seen that for major efforts to take place at the farm, education, training, and common sense must be factored in and continually emphasized for significant reductions in the percent of off-target movement to occur. Buffer zones for some pesticides will also be a viable regulatory tool, but this may prove to be a severe economic hardship for small, family-farm tree crops growers since they can't replant annually.

Ozkan[24] provides an excellent example of a user guide on how to reduce/mitigate spray drift. These guides still must be read, understood, and guidelines therein must be put into practice. Aerial spraying continues to occupy university and federal extension efforts via "Fly-Ins" and Spray Calibration Clinics. Aerial spraying will continue to dominate regulatory attention because of the visual image of losses engendered by aerial sprays. However we choose to reduce drift, it still remains a complex multidimensional problem for policy makers to define what it is, how much is too much, and finally to devise the specific tactics which are especially appropriate for the combination of high risk sites with the use of high hazard (environmental/human/non-target organism exposures) actives currently labeled for such use patterns.

Regardless of the type of application, however, users and manufacturers of crop protection agents (CPAs) still have the responsibility to utilize these CPAs in the most effective way and in a manner that mitigates off-target movement of product. As far back as 1983[10], researchers had suggested the need for grower guides to estimate drift hazards while "on-the-go." These opportunities for a user-friendly, cost and time effective system of assessing spray drift hazards remains untapped today. U.K. programs have focused on a

Table 3. Summary of Recommended Procedures for Reducing Off-Target Movement of Pesticides[a]

Recommended procedure	Example	Explanation
Select a nozzle type that produces coarse droplets.	Use Raindrop, wide-angle full cone, or flooding nozzles.	Use the largest droplets possible while providing necessary coverage. Larger droplets cannot be carried downwind as easily as smaller ones.
Use of the lower end of the nozzle's pressure range.	Use 138 to 275 kpa for Raindrop and less than 172 kpa for other nozzle types.	Higher pressures generate many more small droplets (less than 100 microns).
Lower boom height.	Use the lowest boom height possible while maintaining uniform distribution. Use drops for systemic herbicides in corn.	Wind speed increases with height. Lowering boom height by a few can reduce off-target drift.
Increase spray volume by selecting larger spray tips.	If normal gallonage is 140 to 187 ℓ/ha, increase to 233 to 280 ℓ/ha.	Larger capacity nozzles will reduce spray deposited off-target.
Spray when wind speeds are less than 16 km/h and when wind direction is away from sensitive plants.	Leave a buffer zone if sensitive plants are downwind. Spray buffer zone when wind changes.	More of the spray volume will move off-target as wind increases.
Do not spray when the air is completely calm or when an inversion exists.	Inversions or calm air generally occur in early morning or near bodies of water.	Inversions reduce vertical air mixing, causing spray to form clouds at the lower air levels, which can move downwind.
Use a drift control agent when needed.	Several long-chain-polymer products are available.	Drift additives increase the average droplet size produced by nozzles.

[a] From Bode, L. E., and and Wolf, R., Proc. Ohio State University Pesticide Application Workshop, 1991. With permission.

Table 4. Mitigation Options for Orchard Airblast Applications

Action	Ranking	Estimated % drift reduction	Comments	Field tests/ action agency
			Short-Term Strategies	
Education	1-3	20%	1. Survey current practices. 2. Train individuals to critically adjust sprayer for planting geometry match/season and tree size. 3. Some question what this will do for drift: more on the target means less off-target.	1. Use of video tapes, etc. to "train the trainees" and applicators. (UNIV) 2. Demonstrate usefulness of hands-on calibration and on-site adjustments.
Modify Edge Practices	1-3	40-50%+	1. Least costly and probably most easy practice to accomplish. 2. Spray *inwards* on outer 4-6 rows. 3. Decreased spray volume (psi and/or liquid volume). 4. Automatic or manual shut-off (sprayer) on last 4-6 rows.	1. Requires training. 2. Field tests needed to verify how far into block on parallel rows (SDTF). No data available on % reduction from either parallel or perpendicular rows. 3. No data on volume change effect on drift +/or decreased pest control efficacy (UNIV/FED/SDTF). 4. No drift data available from auto controls.

Table 4. Continued.

Action	Ranking	Estimated % drift reduction	Comments	Field tests/ action agency
Pro-action pilot program to induce grower drift reduction actions	1-3	40-50%	1. Follows evaluation of modified edge practices.	1. Needs survey across U.S. tree crops with volunteer growers. 2. All participants to map all their orchards, identify vulnerable sides, and prepare Farm Management Plan to achieve reduced drift. 3. Cost/benefits assessment needed on these practices. 4. Pro-action Plan by UNIV/FED via EPA/USDA support grant.
Restriction of ai/ applications	4	15-20%	1. Restrict by AI and time of day [not 10:00 AM - 4:00 PM]. 2. Support data may be available from SDTF? 3. Can [illegible]o restrict AI, e.g., safe ones only for those high risk areas.	1. Test effects of day/night spraying (SDTF). 2. Evaluate real time drift monitor or in-depth development needs and assessment (UNIV/FED). 3. Require label changes - who decides which AI and what is high risk area/site, i.e., urban, water, etc.

Table 4. Continued.

Action	Ranking	Estimated % drift reduction	Comments	Field tests/ action agency
			Mid-Term Strategies	
Sensors	5	20-25%	1. This technology should be accelerated. EPA/USDA to fund grants? 2. Currently, it is not cost-effective for small growers.	1. Effectiveness of technology must identified and integrated into the Edge Practice Strategy. 2. Tax incentives are mandatory if majority of horticultural users are upgraded to utilize this practice (equipment) (FED).
Buffer zones	6	15-20%	1. Effectiveness would depend upon width of buffer, size of tree and/or block, and sprayer. 2. Data may be available from SDTF DB?	1. Width of buffer zone may cause undue crop loss (smaller blocks/farms). 2. Tree intercepts may be preferable to open ground [see windbreak discussion].

Table 4. Continued.

Action	Ranking	Estimated % drift reduction	Comments	Field tests/ action agency
Tower and/or tunnel sprayers	7	10-80%	1. No documentation of % reduction, although visual evidence suggests an increased target placement efficiency; hence, a reduced off-target movement.	1. Tower sprayers need drift reduction validation [SDTF, UNIV].
			2. Tunnels useful only for 8' trees/vines [e.g., <5% apple acreage and <u>no</u> other tree crops) but all vine crops.	2. Tunnels currently being investigated. Concentration of AI on equipment is of concern. Correct match of speed and tunnel length is critical - otherwise <u>excessive</u> drift results. Data now being accumulated needs to be summarized [UNIV/FED].
			<u>Long-Term Strategies</u>	
Narrow droplet spectra	8	10-15%	1. Possible with "dialable sprayers". How to easily change nozzles "on the go!" is still a question.	1. If documentation shows validity, then training in adjustments is needed.

Table 4. Continued.

Action	Ranking	Estimated % drift reduction	Comments	Field tests/ action agency
			2. Documentation of drift reduction benefit estimated but not proven - DB may show some evidence.	2. Spray adjuvants can move spectra upward - but no real field data on (1) drift reduction, (2) changes in biological effect (SDTF).
				3. Test alternatives in wind tunnels and in predictive models [SDTF/UNIV/FED].
WIND BREAKS	9	20-40%+	1. Not for pecans or other high tree crops.	1. USDA-SCS has a program of windbreak technology. Use it to investigate the potential of this strategy for drift reduction in horticultural crops [UNIV/FED].
			2. Practicality of the strategy is in question. Will it create other problems?	2. Australia and UK both have studies underway to be reviewed as to its usefulness.
				3. No data exists on artificial net barriers for high risk areas. Wind tunnels and FLUENT, etc., programs could be utilized to provide preliminary data [UNIV/FED].

Table 4. Continued.

Action	Ranking	Estimated % drift reduction	Comments	Field tests/ action agency
Overhead or chemigation	10	20%+	1. Medium to long term project.	1. Use of large drops would reduce off-target movement. 2. Limited to certain sites/crops, but has irrigation potential as well.
Use of helocopters (aerial)	11	10-20%	1. Needs more data. 2. Exchanges the drift problem for a practice that already has a "visual" image question.	1. Other practices will be more effective at drift reduction in a shorter time scale. Still remains an option for very sensitive sites and extremely high trees.
Dwarfing root stock	12	40-50%	1. No dwarf capability except on apples. Even then, tunnel sprayers are practical only for <5% of all apple acreage and no other tree crops. 2. Use of tunnel equipment on grapes is feasible immediately (as it was in the 40's!).	1. Long range breeding program which actually may produce pest tolerant CV's before dwarfing rootstocks become practical.

better informed applicator, e.g., the British Crop Protection Council (BCPC) Nozzle Selection Guide[2] outlines how to select a nozzle for the intended activity. More recently, the BCPC has taken this a step further by reviewing the potential of this same guide to index the driftability of nozzle/pressure selections and another program focused on an index of risk from the delivery system in general. Again, greater attention is being placed at educating the user that there are ramifications to every nozzle choice, and that one should be aware of these limitations in order to optimize the pesticide application process.

Chemical companies still report that over half of drift complaints involve application procedures known to be "off-label." Thus, education, stewardship, and drift management planning are the responsibility for all users of CPAs. The manufacturer will have an increasing role in this education because of the "cradle-to-grave" philosophy engendered by the U.S. Environmental Protection Agency (USEPA). Consequently, actions by state/federal/private partners towards a common goal of environmental stewardship via resource management tactics will place agriculture in a more favorable light (by the public), and increase the confidence in future agricultural and policy goals.

IV. POLICY IMPLICATIONS AND FUTURE OF DRIFT MITIGATION

Recent trends in pesticide policy suggest an increased concern about pesticide drift and our ability to understand its risk, costs, and complexity in order to accurately predict and devise mitigation strategies. Among the significant trends are (1) legislative mandates, e.g., 50% reduction in 10 years, as developed in some European countries (and predicted as an oncoming event in the U.S.); (2) increased liability, e.g., neighbors, public roads and/or water are now next to our sprayed crop lands; and (3) a general environmental movement which demands more accountability of pesticide users and manufacturers.

It is somewhat ironic that the inefficiency of the pesticide application process results in a focus on off-target movement and how to reduce spray drift, while the national effort to optimize and improve the efficiency of delivery of actives to specific target sites remains relatively low key (with minimal national funding) in spite of governmental edicts on a pesticide reduction philosophy; e.g., the need to get more growers on integrated pest management (IPM) and the overall environmental concerns about current pesticide use with >65% as herbicides. With conservation tillage on the increase (thus requiring post-emergence herbicides), a greater attention to labeling instructions and proper user guidelines of maximizing the efficiency of post-emergent herbicide spraying seems opportune.

With increased legislative programs aimed at reduction of pesticides in the environment, strategies to reduce drift must also have increased understanding of "how, what and why" in order to predict and, hence, deliver successful user tactics. New computer models and specific technologies involving aerial drift mitigation via upwind wing releases to minimize exposure to environmentally sensitive areas should prove helpful. In addition, the use of scanning laser systems, e.g., ARAL [Atmospheric Environmental Services Rapid Acquisition Lidar] and ecotoxicological studies of non-target exposure under better defined drift scenarios will allow progress to be made without severe legal constraints. However, in the end, the tradeoffs between the use of CPAs for better defined benefits must be integrated with a strategy to mitigate the drift consequences of that use pattern. Externally applied CPAs (as opposed to biotechnology strategies, e.g., gene insertion, etc.) still offer the greatest immediate opportunities to easily adjust to highly variable pest pressures. Coupled with increased knowledge of varietal tolerances and accurate pest assessments, hydraulic atomization of CPAs will continue to offer economic, achievable advantages.

Aerial application of pesticides will continue to be under greater scrutiny by policy makers/legislators because of a highly visible inefficiency. Mickle[21] reports a generic approach to minimizing the impact of spray drift to non-target species in Canada and proposes regulatory guidelines for assessing spray drift (Figure 10). One relevant question is what does 3% (of released AI spray) mean biologically at 50 or 100 m from target site? Buffer zones do not necessarily address the uniqueness of site specificity, user resources, product toxicity, or ecosystem sensitivities. Mickle[21] also presents a draft buffer zone guide (Figure 11) intended for use patterns of biologicals, which highlights the need for product specificity information when developing mitigation options in the registration process.

Ground application scenarios might well involve a more holistic systems approach to drift mitigation by incorporating multiple approaches for drift risk reduction such as nozzle selection, equipment utilization optima, evaporation retardants, and farm management plans for high risk sites (including buffer zones, application, and AI restrictions). These tactics will include a recognition of site-specific needs and use of spray advisory indexes, all of which will require increased decision making (information) and training (Table 4).

An innovative industry will benefit from having the foresight to effectively address the issues of accountancy of toxin placement in a crop environment. Assessing and implementing an effective strategy of drift reduction, even with the use of the "green" biopesticides is essential. It will insure that the industry has not only an understanding of a "perceived concern," but has taken appropriate good stewardship tactics to take advantage of the opportunities that public discussions offer for assuring a competitive U.S. agriculture in the 1990s and beyond.

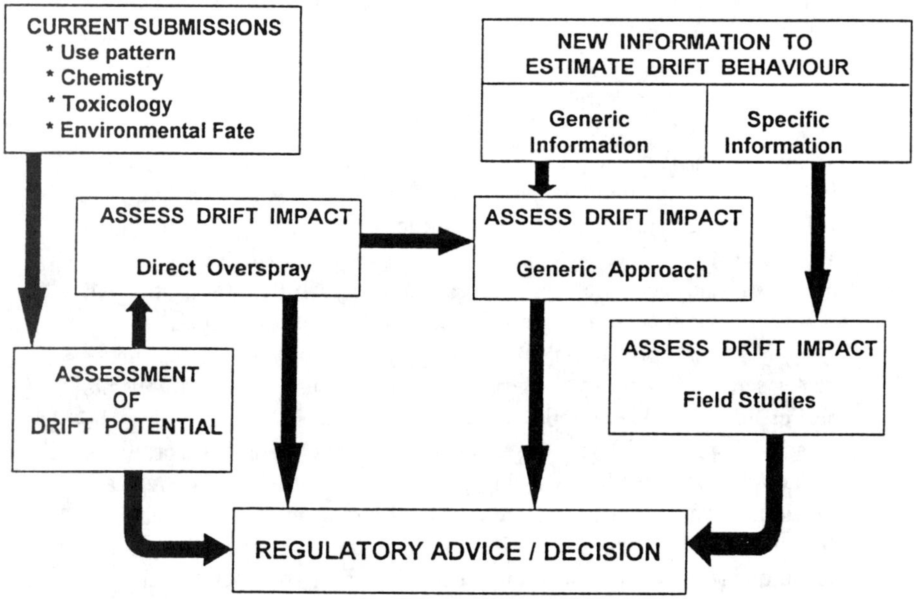

Figure 10. Regulatory scheme for biorationals relating assessment of drift behavior in environment.[21] (From Mickle, R. E., Biorational Pesticides Formulation and Delivery, American Chemical Society, Washington, D. C., 1994. With permission.)

BUFFER ZONES
BIOLOGICALS

AREA OF CONCERN	BUFFER ZONE RANGE
HABITATION	
Residential	600 m
WATER	
Potable	50 - 300 m
Lakes	0 - 50 m
Municipal Intake	10+ - 3200 m
Fish Hatcheries	0 - 50 m
Fish Bearing Waters	0 - 50 m
SENSITIVE CROPS	
Berry Patches	0 m
Organic Farms	Specific Buffers
ENDANGERED SPECIES	0 - 500* m

* Eagle Nesting

Figure 11. Proposed buffer zones for biologicals on basis of perceived risk to nearby areas.[21] (From Mickle, R. E., Biorational Pesticides - Formulation and Delivery, American Chemical Society, Washington, D. C., 1994. With permission.)

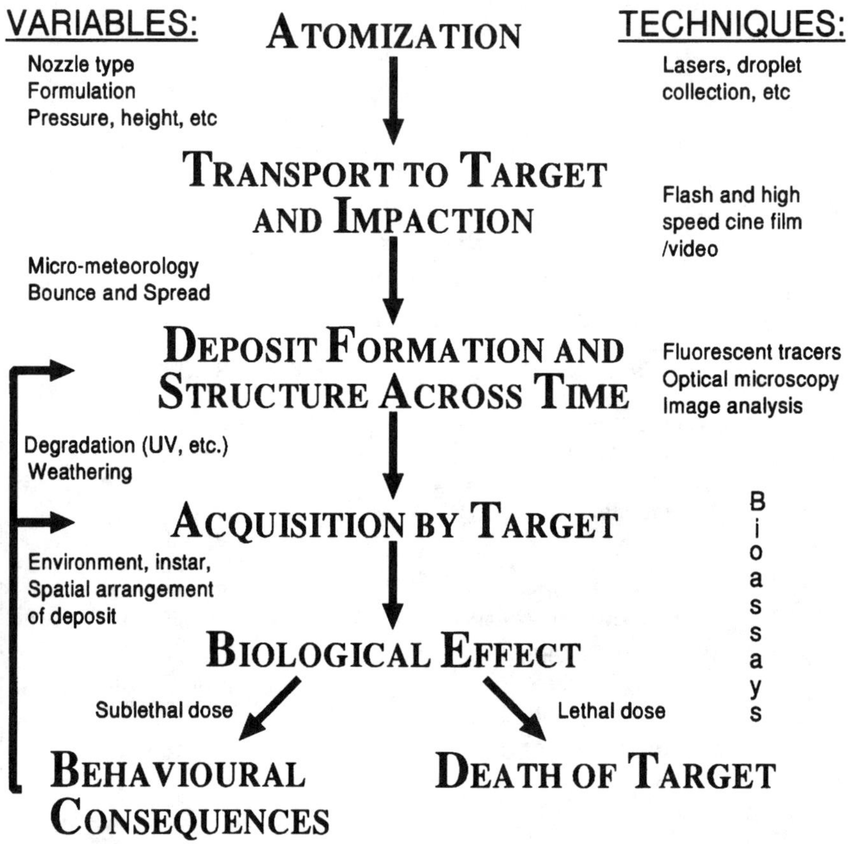

Figure 12 Schematic of the various stages involved in the transport of pesticide a.i. from atomiztion to biological effect.

REFERENCES

1. Adams, A. J., Chapple, A. C., and Hall, F. R., Droplet spectra for some agricultural fan nozzles, with respect to drift and biological efficiency. Pesticide Formulations and Application Systems, Vol. 10, ASTM STP 1078, Bode, L. E., Hazen, J. L., and Chasin, D. G., Eds., American Society for Testing and Materials, Philadelphia, 1990.

2. BCPC, Nozzle Selection Handbook, British Crop Protection Council, Bracknell, U.K., 1988.

3. Bilanin, A. J., Teske, M. E., Barry, J. W., and Eckblad, R. B., AGDISP: The aircraft spray dispersion model, code development and experimental validation, Trans. of ASAE, 32(1), 327, 1989.

4. Bode, L. E., Butler, B. J., and Goering, C. E., Spray drift and recovery as affected by spray thickener, nozzle type, and nozzle pressure, Trans. ASAE, 19(2), 213, 1976.

5. Bode, L. E. and Wolf, R., Recommendations for Reducing Off-Target Movement of Pesticides, Proc. Pesticide Application Workshop, The Ohio State University, Columbus, OH, 1991.

6. Chapple, A., Downer, R., and Hall, F., Effects of Pump Type on Atomization of Spray Formulations, Bode, L. E., and Chasin, D., Eds., ASTM STP 1112, American Society for Testing and Materials, Philadelphia, 193, 1992.

7. Chapple, A. C., Downer, R., and Hall, F., Effects of spray adjuvants on swath patterns and droplet spectra for a flat-fan hydraulic nozzle, Crop Prot. 12,(8), 579, 1993.

8. Dorr, G. and Pannell, D., Economics of improved spatial distribution of herbicide for weed control in crops, Crop Prot. 11, 385, 1992.

9. Downer, R., Chapple, A. C., Wolf, T. M., Hall, F. R., and Hazen, J. L., A new dry drift reduction additive with improved shear stability, Proc. IUPAC, Vol, 3, Washington, D. C., 698, 1994.

10. Elliot, J. G. and Wilson, B. J., Eds., The Drift of Herbicides, BCPC Publications, Croydon, U.K., 1983.

11. Fox, R. D., Reichard, D. L., Brazee, R. D., and Hall, F. R., Downwind residues from air spraying of a dwarf apple orchard, Trans. ASAE 33(4), 1104, 1990.

12. Fox, R. D., Reichard, D. L., Brazee, R. D., Krause, C. R., and Hall, F. R., Downwind residues from spraying a semi-dwarf apple orchard, Trans. ASAE 36(2), 333, 1993.

13. Gilbert, A. J. and Bell, G. J., Evaluation of the drift hazards arising from pesticide spray application, Aspects Appl. Biol., 17, 363, 1988.

14. Green, G., Ed., Proc. Symp., Aerial Appl. Pestic. For., AFA-TN-18, NRC No. 29197, National Research Council of Canada, Ottawa, 387, 1987.

15. Hall, F. R., in Handbook of Pest Management in Agriculture, Pimental, D., Ed., CRC Press, Boca Raton, FL, 1991.

16. Hall, F., Chapple, A., Downer, R., Kirchner, L., and Thacker, J., Pesticide application as affected by spray modifiers, Pest. Sci., 38, 123, 1993.

17. Hall, F., Kirchner, L., and Downer, R., Measurement of evaporation from adjuvant solutions using a volumetric method, Pest. Sci., 40, 17, 1993.

18. Hess, C. E., Agricultural technology and society, Proc. Ohio State Univ. Battelle Endowment Technol. Hum. Affairs, 1, 1, 1987.

19. Knoche, M. Effect of droplet size and carrier volume on performance of foliage-applied herbicides, Crop Prot., 13(3), 163, 1994.

20. Maybank, J., Yoshida, K., and Grover, R., Spray drift from agricultural pesticide applications, APCA J., 28, 1010, 1978.

21. Mickle, R. E., A generic approach to minimize impact on non-target species in Canada, in Biorational Pesticides - Formulation and Delivery, American Chemical Society, Washington, D. C., 1994.

22. Miller, P., Spray drift and its measurement, in Application Technology Matthews, G. and Hislop, E., Eds., CAB Int., 101, 1993.

23. Nordby, A. and Skuterud, R., Effects of boom height, working pressure and wind speed on spray drift, Weed Sci. 14, 385, 1975.

24. Ozkan, E., Reducing Spray Drift, OCES publication, Ohio State University Press, Columbus, OH, 15 pp., 1991.

25. Payne, N., Spray deposits from aerial insecticide spray simulant applications to a coniferous plantation at low and high speeds, Crop Prot., 13(2), 121, 1994.

26. Payne, N. J., Helson, B. V., Sundaram, K. M. S., and Fleming, R. A., Estimating buffer zone widths for pesticide applications, Pestic. Sci., 24, 147, 1988.

27. Picot, J. J. C., Kristmanson, D. D., and Basak-Brown, N., Canopy deposit and off-target drift in forestry aerial spraying: the effects of operational parameters, Trans. ASAE 29(1), 90, 1986.

28. Pimentel, D. and Levitan, L., Pesticides: amounts applied and amounts reaching pests, Bioscience, 36, 86, 1986.

29. Reichard, D. L., Zhu, H., Fox, R. D. and Brazee, R. D., Wind tunnel evaluation of a computer program to model spray drift, Trans. ASAE 35(3), 755, 1992.

30. Reichard, D. L., Zhu, H., Fox, R. D. and Brazee, R. D., Computer simulation of variables that influence spray drift, Trans. ASAE, 35(5), 1401, 1992.

31. Riley, C. M. and Wiesner, C. J., Off-target pesticide losses resulting from the use of an air-assisted orchard sprayer, in Pesticide Formulations and Application Systems, Vol. 10, Bode, L. E., Hazen, J. L., and Chasin, D. G., Eds., ASTM STP 1078, American Society for Testing and Materials, Philadelphia, 1990.

32. Salyani, M. and Cromwell, R. P., Adjuvants to reduce drift from handgun spray applications, in Pesticide Formulations and Application Systems, Vol. 12, Devisetty, B. N., Chasin, D. G., and Berger, P. D., Eds., ASTM STP 1146, American Society for Testing and Materials, Philadelphia, 1993.

33. Salyani, M. and Cromwell, R. P., Spray drift from ground and aerial applications, Trans. ASAE 35(4), 1113, 1992.

34. Spillman, J. J., Spray impaction, retention and adhesion: an introduction to basic characteristics, Pestic. Sci., 15, 97, 1984.

35. Walklate, P. J., A simulation study of pesticide drift from an air-assisted orchard sprayer, J. Agric. Eng. Res. 51, 263, 1992.

36. Wolf, T. M., Grover, R., Wallace, K., Shewchuk, S. R. and Maybank, J., Effect of protective shields on drift and deposition characteristics of field sprayers, Can. J. Plant Sci. 73, 1261, 1993.

37. Young, B. W., A method for assessing the drift potential of hydraulic nozzle spray clouds, and the effect of air assistance, in Air-Assisted Spraying in Crop Protection, Lavers, A., Herrington, P., and Southcombe, S. E., Eds., Lavenham Press, U.K., 1991.

Chapter 13

Enhancing Uptake and Translocation of Systemic Active Ingredients

Roger J. Field and Farhad Dastgheib

CONTENTS

1-8493-7678-5/96/$0.00+$.50

I. INTRODUCTION

The formulation of pesticides is a critical issue in ensuring improved efficacy, more effective targeting, and for achieving significant reductions in environmental impact. While this chapter is concerned principally with efficacy, the wider issues of restricting entry of active ingredient and co-formulants into non-target organisms or the environment are also significant considerations. It is simply not appropriate to consider efficacy in isolation from environmental and economic issues that are critical to the viability of all primary production systems.

Detailed studies of the effects of formulation on the efficacy of pesticides are poorly documented considering the total number of commercial products that are available. The proprietary value of information relating to formulations has precluded its open publication. This has led to emphasis on reports that detail either the effect of added adjuvants or to rather simplistic comparisons between different chemical forms of an active ingredient. The latter studies are typically carried out in the absence of reference to other formulation components or are based on minimal formulations that would not be commercially useful. Most commercial formulations components, other than the active ingredient and major adjuvants like surfactants, have not been the basis of major study and have largely been ignored. Recent reviews have inevitably tended to make generalizations about formulation issues, due to the absence of specific information.[42,109,157] The present review may well be criticized on the same basis, but it does present a critical assessment of those formulation factors that influence the efficacy of the transfer processes that take active ingredients from the point of deposit of a droplet to the site of action. This requires consideration of both pesticide formulation factors and also the anatomy and physiology of target organisms. The key processes of uptake and translocation have often been viewed from the narrow focus on the pesticide delivery system rather than in the context of the overall physiological behavior of the target organism. These are crucial interactions if two of the overall objectives are to achieve highly reproducible efficacy at low rates of pesticide use.

Notwithstanding the earlier comments, the most significant advances in improving pesticide performance are in manipulation of formulations, principally by the introduction of novel adjuvants. Adjuvants are a very large and heterogenous group of substances representing diverse and complex chemical types. They are defined as materials which can enhance the action of a pesticide by facilitating or modifying the characteristics of spray formulations.[69,130] In this chapter the term Adjuvant will denote both tank-mix products and inert ingredients incorporated in the concentrate by the pesticide manufacturer. The importance of adjuvants can be gauged by the fact that pesticides formulated in the absence of adjuvants may exhibit as little as 10% of their potential biological activity.[215] Inclusion of adjuvants in herbicide formulations has become an almost universal practice.[70,84,129] The range of adjuvant types is considerable but typically they act as wetting agents, penetrants, spreaders, co-solvents, coupling agents or deposit builders, hydroscopic agents or stickers, stabilizing agents acting as emulsifiers or dispersants, and activators.[130]

In most cases of adjuvant use their mode of action is not well understood owing to the complexity of the chemical, environmental, and biological systems in which they operate. To understand how adjuvants act, critical factors that impinge upon the efficacy of pesticides need to be considered. For example, it is unreasonable to deal specifically with spray application procedures without reference to the concentration and distribution of the active ingredient in the formulation. Thus, following application of a herbicide, foliar retention and uptake into living cells is a critical step before any biological activity can occur. Typically the herbicide must be able to react with a suitable receptor or biologically active site and this will frequently require translocation to specific sites of action. These three main processes all depend upon the physical and chemical properties of the herbicide, its formulation, the plant species, and specific climatic and edaphic conditions at the time of application.

This chapter will concentrate on the effect of formulation factors on the uptake and translocation of pesticides in plants, drawing on examples from a range of active ingredient and adjuvant types, but with an overall emphasis of describing the general principles involved and the strengths and weaknesses of the current understanding of the mechanisms.

II. FORMULATION CHARACTERISTICS

A. PESTICIDE TYPES

Given the wide range of chemical structures of pesticide active ingredients, it is inappropriate to attempt to classify them on that basis alone. The more appropriate approach, in the context of this review, is to consider broad categorization based on either physicochemical characteristics or mode of action. The overall objective of delivering pesticide active ingredient to a distinct site of action may be unrelated to its mode of action, or for that matter the nature of the target organism. For example, the need for systemicity for a herbicide targeted at the control of a perennial weed requires a similar delivery system to a fungicide for control of a root disease or a foliar-applied systemic aphicide. The features that the three pesticides have in common are an ability to enter the plant system and translocate to points distant from the site of application. It can be reasonably argued that the physicochemical characteristics required for this to be achieved may well be similar even though the mode of action of the pesticides are clearly very different.[22] Additionally, the co-formulants in the three formulations may be very similar.

Considerable progress has been made in determining a physicochemical basis for understanding the uptake and transport of herbicides in plants.[21,22] Thus Bromilow and Chamberlain[21,22] have produced a simple model indicating the properties required by non-ionized compounds and weak acids for the various types of transport that range from non-systemic, largely immobile compounds to those that demonstrate maximal phloem mobility. While it is understood that the characterization of types of transport is essentially arbitrary, the current approach is helpful in predicting performance and is ahead of the crude attempts that have been made previously that were based solely on broad chemical description. The latter approach simply emphasized that highly lipophilic active ingredient formulations, such as esters, show quite different behavior to hydrophilic, polar formulations such as amine salts.[27,29,145,146,147]

There are huge variations in systemicity of herbicides that can be related to the octan-1-ol/water partition coefficients (K_{ow}).[23] Thus, compounds with a log K_{ow} values greater than 3, such as trifluralin [2,6-dinitro-*N,N*-dipropyl-4-(trifluoromethyl)benzenamine], and diflufenican [2',4'-difluoro-2 (ααα-trifluoro-m-tolyloxy) nicotinanilide] are non-systemic, while compounds with values less than 0, such as maleic hydrazide [1,2-dihydro-3,6-pyridozinedione] and glyphosate [N-(phosphonomethyl) glycine], are highly systemic in both phloem and xylem and are referred to as being ambimobile.[23] However, these broad categorizations refer to transport within the plant and do not include the initial phases of uptake into the symplast of leaf and stem tissues. For example, because of their high lipophilicity and the nature of cuticular barriers, compounds demonstrating poor translocation are likely to more readily transfer across the cuticle and into leaf tissues than those compounds that are more hydrophilic. Thus, while a broad classification of pesticides based on physicochemical measurements has utility, it is still critical to analyze performance in the overall system by determining performance in the key subparts of the overall process. It is therefore not surprising that those herbicides that are classified as intermediate in relation to their oil-water partition coefficients are recognized as performing adequately in all subphases of the overall transfer process.

A number of pesticides can be formulated so as to vary their physicochemical characteristics and enhance their biological performance. Some common herbicides, such as those in the chlorophenoxyacetic group, can be formulated either as esters or as amines. Such variations offer an opportunity for selectivity and for targeted use.

A major challenge is to understand the interrelationship between the broad physicochemical characteristics of active ingredients and specific formulation requirements that will optimize delivery in a specific biological system. It is only where the inherent toxicity of co-formulants may interfere with the mode of action of the active ingredient that specific care should be taken in overall formulation. Although the correct formulation can result in great increases in efficacy of agrochemicals, it is the physicochemical characteristics of active ingredients which are the main determinants of their uptake and translocation. As will become apparent in succeeding sections, the greatest scope for enhancement is in promotion of foliar uptake, rather than enhancement of translocation.

B. ADJUVANT TYPES

There is a rather unsatisfactory taxonomy for adjuvants with classification being based on function, as described above, chemistry, time of addition, or perhaps a combination of the three. The most prominent group of adjuvants is surfactants which while seen as distinct from crop oils and perhaps humectants, may have similar effects. The similarities extend to the option to include such adjuvants in base formulations or to be added to tank mixes immediately before application. In the context of the present chapter it is necessary to describe the chemical and physicochemical characteristics of key groups of adjuvants in order to understand their possible impact on the delivery of active ingredient in biological systems.

1. Surfactants

Surfactants possess surface modifying properties and are able to reduce the surface energy of solvents by aggregating to the surface.[215] They operate at the interface between hydrophilic and hydrophobic phases to improve wetting, spreading, and retention.[11] In general, a surfactant molecule consists of two parts; a hydrophobic or lipophilic (apolar) part and a hydrophilic (polar) part. It is the combined hydrophilic and hydrophobic nature of these molecules, as well as their interactions with adjoining molecular groups, that determines their emulsifying, dispersing, spreading, wetting, solubilizing, and other surface-active properties.[69]

The wide range of chemistry of surfactants makes precise grouping and structure-activity relationships very difficult to define. The most common classification is on the basis of the ionic charge on the hydrophilic portion of the molecule.[69,109,110]

Nonionic surfactants are the most common type of surface active agents, deriving their hydrophilic characteristics from nonionizable groups such as phenolic and alcoholic hydroxyls, carbonyl oxygens of esters and amides, ether oxygens, and analogous sulphur-containing configurations. Their nonionic nature is often advantageous in formulations because of their lack of reactivity with ions present in hard water (e.g., calcium, magnesium, or ferric ions) and their chemical compatibility with many other chemicals.[94]

Cationic surfactants ionize in water such that the hydrophilic group becomes positively charged. Primary, secondary, tertiary, and quaternary amino groups and ammonium cations are the most common types of cations formed by these surfactants.

Anionic surfactants are ionizable in water to yield a negative ion. Generally, these anions are derived from sulfonic or carboxylic acids and sulfuric esters, for example, sodium stearate, sodium lauryl sulfate, and alkyl phosphoric ester. Such polar groups all lend a high water solubility to the molecule. These are commonly used either alone in herbicide formulations or in blends with nonionic surfactants.

Amphoteric surfactants exhibit both anionic and cationic properties depending on the pH.[11,130,215]

The hydrophilic-lipophilic balance (HLB values) for each surfactant is also used to categorize their applicability as emulsifiers, detergents, and/or activators. The HLB value confers a specific surfactant characteristic and can be broadly classified as follows:[130]

HLB 4-6 water in oil emulsifier
HLB 7-9 wetting agent
HLB 8-18 oil in water emulsifier
HLB 10-18 solubilizer

While an optimum HLB is believed to exist for the performance of a surfactant in aiding penetration of a specific pesticide, the relationship tends to be adequate only for trends where a single homologous series of surfactants is being studied. Different homologous series are unlikely to have the same optimum HLB values, i.e., the system is not absolute.[10,215] The index of HLB values best suits nonionic agents and is of little value for anionic agents. Although the HLB system has been widely used to select emulsifiers for pesticide formulations and for selecting penetrant aids for certain herbicides, over-reliance on such concepts is questioned.[215] In addition, the presence of water in the spray formulation causes hydration and/or hydrogen bonding between parts of the surfactant molecule which may modify the HLB value.[118]

Surfactants tend to exist in water solutions as individual molecules at relatively low concentrations (< 0.2 g L^{-1}). As the concentration is raised and reaches a critical level, molecules aggregate into clusters or micelles.[215] The critical micelle concentration (CMC) or concentration range tends to occur near the point of inflexion in plots of surface tension versus surfactant concentration. Thus, the CMC range is reached when surface tension reaches a minimum value, beyond which there is little change.[152,153] Typically, the CMC occurs at about 0.1 g L^{-1} although this varies widely with solution, surfactants, temperature, and many other factors.

The number of surfactant molecules per micelle, called the aggregation number, largely depends on surfactant structure and the nature of the surfactant hydrophile and hydrophobe. The most common surfactants with organic structures, have typical aggregate numbers ranging from 20 to 150. Organosilicone surfactants may have smaller aggregate numbers of around 5 to 31 and some do not form micelles.[6,172,]

The mean molar ethylene oxide (EO) content is a major factor in determining surfactant performance. Thus, significant differences in herbicide toxicity have been noted with surfactant hydrophile content, with an optimum EO range usually being observed for maximal activation.[97] The importance of EO content for surfactant performance, and the nature of specific interaction with particular pesticides, is considered in a later section in the context of the predictive model proposed for optimizing the performance of formulations.[97,98,200]

2. Oil-Based Adjuvants

The collective group of oil-based adjuvants is very heterogeneous with the majority being petroleum-based products rather than vegetable-based oils. The taxonomy is further complicated by common reference to crop oils which in most cases are petroleum products rather than of vegetable origin. Irrespective of their origin, oils that are used as adjuvants include only refined products, and they exhibit specific effects that are dependent upon such variables as pesticide type and target species specificity. Oil-based adjuvants typically consist of a mixture of a phytobland oil and emulsifiers which are capable of forming stable oil-in-water emulsions, thereby providing a basis for dispersion of pesticides in solution.

The performance of oil-based adjuvants in reducing vapor loss of pesticides and ameliorating the adverse effect of climate is now well acknowledged.[110] Investigations have tended to emphasize the advantage of adding oil-based adjuvants to enhance application, rather than a thorough analysis of their effects on enhancing pesticide transfer into and within the plant.[3,7,123,131] However, there is little doubt that oil-based adjuvants may enhance uptake of active ingredient into the plant as determined by a number of specific chemical factors and their interaction.[122,135,136,211,212]

3. Other Adjuvants

It is beyond scope of this review to consider the significance of minor adjuvants on enhancing the uptake and translocation of pesticides. With perhaps two exceptions, mentioned below, the other adjuvant classes have more effect on the delivery of pesticides to the target than on the biological processes of uptake and translocation.

Humectants are designed to prolong the period of hydration of pesticide deposits, thus enhancing the opportunity for uptake of active ingredient into target organisms. Humectants are typically materials such as glycerols, ethylene and propylene glycol, mixtures of fatty acids, and polyhydric alcohols. The most pronounced effects of humectants are with water soluble pesticides, although performance is often variable and highly dependent upon relative humidity.[38,125,128]

There has been great speculation about the advantages of adding inorganic salts, such as ammonium sulphate, magnesium sulfate, and diammonium hydrogen phosphate to herbicide formulations to improve the efficacy of polar compounds such as glyphosate and amine formulations of 2,4-D [(2,4-dichlorophenoxy)acetic acid].[90,110] No universally accepted explanation of why the addition of inorganic salts should enhance herbicide performance has been forthcoming. Differenct species showed differential response to the enhancement of glyphosate activity by diammonium sulfate.[134] The antagonistic effect of various cations and anions and the characteristics of residual deposits on leaf surfaces from the herbicidal sprays containing various salts have been investigated.[137,138] It is clear that simple in-tank changes to solution pH or an interaction with other adjuvants is not solely responsible for the observed improvements in field efficacy.[39,134,208,209]

III. FOLIAR UPTAKE AND TRANSLOCATION PATHWAYS

A. UPTAKE PATHWAYS

Following retention of herbicide deposits on the surface of leaves and stems there is a requirement for transport to the site of herbicide action, which is never at the point of contact on the plant surface. Even contact herbicides, such as paraquat (1,1'-dimethyl-4,4'-bipyridinium ion), are required to enter living cells to initiate plant damage and herbicide activity. One of the constant factors in achieving herbicide efficacy is the structure of the target plant, although physiological status may vary depending upon prevailing growth conditions and the stage of plant development. An analysis of plant structure and the performance of associated transport processes is critical to understanding the barriers to the distribution of herbicides within the plant and has been well covered in recent reviews.[22,55] The interaction between species-specific transport systems and the physicochemical properties of individual pesticides leads to a number of unique outcomes with influence on both efficacy and selectivity.[22]

1. Cuticle

The outer surface of plant shoots is covered by a cuticle. It must be traversed by diffusion of aqueous or liquid phases if herbicide active ingredient is to enter the underlying living cells. The cuticle is an effective barrier to transport both into and out of the leaf, and it is critical in ensuring that the plant maintains normal physiological function.[157] The degree of impermeability of the cuticle varies between species and is dependent upon plant growing conditions and plant age. The semi-permeable nature of the plant cuticle allows the entry of pesticides under highly specified conditions, formulations being a major factor.

The cuticle is a complex structure, primarily composed of a cutin matrix. The matrix is made up of insoluble high molecular weight polyesters composed of long-chain substituted aliphatic acids, combined with hydroxyl, aldehyde, ketone, epoxide, and unsaturated groups.[95,109,157] The cutins are typically monomers having chain lengths of C16 to C18, or mixtures thereof, the composition of which is dependent upon the plant species and to some extent the growth environment.[95] Embedded in the cutin matrix are epicuticular and cuticular waxes that determine both the surface topography and the physicochemical properties of the leaf surface. The orientation and configuration of these waxes varies between plant species, conditions for plant growth and the age of particular plant organs. The crystalline waxes typically form filaments, plates, ribbons, rods, or dendrites. They are highly diagnostic for a particular species and can be recorded by scanning electron microscopy.[8] These epicuticular waxes have a variable composition which includes long-chain aliphatics, pentacyclic triterpenoids, sterols, and flavanoids. In contrast, the cuticular waxes are typically composed of short-chain fatty acids of variable carbon length, but typically C16 and C18. Detailed studies of the chemistry and morphology of these cuticle components can be found elsewhere.[95,109,157]

One key characteristic of the cuticle is that it is chemically and physically heterogeneous. This provides for variable avenues for the penetration of externally applied compounds. In addition, the cuticle is fully connected to the underlying epidermal cells; the idealized cuticle structure includes carbohydrate fibers extending from the underlying epidermal cell wall and middle lamella into the cuticle.[150,157] These carbohydrate fibers provide a possible hydrophilic pathway through the cuticle while the lipophilic pathway is through the epicuticular and cuticular wax. Therefore, the heterogeneity of the cuticle provides for a complex array of pathways with the overall efficacy of the uptake process being dependent upon the physicochemical characteristics of both the active ingredients and adjuvants contained in a particular formulation.

Conjecture about the structure of the plant cuticle and its connection to epidermal cells adds to the difficulty in providing a simple and universally held view of the mechanism of transfer of active ingredient across this barrier. As the cuticle is not living, the diffusion of specific compounds is the most likely mechanism for transfer and should obey Fick's first law.[143,157] The rate of diffusion will be dependent upon the physicochemical composition and thickness of the cuticle and the physicochemical characteristics of the diffusing compound. Thus, thin cuticles, which are typically found on young plant organs or those growing in a protected environment, will be more permeable than thick cuticles.[157] However, it has been noted that cuticular structure and composition are so variable between species that these characteristics may override the simpler effects of thickness.[157] If transfer obeys the laws of diffusion then the magnitude of concentration gradients of active ingredients will be a major factor determining the rate of transfer. The physical form of pesticide deposits on the leaf surface and their contribution to a pool of the active ingredient will thus be as important as the rate at which active ingredient is removed from the inner surface of the cuticle. While it may be satisfactory to analyze the overall rate of transfer of active ingredients across a cuticle, there is a requirement to understand the precise nature of transcuticular movement.

In a broad sense, two major transcuticular pathways have been identified, although both are likely to be discontinuous. The wax components of the cuticle may offer preferential transport routes for apolar, lipophilic materials. In contrast, the cutin matrix and the associated carbohydrate projections from epidermal cell walls provide a more polar route for hydrophilic compounds. Polar compounds traversing through the latter route will still be required to pass through the epicuticular wax at the surface, and this may be the major limiting factor in the transfer of these compounds. Reference has been made to hydration of the cuticle which, while not proven, is likely to be associated with carbohydrate fibers and possible air spaces in the cutin matrix. Thus, it is possible that highly polar compounds may successfully move across the bulk of the cuticle provided they can traverse the initial epicuticular wax barrier. Such transport routes may be tortuous because of the discontinuous nature of cuticular wax and carbohydrate fiber deposition and location. Pesticide formulations containing multiple components, including two or more active ingredients, may utilize quite different transcuticular pathways. Each component may have a characteristic rate of transfer that will be both species-specific and dependent upon plant growth characteristics, physiological status, and environmental conditions.

The role of stomata in the transfer of active ingredient across the cuticle has always been controversial. It has been suggested that the guard and accessory cells of the stomata may be preferential entry sites associated with reduced surface fine structure or the existence of ectodesmata.[34,142,166] Even under conditions where no direct infiltration of solution into stomata was observed, open stomata may enhance total uptake of active ingredient compared to closed stomata.[56] A more desirable mechanism is that pesticide spray solutions enter through open stomata and flow directly into substomatal cavities, which are virtually free of epicuticular wax.[205] The major limiting factor in direct stomatal infiltration is the surface tension of applied solutions. Aqueous solutions with surface tensions approaching that of pure water (72.25 mN m^{-1} at 25°C) will not penetrate stomatal pores.[169] However, reduction in solution surface tension, typically by adjuvant addition, to values below 30 mN m^{-1} should facilitate stomatal entry in most species.[62,63,169] Several techniques have been used to demonstrate stomatal entry, which include the use of fluorescent dyes, radiotracers, and particulate suspensions.[48,62,63]

If stomatal penetration were to occur to a significant degree it would reduce the critical rain-free period required after spraying and thereby improve the flexibility of pesticide application timing.[63] The rapid entry of pesticide into substomatal cavities would provide a rain-fast reservoir of active ingredient that might enhance pesticide efficacy in the event of postspray rainfall.[62,63,140,141] Entry of solution into the sub-stomatal cavity does not avoid the passage of active ingredient across the cuticle which typically lines the substomatal cavity.[124,150] The high relative humidity in the substomatal cavity would reduce drying and volatilization losses. Further, photodegradation would be reduced and uptake increased because the cuticle is more easily penetrable than the leaf surface cuticle.[11,156]

The basal cells of trichomes appear to be preferential sites of entry, as do regions overlying veins and the anticlinal cell walls.[111,119,214] It has been speculated that rapid uptake in such regions may be due to an increased presence of polar pathways associated with an abundance of ectodesmata-type structures.[72,120,168]

2. Leaf Tissues

Analysis of the overall pathways of pesticide movement from leaf surfaces to sites of action is dominated by considerations of trans-cuticular uptake and long-distance translocation. Movement across leaf tissues, principally the epidermis and spongy and palisade mesophyll layers, has not received significant attention, other than descriptions of apoplastic and symplastic pathways.[42,55] Recent reviews have concentrated attention on either the transcuticular or long-distance transport of pesticides.[55,109]

There is scant recognition of the complexity of leaf tissues and the significance that their physical arrangement and their physiological status may have on the selective transfer of foliar applied pesticides to long distance transport tissues. The apoplastic transport route through cell walls and intercellular spaces, would involve the diffusion of active ingredient and be subject to some of the same physical limitations that are imposed on trans-cuticular movement. It is proposed that the symplastic route across mesophyll tissues could be inherently more rapid, perhaps depending on the presence of appropriate carrier mechanisms but certainly upon the ability of active ingredient to traverse at least one cell membrane. Coupland[42] has identified four transport pathways for the movement of active ingredient from epidermal cells. Active ingredient may remain in aqueous solution and move along an apoplastic route directly into the transpiration stream. The second and third options may depend on the physicochemical characteristics of the pesticide. They may involve either partitioning into the lipid phase of the plasmalemma and from there movement into the cell, or active ingredient may be transported directly into the cell by naturally-occurring permeases, which are proteins that are resident in the plasmalemma and which facilitate the absorption of organic solutes into the cell.[42] An alternative to the latter process may be the involvement of co-transporters, with the uptake of active ingredient being linked to H^+-pumps. The fourth proposal is that the fluid-phase endocytosis may be involved, allowing extracellular material, usually in solution, to be taken up by invagination of the plasmalemma. This is a nonselective mechanism and is therefore of no value in predicting pesticide activity or in the design of molecules to utilize the process. However there are refinements, and receptor-mediated endocytosis may occur that involves a specific ligand-receptor interaction, and there is potential for this being a selective process for pesticides to move into plant cells.[42,100,101]

B. PHLOEM TRANSLOCATION

The presence of phloem tissue in leaf veins and the intimate association of veinal tissue with mesophyll cells, provides for an effective capture and transfer of the products of photosynthesis. The dominant product is sucrose, and considerable knowledge has been gained of the mechanisms involved in the transfer and loading of sucrose into the sieve element-companion cell complex and subsequent long-distance transport in sieve tubes.[82,93,213] These processes have been reviewed in the context of the translocation of foliage-applied herbicides.[42]

The pathways involved in the transfer of sucrose and other metabolites from mesophyll cells to the sieve element are relevant to the routes taken by pesticides. The apoplastic route involves metabolites being taken up from the apoplast surrounding the sieve element-companion complex. This would involve a continuous pathway for those pesticides that move across the epidermis and mesophyll tissues in the apoplast, or a reentry into the apoplast for those that have taken a symplastic route. Currently, the loading of phloem appears to be from an apoplastic source although the involvement of direct plasmodesmata connections and a symplastic continuum is still a consideration.[42,103,163] One implication of phloem loading from an apoplastic metabolite source is the requirement for the transfer across at least one cell membrane before entry into the sieve tube is achieved. The identification of a specific sucrose transporter, which is a permease that can specifically bind sucrose, is of major significance in understanding the processes that may be involved in the loading of pesticides.[162,216] Phloem loading has often been identified as a limiting factor in the translocation of endogenous solutes and must clearly be of considerable importance to the translocation of pesticides, particularly given the risk that the latter compounds may damage specific cellular structures and functional mechanisms.

One of the major limitations to the understanding of the loading of pesticides into phloem cells has been the difficulty in identifying technological approaches that define the precise pathway. Typically, phloem cells are surrounded by other vascular and support tissues associated with leaf veins which in turn are embedded in mesophyll tissue. Probing individual sieve tubes is extremely difficult and for the most part has relied on artificial experimental systems, frequently utilizing explants of whole plants. The use of aphid (*Aphis* sp.) mouth parts to remove the contents of sieve tubes has permitted the phloem loading capability of individual pesticides to be compared, some aspects of the mechanisms to be determined, and comparisons made with the relative longitudinal transport of the same pesticides.[44,45,64,65] It has been shown that the relative phloem loading of pesticides may not be directly related to their long-distance transport capability.[64,65] The herbicide maleic hydrazide was more readily loaded into sieve tubes than was 2,4,5-T [2,4,5-trichlorophenoxy)acetic acid], although the latter had greater longitudinal movement in the phloem.[64,65] Whether these results relate to the availability of specific carriers at the point of loading or to other factors has not been fully determined.[44,64] The majority of phloem loading studies with aphids have involved mature stem tissues rather than leaves. However, it is reasonable to suggest that the processes of phloem loading may not vary greatly between tissues, and that the apoplastic and symplastic pathways described for phloem loading in willow (*Salix viminalis* L.) may be similar in leaf tissue. One of the critical findings of these studies is that pesticides may be loaded into sieve tubes against a concentration gradient, implying that the process is carrier driven, although undoubtedly highly specific, and that such processes explain the often rapid translocation demonstrated by pesticides following application to a leaf surface.[64]

The most significant studies on phloem loading have involved the chlorophenoxyacetic acid herbicides. Because these compounds readily dissociate, it has been proposed that their distribution between compartments in the leaf may be determined by the pH in the compartments that are separated by a cell membrane. This is the basis of the "ion trap" effect in which weak acids are accumulated in plant cells of high pH.[22] Typically phloem sieve tube contents have a relatively high pH. They will therefore tend to preferentially accumulate associated weak acids such as the chlorophenoxyacetic acid, sulfonylurea, and cyclohexanedione classes of herbicides, as compared to their accumulation in mesophyll cells. Thus, the phloem loading of pesticides that are weak acids may rely more on the concept of the ion trap effect than on specific carrier-mediated phloem loading.[22,46,47,159,164]

A more recent approach has been to develop both mechanistic and mathematical models to relate the specific physicochemical properties of pesticides to their ability to load into phloem cells.[42,85,102,112] The phloem systemicity of a pesticide is related to concentration of active ingredient in phloem sap and the apoplastic concentration at the point of treatment. The concentration factor is dependent upon the physicochemical properties of the active ingredient as well as a range of plant parameters. Coupland[42] has provided an excellent summary of this model. His model differs from other approaches in identifying the phloem loading step as critical to determining the magnitude of eventual distribution of active ingredient in the target plant.

While phloem loading is likely to be a major limitation to the translocation of pesticides it is not the sole factor. Some workers have proposed that most xenobiotics can enter the phloem quite freely.[22,210] It is argued that the more important factor for long distance transport in the phloem is retention within the sieve tube and that compounds that are not readily unloaded will show greater systemicity.

The distribution of solutes that enter the phloem system is described in terms of the movement from regions of high assimilate supply (sources) to regions of assimilate demand (sinks).[55] The direction of phloem translocation is determined by the relative strength of sources and sinks and may change as the physiological characteristics of single organs vary, which is typically determined by such factors as plant age, specific growth and development, and environmental manipulation. Developmentally mature organs, including expanded leaves, typically act as sources for developing organs. In the context of the present review it is not necessary to elaborate on the proposed mechanisms of phloem transport, but it is critical to indicate that the pattern of assimilate distribution will determine the fate of co-translocated pesticides. The polar nature of the phloem transport environment, with metabolites moving in a continuous aqueous stream along active sieve tubes, is a major determining factor in the transport of specific pesticides. Highly polar pesticides such as glyphosate are phloem mobile and readily translocated to growing tissues both below and above ground.[43,53] This does not mean that the co-translocation of a mobile herbicide such as glyphosate and associated assimilates occurs at either the same velocity or at that same mass transfer. The inherent leakiness of the phloem system, and lateral transport out of the system to adjacent sink tissues, will result in significant variations in the transport characteristics of specific solutes. There are extreme examples where the physicochemical characteristics of a specific pesticide significantly limit its ability to be translocated effectively in the phloem system, for example the herbicides acifluorfen[92] (5-[2-chloro-4-(trifluoromethyl)phenoxy]-2-nitrobenzoic acid), bentazon[104] (3-(1-methylethyl)-(1*H*)-2,1,3-benzothiadiazin-4(3*H*)-one 2,2-dioxide), diclofop-methyl[88] (methyl2-[4-(2,4-dichlorophenoxy)phenoxy]propanoate), and isoproturon[2] [3-(4-isopropylphenyl)-1,1-dimethylurea].

The hypothesis that nonionized compounds and weak acids are translocated in relation to their physicochemical properties has been promoted.[21,22,163] A similar approach has been developed by Kleier.[112] However, the relationships are not necessarily straightforward. It would be unreasonable to derive a simple formula, based on physicochemical properties of pesticides, and use it as a basis for predicting the transfer of a pesticide from a leaf surface to a distant site of action, a process which may involve transfer across the cuticle and leaf tissue, then long-distance transport in the phloem. The total process is complex, with each transport component possessing unique requirements for chemical characteristics, including the poorly documented relationship between the pesticide and its effect on the physiological processes of the plant. In a gross way the latter is often represented by the extent of pesticide metabolism, but it undoubtedly also includes many subtle effects that contribute to the overall mode of action of the pesticide. In addition, there is some difficulty in interpreting and comparing data on the transport of herbicides owing to significant variations in the method of calculation. This important issue will be addressed in the following section.

1. Quantification of Phloem Transport

There are several methods for presenting radiotracer data on the translocation of pesticides in plants. Translocation can be expressed in terms of:

- The total amount of radioactivity applied to the plant
- The percentage of the radioactivity absorbed by the plant, (this requires a definition of absorption), or

- The percentage of the radioactivity recovered from the plant.

In most radioisotope studies it is customary to use only one method for presentation of data, and it is often difficult to compare the results from different reports when the bases for presenting data are different. Care needs to be exercised in the interpretation of translocation data. While translocation as a percentage of the applied dose provides useful information from a practical viewpoint, by indicating what relationship there is between the applied dose and the mass of pesticide reaching the site of action, it should not be confused with data that are based on radioactivity recovered from the plant, where a more effective estimate of the mobility of a particular pesticide can be determined. The latter measurements are of greatest use in trying to establish relationships between the physicochemical characteristics of a pesticide and their physiological behavior in the plant. Such studies allow estimates to be made of the relative ranking of pesticides in terms of their phloem mobility.[23,41]

Analysis of translocation data calculated on the basis of the amount of herbicide applied to a leaf surface can lead to erroneous claims for high phloem mobility, when in reality it may be another part of the overall transport process, such as transcuticular transfer, that may be the major differentiating factor between the performance of two pesticides or two formulations of a single pesticide. Even a crude review of the methods of quantifying translocation indicates that workers often make a large number of assumptions and have a tendency to oversimplify the complex system of transfer of pesticide active ingredient from a point of retention on a plant surface to the site of action. A key point that is frequently overlooked is the importance of effective recovery of radioactivity. Thus, translocation of chlorsulfuron (2-chloro-*N*-[[(4-methoxy-6-methyl-1,3,5-triazin-2-yl)amino]carbonyl]carbonyl] benzenesulfonamide) in wheat (*Triticum aestivum* L.) cultivars, as a percentage of applied herbicide was variable if no account was taken of the efficiency of recovery of radioactivity.[50,51] Corrections for recovery resulted in no cultivar differences in chlorsulfuron translocation.

C. FORMULATION ENHANCEMENT OF PESTICIDE PERFORMANCE

1. Adjuvant Action in Pesticide Uptake

It is clear that the current level of scientific investigation of pesticide transport partitions the system into two basic processes: uptake and translocation, with only scant reference to the complex compartmentalization that may occur in leaf tissues. This is particularly the case in investigations of formulation or adjuvant effects on the total transport of a particular pesticide. The following sections deal with the enhancement or modification of pesticide transport associated with the addition of specific adjuvants, or where known, detailed changes to other formulation components. One of the major limitations is the absence of detailed research on the performance of specific formulations. The commercial value of withholding knowledge on formulation composition has precluded significant publication in this area. The majority of published information reports on minimal formulations that are unlikely to be commercially available. Thus, it is common for studies to report the advantage of surfactant addition to an active ingredient but purely at an experimental level. In other cases adjuvant additions to fully formulated active ingredients have been investigated but are difficult to interpret because of the lack of knowledge of the composition of base formulation. This precludes any consideration of detailed interactions between added adjuvants and core formulations components.

It is generally accepted that formulation modification, specifically surfactant addition, can enhance the efficacy of pesticides principally by modifying conditions for retention and absorption.[11,108] Surfactants are not generally believed to affect directly the translocation of pesticides in the plant.[9,41] This is despite evidence that suggests that surfactants can enter the symplast of the leaf following foliar uptake, even though they may be present in very low concentrations and/or become partially degraded.[71,96,148,149,201,202] There is clearly a major interaction between pesticide uptake and translocation that is important in determining the optimum rate of surfactant for each pesticide-plant system. The objective should be to establish the maximum amount of pesticide that can be both absorbed and translocated to allow accumulation of an effective dose at the site of action.[69,220]

The role of surfactants and other surface active agents in modifying pesticide uptake is complex and poorly understood. The mechanism of action has been attributed to a number of factors. The following list includes consideration of historical and current explanations of surfactant mode of action.

1. Increasing the area of contact of droplets with the leaf.[48]
2. Increasing spray retention at sites favorable for uptake.[99]
3. Acting as a humectant to keep spray droplets moist for long periods to lengthen the time for uptake.[48,99,108,177]
4. Modifying the external barrier of leaf wax by allowing dual solubility. The hydrophobic and hydrophilic components of surfactants allow solubility in both the wax regions and the aqueous cutin, respectively. This may also be important in penetration of the outer cell membrane where dual solubility (lipoidal and proteinaceous layers of the membrane) is likely to be important.[35,139,144,219]
5. The hydrophilic pathway of entry may be enhanced by surfactant addition. It has been proposed that surfactant molecules penetrate and line themselves in monolayers into areas of imperfection or entry. With their hydrophobic ends aligned in or on the cuticular

waxes, the protruding hydrophilic ends are then able to provide 'hydrophilic channels' for water-soluble pesticides to diffuse through into the cell wall region of the epidermal cells.[188]

6. Some surfactants increase the permeability of the leaf cuticle and plasma membranes.[87,,108,153,190,207] The enhanced activity of some herbicides is therefore believed to be a result of the surfactant-induced breakdown in the cell's normal regulation of permeability. Protoplasmic streaming was also found to be inhibited by some surfactants.[87] Herbicide formulations containing an organosilicone surfactant, Silwet L-77®(trisiloxane polyethoxylate), were found to increase electrolyte leakage from gorse (*Ulex europaeus* L.) much more from bean (*Vicia faba* L.) leaves[196] than formulations without surfactant.[79] However, the same organosilicone surfactant was found to be less phytotoxic than Triton X-45® (polyethylene glycol *p*-isooctylphenyl ether) or Agral® (nonylphenoxypolyethyoxyethanol), as displayed by lower evolution of ethylene.
7. Surfactants may interact with the active ingredient and the plant cuticle to form a specific complex. This is thought to affect foliar uptake more than the lowering of surface tension by surfactants.[106,108,153,157] Surfactants may change the viscosity of cuticular waxes by mixing or co-micellizing with amphipathic acids and alcohols of the wax.[157]
8. Surfactant-induced penetration may involve enhancement of movement of active ingredient along cell walls in the region of the wall/cytoplasm interface by the lowering of interfacial tensions.[48,99,108,153]
9. The rapid, surfactant-induced inflow of pesticide deposits into stomatal cavities has been shown to be promoted by specific surfactants, notably the organosilicone group.[32,63,192]

a. Transcuticular Uptake

Some recent research has analyzed the transcuticular pathways in attempting to understand the behavior of both surfactant and active ingredient, in contrast to simply analyzing the adjuvant-enhanced behavior of active ingredient. Thus Schönherr and co-workers have suggested that foliar uptake of pesticides and activation by adjuvants can be successfully optimized only if: (1) all relevant resistances in the diffusion path are characterized separately and quantitatively; (2) the effects of adjuvants on different resistances are known: and (3) if the factors determining the resistance can be related to physicochemical properties of cuticles, active ingredients, and adjuvants.[171] While it is possible to describe the broad parameters for the kind of information required to describe a transcuticular model, there is still limited understanding of the dynamics of the process.

After deposition on the surface of the cuticle pesticide droplets achieve more effective contact with the surface if surfactants are present. Surfactant-induced reductions in droplet contact angle are well documented, but perhaps of greater importance is the extent to which the inclusion of surfactants facilitate the active ingredient in contacting all the micro-relief detail associated with the deposition of epicuticular wax.[171] The organosilicone surfactants reduce solution surface tension to such a low value (< 30 mN m^{-1}) that no distinct droplets remain on the surface.[28,83,192] The excellent surface contact promoted by organosilicone surfactants is a key feature in their mode of action.[63,197] It is clear that there are no simple relationships between effective contact area of a deposit and the rate of transfer of an active ingredient through the cuticle.[199] It is probably unreasonable to expect a simple relationship given the substantial variation in physicochemical properties of active ingredients and surfactants. Highly lipophilic active ingredients, that would presumably follow an apolar pathway across the cuticle, show enhanced uptake when formulated with nonionic surfactants with low ethylene oxide (EO) content (5-6), which typically have high surface activity and promote high surface contact. In contrast, nonionic surfactants with higher EO content (15-20) typically have poor droplet spreading capability but are most effective with polar active ingredients.[200] Unfortunately, this generalized model that relates surfactant EO content to the lipophilicity of active ingredients appears to break down for compounds of intermediate polarity.[200]

The claim that surfactants may erode cuticular surfaces by dissolving or disrupting epicuticular and cuticular wax is not established.[97,109] However scanning electron microscopy has revealed surfactant precipitation near the edge of a pesticide deposit, although its significance is not clear.[109,194] A related phenomenon may be the ability of surfactant to promote the dissolution of the deposit by solubilization of the active ingredient into surfactant micelles.[97,203]

While solubilization probably makes a contribution to the movement of pesticides to sites of absorption, there is little evidence to suggest that this is a key role for surfactant-enhanced uptake of active ingredient.[199]

Retention of moisture in pesticide deposits by the humectant action of surfactants may be beneficial in promoting uptake. The humectant effects of surfactants are difficult to establish if the surfactant has additional effects associated with cuticular penetration. However, it has been shown that high EO content surfactants that do not penetrate a cuticle may promote the uptake of polar active ingredients.[199] A related issue is the extent to which hydration of pesticide deposits is a requirement for continued foliar uptake. The role of surfactants in this situation is unclear. Their addition modifies the physical appearance of the active ingredient, generally leading to the formation of amorphous and film-like deposits rather

than pure crystalline forms.[121] Recent research suggests that foliar uptake for an amorphous deposit may or may not be different from a crystalline form of an active ingredient.[199]

The mechanism of surfactant-induced transfer of pesticide active ingredient across the cuticle is not established, but it could involve one or more of the following processes, as described by Stock and Holloway.[199]

1. Changes in solubility relationships and partitioning processes that are favourable to transfer.
2. Reduction in the resistance of the cuticle to diffusion.
3. Activation of specific polar or apolar routes through the cuticle.

A detailed analysis of the above three possibilities is contained elsewhere,[199] but there is not a strong case for a single dominant process and many of the current investigations are inconclusive.[171,199] One of the intriguing questions that relates to all three propositions is the extent to which a surfactant is able to enter and transfer across the cuticle and its relationship with particular active ingredients that may be undergoing some form of co-penetration. The use of radiolabeled surfactants has demonstrated their movement into and across the cuticle, and they therefore have the opportunity to either pre-penetrate or co-penetrate with the active ingredient.[97,109] More recent work by Stock and Holloway has challenged the validity of the assumption that surfactants may act as a co-penetrant with the active ingredient.[199] A detailed series of experiments using radiolabeled surfactants established that interactions between surfactants and active ingredient may occur during their mutual uptake, but the precise mechanisms involved were not discerned. It was envisaged that there may be three main types of interaction:

1. A direct association between surfactant monomers and molecules of the active ingredient.
2. Transport of active ingredient within surfactant micelles.
3. An indirect action between the two components at a common site during foliar penetration.

The authors did not consider that the first two proposals were plausible because activator surfactants, which are typically nonionic, would be unlikely to form strong physical or chemical bonds with active ingredients. In addition, given the large molecular dimension of some surfactant micelles, it is unlikely that they would be involved in the transcuticular transport process. The third process, an indirect action, proposes an effective preconditioning effect of the surfactant on the cuticle and is a more reasonable proposition. However, in experiments where leaves were preconditioned with a surfactant, the presence of the surfactant in the cuticle or other internal tissues of the leaf was not sufficient to induce uptake of herbicide active ingredient from the leaf surface.[199] While this result could be interpreted as supporting a co-penetration hypothesis, it could also indicate that the results were highly surfactant-specific. Thus, an alcohol ethoxylate

surfactant with an EO content of 15 was likely to exert its effect mainly on and within the plant cuticle, unlike surfactants with a lower EO content where action would depend on a continuous rapid flux through the cuticle for promoting the transfer of active ingredient.[198,199]

Recent research has attempted to investigate the transcuticular uptake of organosilicone surfactant oligomers in the presence and absence of the polar herbicide glyphosate.[226] The surfactant oligomers were identified by gas chromatography. The results showed that the mode of action of the organosilicone surfactant Silwet L-77® was dependent on both surfactant concentration and interaction with glyphosate.[226] In the presence of glyphosate the uptake of Silwet L-77® was primarily by stomatal infiltration, while at the lower concentrations (2 g L^{-1}) there was significant transcuticular diffusion. Uptake of glyphosate when applied alone was low; stomatal uptake occured following the addition of Silwet L-77®. When Silwet L-77® was applied alone there was no evidence of oligomer discrimination while with the addition of glyphosate, particularly at low surfactant concentrations (1 g L^{-1}), some oligomer discrimination did occur. The authors propose that this may be the result of "cuticular sieving" based on differential size and lipophilicity of the oligomers, or as the result of an interaction with the herbicide on the leaf surface.[226]

It is clear that the magnitude of surfactant-active ingredient interaction varies in relation to the physicochemical properties of both chemical entities and is influenced by their concentration and the nature of the target species. There are interactive and noninteractive mechanisms operating which may both lead to promotion of active ingredient uptake, although at this stage the precise nature has not been elucidated.[198] Current analyses emphasize the difficulty of identifying specific sites of action. In situations where penetration of a surfactant is required to activate uptake of an active ingredient, the key sites could be located in either the cuticle or the walls of epidermal cells, and perhaps even in mesophyll and vascular tissues of the leaf. A great deal of store has been placed on the evidence produced from work with isolated cuticles.[37,158,174,175,176] While these results confirm the differential permeability of isolated cuticles, the experimental systems are by definition artificial and the relevance of these studies to the interpretation of events in intact plant systems has been questioned.[198] Isolated cuticles have been used to examine the effect of surfactants on the permeability of water and solutes.[81,170] They concluded that the rate-limiting step in transcuticular transport was the presence of cuticular lipids.

The foregoing statements suggest that it would be inappropriate to make generalizations about the performance of particular surfactant - active ingredient combinations based simply on physicochemical characteristics. However, it is equally apparent that some prediction of performance is required, and this has led to the development of a predictive model based on a response surface approach.[74,97] The model attempts to predict that the uptake of lipophilic active ingredients (log octanol-water partition coefficient *(P)* > 3.0) would be enhanced best by surfactants of low EO content, and that hydrophilic active ingredients (log *(P)* < -3.0) enhanced by surfactants of high EO content. Active ingredients which have intermediate lipophilicity are expected to demonstrate little preference for surfactant EO content in maximizing their foliar uptake. The model takes account of the concentration of surfactant added to the formulation and the nature of the target species. The advantage in the development of the response surface model is that it can be tested and its limits determined. Thus, the performance of a highly polar herbicide, glyphosate, was investigated relative to the importance of surfactant hydrophobe composition, EO content, and concentration on the promotion of foliar uptake.[74] While the original model had been shown to be effective in predicting the uptake of polar, neutral compounds such as 3-*O*-methylglucose, it was found to be equally applicable to glyphosate which has a highly ionic character. In both cases, the uptake is facilitated by the inclusion of high EO content surfactants in the formulation. Significantly, the model is able to predict the performance of a range of surfactants including C_{13}/C_{15} aliphatic, C_{13}/C_{15} primary aliphatic amines and nonylphenol classes of surfactants.[74] Recent results from Holloway and co-workers have demonstrated the applicability of the response surface model to lipophilic active ingredients.[98]

The development of generalized predictive models that can determine the performance of a wide range of surfactant-active ingredient combinations is a major challenge given the wide range of physicochemical interactions that may occur between particular surfactant-active ingredient partners.[86,97] Highly lipophilic, nonpolar active ingredients require solubilization into surfactant micelles.[109] Solubilization is influenced by the structure of the surfactant and active ingredient, as well as external conditions, particularly temperature. The site of solubilization in the surfactant micelle is a key factor and demonstrates the structural specificity required.[203] For example, highly insoluble active ingredient is typically located either in the core of the micelle or within the hydrophobic portion of the micelle, and solubilization should thus be enhanced by an increase in the alkyl chain length of the surfactant. Where solubilization occurs in the hydrophilic portion of the surfactant, it increases in accordance with the size of the hydrophilic group and may be influenced by polarity, chain branching, and molecular size and configuration.[109,203] Highly hydrophilic, polar compounds coexist with surfactant molecules without incorporation into micelle structures.

The overall conclusion is that there is a wide disparity in the uptake of surfactants into the cuticle and subsequent movement into leaf tissues. The EO content appears to be a critical feature; uptake into foliage is more rapid when the hydrophyl chain is short (EO content of 6-7).[191] Uptake studies with a series of alkylethoxylate and nonylphenol surfactants have confirmed that uptake characteristics vary with hydrophyl chain length and are species specific.[97,185,198]

b. Stomatal Uptake

Since the introduction of selective pesticides that require systemic action, there has been speculation that entry into leaf tissue could be facilitated by stomatal entry. The parameters required to permit stomatal entry of solution were determined over 20 years ago, the key requirement being solution of surface tension below 30 mN m^{-1}, although this was dependent upon the specific surface.[169] Achievement of low surface tension, approaching 20 mN m^{-1}, in formulated pesticides was not achieved until the availability of fluorocarbon and organosilicone surfactant classes. Of these only the organosilicone surfactants are both environmentally and commercially acceptable. The development of organosilicone surfactants from their earliest uses in the promotion of micronutrient uptake in trees has been reviewed recently.[192]

There is a significant effect of organosilicone surfactants on droplet adhesion to leaf surfaces. This is associated with a significant reduction in solution surface tension and in particular the dynamic surface tension of droplets prior to impact with the leaf surface. The role of dynamic surface tension in the droplet impaction process has now been demonstrated definitively.[197] The ability of organosilicone surfactants to wet low energy surfaces has been interpreted in terms of the molecular processes and the structure-performance relationships which govern the behavior of this surfactant class.[6] The generalized chemical structure of organosilicone surfactants is a T-shaped conformation, with a fully methylated siloxane backbone from which one or more polyether tails are suspended.[192] The hydrophilic portion of the molecule is similar to that of most conventional surfactants and comprises a chain of EO units. The unique chemical structure of organosilicone surfactants contributes to their special wetting properties.[6] The surfactants have an ability to lower the liquid-air surface tension to extremely low values, have fast kinetics of adsorption at the liquid-air and solid-liquid interfaces, have high affinity for low energy surfaces, and form a favorable orientation and structure with adsorbed molecules.[6,83] The overall effect has been attributed to a form of molecular zippering at the solid-liquid interface.[6] Conventional surfactants tend to lie flat on the surface exposing hydrophobic patches, which impedes spreading; in contrast, both the hydrophobic and polyether groups of organosilicone surfactants maintain a specific orientation that promotes the advance of the droplet edge.[6]

The addition of the organosilicone surfactant Silwet L-77® to polar formulations of herbicides such as glyphosate and triclopyr ([(3,5,6-trichloro-2-pyridinyl)oxy]acetic acid) enhanced the rate of active ingredient uptake and provided the first indication of surfactant-induced stomatal uptake of pesticides.[14,27,28,62,63,195,196] The conditions required for stomatal uptake are precise and there is requirement not only for a stomatous surface but also that the stomata should be open.[32,62,63] Closure of the stomata by the use of abscisic acid or reducing water availability does not result in organosilicone surfactant-facilitated uptake.[27,32,226] Assuming open stomata, stomatal entry may occur provided the stomatal pore wall morphology is such that the contact angles formed between solution and the pore walls are conducive to a positive pressure difference, which induces liquid movement into the substomatal cavity.[28,169] The solution surface tension is a key factor; it needs to be less than the critical surface tension value ($\gamma_{crit.}$) for the plant surface and solution combination.[28,169] There has been no major analysis of the effect of surface pore diameter on rates of solution infiltration. However, in one study it was shown that total pore area per unit of leaf surface was a more critical determinant than individual pore surface area in predicting the magnitude of stomatal uptake of a herbicide.[27] Thus, field bean was a more effective model system to replace gorse spines than was dwarf bean (*Phaselous vulgaris* L.). The typical pore area for field bean and gorse was 155.3 and 26.9 μm^2, respectively, although stomatal frequency varied from 52.7 pores mm^{-2} in field bean to 448.7 pores mm^{-2} in gorse. In contrast the pore area for dwarf bean was 11.8 μm^2 and the stomatal frequency 82.4 pores mm^{-2}.[27] These results add weight to the proposition that stomatal infiltration is determined by a number of physical parameters, both in the formulated solution and in the biological system under test. These findings also relate to the nature of the leaf surface deposit which, with the inclusion of an organosilicone surfactant, is typically a thin film without definition of a droplet or a measurable contact angle. The relationship between droplet spreading and stomatal pore area requires further investigation, particularly given the nature of the organosilicone surfactant-induced antagonism of herbicide uptake in some species, as discussed below.

In the case of organosilicone surfactants there is a very large disparity between proposed CMC values, which are typically at 0.1-0.5 g L^{-1}, and optimum rates for promotion of stomatal infiltration which are close to 2 g L^{-1}.[28,192] This has led to the development of an "adsorptive dilution hypothesis" in which there is a migration of surfactant from the bulk solution to the expanding droplet surfaces as the droplet spreads.[28] This dilutes the surfactant concentration within the bulk solution, which can in turn increase the surface tension, resulting in reduced efficacy at concentrations in the range 0.5 to 1.5 g L^{-1}. In a model system using field bean the uptake of a triclopyr-triethylammonium formulation containing 1 g L^{-1} Silwet L-77® was presumably reduced because adsorptive dilution had reduced the effective concentration of surfactant in the bulk solution to 0.48 g L^{-1}.[28] This surfactant concentration was close to the calculated CMC, but was too low to promote stomatal infiltration and actually reduced uptake of the herbicide. At 1 g L^{-1} the surface tension of the formulation was close to 22 mN m^{-1}, which should have been sufficient to permit stomatal infiltration. These are critical observations in determining the efficacy of organosilicone surfactants.

Organosilicone surfactants are most effective in promoting the stomatal infiltration of herbicide formulations that contain polar active ingredients. Thus there are many reports of the promotion of uptake of herbicides formulated as amine salts, glyphosate[14,62,63,75,78,80,192,225] and triclopyr.[27,28,30,32] Addition of organosilicone surfactants to emulsifiable concentrates does not promote stomatal infiltration, generally because such formulations do not satisfy the requirement for solution surface tension of less than 30 mN m^{-1}, because of co-formulant interference with the physical properties of organosilicone surfactants.[27]

In some situations where solution surface tension requirements are met and plant surfaces have open, functional stomata, there is an apparent antagonism to herbicide uptake inclusion of an organosilicone surfactant in the formulation.[66,76,77,107] The uptake of glyphosate by a number of grass species, including colonial bentgrass (*Agrostis tenuis* Sibth.), orchardgrass (*Dactylis glomerata* L.), quackgrass (*Elytrigia repens* Beauv.) and dallis grass (*Paspalum dilatatum* Poiret.), which all have stomatous adaxial surfaces was reduced by the addition of 1 g L^{-1} Silwet L-77®.[66] This was in contrast to positive effects on uptake found in perennial ryegrass (*Lolium perenne* L.) under the same conditions. The antagonism was not attributed to droplet spreading or drying. A detailed investigation with dallis grass showed that while droplet drying *per se* was not a factor, the addition of the humectant glycerin at 30-60 g L^{-1} overcame the organosilicone antagonism and promoted the foliar uptake of glyphosate.[66] The results emphasized that there is poor knowledge of the role of humectants and no information on their interaction with organosilicone surfactants. A detailed investigation of organosilicone surfactant-induced antagonism of glyphosate in wheat revealed the importance of the chemical structure of the surfactant. Glyphosate uptake was influenced by: (1) the ratio of the ethylene oxide hydrophile to the siloxane hydrophobe, (2) the mean EO chain length of the surfactant and (3) the chemical nature of the terminal group on the EO chain.[76,77] It was concluded that antagonism was substantially reduced by increasing the polarity of the organosilicone surfactant molecule. Thus, an increase in the size of the hydrophobic backbone from 3 to an average of 3.7 siloxane units enhanced both the rate and total amount of glyphosate taken up. However, interpretation is complicated because when the hydrophobe increased there was a concomitant rise in total EO content of the molecule (from 8 to a mean of 13.6), which may have been substantially responsible for the increased performance and efficacy.[77] The latter study confirmed earlier findings that glycerin could overcome antagonism, and it suggested that this was associated with enhanced stomatal flooding of solutions containing Silwet L-77® and glycerine.[66,67,77] Glycerine's role may be to assist in maintaining low interfacial tensions of the organosilicone surfactant at the liquid-air or liquid-solid surfaces of the droplet, thus overcoming the increased surface tension by adsorptive dilution which may preclude stomatal entry.[26,28,77]

The development of simple models to describe surfactant performance is a major advance, although largely based on the physicochemical characteristics of formulation components.[32,86,97,127,161] It is arguably more important to describe herbicide fluxes in terms of simple biological models, which allow mechanistic explanations of the total system.[27,32] For example, the use of the Michaelis-Menten formalism to describe the uptake of the herbicide triclopyr-triethylammonium in the presence of the organosilicone surfactant Silwet L-77® has put appropriate weighting on the biological contribution to the total uptake process.[27,32] Using a model system of field bean leaves and uptake over a 24-h period, the Michaelis-Menten formalism was found to describe the data well and enabled a mechanistic explanation of the effects of Silwet L-77® on triclopyr uptake.[32] Silwet L-77® significantly reduced the time to reach half the maximum uptake (k) of herbicide (by 2.8 times) by providing a stomatal entry route into the plant. However, the organosilicone surfactant did not significantly increase the total amount of active ingredient entering the plant after 24 h ($V_{\max}$). The relatively short time for the Silwet L-77® treatment to achieve V_{max} (1-3 h) has special significance for reducing the critical rain free period after spraying.[62,63] The control formulation without surfactant required longer (17-24 h) to reach the same asymptotic value as the Silwet L-77® formulation for $V_{\max}$ uptake. In Michaelis-Menten kinetic terms, the effect of uptake induced by the presence of Silwet L-77® is one of competitive inhibition. The situation involves two treatments having different values of k but equal $V_{\max}$ values, as demonstrated by Eadie-Hofstee linear plots.[32,143] Values for $V_{\max}$ were statistically the same but the time to reach half the maximum uptake (k) was significantly decreased when Silwet L-77® was included in the triclopyr formulation. In Michaelis-Menten theory, k

was decreased by the presence of a competitive inhibitor, Silwet L-77®.

The suitability of Michaelis-Menten theory to the triclopyr uptake data suggests the involvement of active and/or facilitated diffusion in the foliar absorption of triclopyr, both with and without the organosilicone surfactant. In the present experimental system it is likely that facilitated diffusion or active transport of triclopyr takes place when it moves across the cellular membranes into the symplasm.[52,64] Provided that triclopyr can be retained within the symplasm and is phloem-mobile, it can be loaded into the conducting sieve elements for long-distance transport.[18] Thus, maintenance of the herbicide flux following stomatal entry is critical, and it involves continuous removal of triclopyr from the leaf system by phloem translocation. The present model does not determine the rate limiting step in the uptake process. This appears to involve carrier-mediated transfer at a cell membrane site located within mesophyll cells or at the phloem loading step.[32]

2. Translocation and Adjuvants

The foregoing sections indicate the difficulty in describing the factors limiting or promoting the long distance translocation of pesticides, and the imperfect knowledge that exists in describing fundamental aspects such as phloem loading. Thus, it is difficult to ascribe a particular contribution to other formulation components. For example, it is not satisfactory to imply that manipulation of formulations by surfactant addition increases the translocation of pesticide active ingredient if the key area of enhancement is uptake, rather than translocation *per se*. Some of the more extravagant claims for improvements in translocation by formulation manipulation are related to the method of data calculation, as described previously, rather than enhancement of a key component of the translocation system. Reports about specific surfactant-enhanced translocation must be viewed with a degree of skepticism if the translocation process is defined simply as the longitudinal transfer of solutes in phloem cells. Thus, recent assertions that organosilicone surfactants enhanced translocation by bringing active ingredient in close proximity to vascular tissues is somewhat erroneous and, in physiological terms, an artifact created by the method of calculation.[192,195,224]

However, it is clear that formulation components, particularly surfactants, are not involved simply in surface phenomena and may actively pass through the cuticle and enter both the apoplast and symplast of the leaf. This is particularly evident for the organosilicone surfactants that enter sub-stomatal cavities and spread laterally along the cell walls of epidermal cells and perhaps mesophyll cells.[27,31,62] The arguments in favor of co-penetration of active ingredient and surfactant are particularly relevant to the involvement of surfactants in enhancing the specific process of translocation.[198,199] The evidence for formulation components actively promoting the process of translocation is circumstantial because of the practical difficulty of locating adjuvants at specific cellular or transport sites. Thus, while the limited availability of radioactively labeled adjuvants imposes one form of limitation, the inability of scientists to develop consistent and acceptable techniques for the cellular localization of such materials, and then to interpret the physiological significance of such findings, is equally limiting.

Many recent reviews on herbicide translocation have not acknowledged the importance of formulation and the inclusion of adjuvants.[42,52,54,110] Other articles have provided limited, essentially nonphysiological information.[55,192] Coupland has provided the most effective review with the inclusion of data that assists in understanding the relation between surfactant addition and the translocation of herbicides.[41] It is concluded that surfactants are generally poorly translocated in plants; this is not necessarily due to ineffective foliar absorption and probably not to the inability to reach vascular tissues.[41,181,184] It is noted that those surfactants that produce significant reductions in solution surface tension (< 30 mN m^{-1}), organosilicone and fluorocarbon types, rapidly enter leaves and diffuse extensively through the apoplast.[62,223] Thus, while access to leaf tissues by surfactants is not a major limiting factor, their inability to transfer across the sieve element plasmalemma and be actively absorbed by sieve elements limits their entry into the long-distance transport system. In the case of the organosilicone surfactant Silwet L-77®, its inability to enter sieve elements has been demonstrated by Coupland, Field, and Loeffler (unpublished). They used aphids to sample the phloem sap of willow plants, using two explant systems; isolated bark strips in which the surfactant solution was applied directly to the cambium surface, and whole stem segments in which surfactants were supplied by xylem perfusion. Two surfactants were tested, Silwet L-77® which was analysed by nuclear magnetic resonance (NMR) spectroscopy utilizing the characteristic signal from the trimethysilyl protons, and Triton X-100 (octylphenol ethoxylate, 9-10 EO units) which was radiolabeled. Phloem samples were collected as aphid honeydew. No radioactivity was measured in either explant system over the range of Triton X-100 concentrations (0.1-2.5 g L^{-1}). No Silwet L-77® was detected in similar experiments over a concentration range of 1-2 g L^{-1}. It was concluded that neither of these two surfactants, which are of contrasting chemistry, entered sieve elements. While this does not remove the possibility of surfactants being involved in translocation by enhancement of sieve tube loading, it does suggest that they do not participate in long-distance transport.

Further corroboration of this hypothesis has been provided by investigations of the direct effect of surfactants on phloem loading using short half life radioisotope studies to determine assimilate movement.[41] These results showed that the organosilcone surfactant Silwet L-77® at 5 g L^{-1} could considerably inhibit assimilate loading and subsequent translocation of solutes. It is postulated that if herbicides show co-translocation patterns with assimilate then their overall long-distance transport in the phloem could be restricted. Such postulations are given additional weight when the components of the overall herbicide transfer process from leaf surface to arrival at a distant site of action are investigated.[27] In this case the concentration of the organosilicone surfactants Silwet L-77® and DC X25152 had a major effect on the translocation of triclopyr butyl ester. Translocation of triclopyr as a percentage of net uptake decreased linearly from 41.4% in the absence of Silwet L77® to 21.2% at 5 g L^{-1}. In contrast, there was a fourfold increase in foliar uptake, from 18.0% in the absence of Silwet L77® to 76.6% at 1 g L^{-1}, which declined to 68.7% at 5 g L^{-1}. It is concluded that rates of Silwet L77® in the range 1-3.2 g L^{-1} are optimum in promoting significant foliar uptake while causing only intermediate levels of damage to the translocation mechanism.[27] A further factor in the apparent poor translocation of surfactants is that they may be rapidly metabolized to immobile products soon after foliar absorption.[41,187,202]

IV. ROOT UPTAKE AND APOPLASTIC TRANSPORT

Of 165 herbicide active ingredients currently registered in the United States, 45% are recommended for soil application. Moreover, many of the foliar-applied compounds can also be absorbed by roots. A major portion of any chemical spray directed to the foliage will eventually reach the soil surface. Thus, properly described as the ultimate sink, the soil is of major concern from the viewpoint of environmental safety.[116] A good understanding of the mechanisms involved in the movement and fate of pesticides in the soil, their uptake and translocation in plants, and the effect of formulation on these processes is essential for improvements in the efficacy of active ingredients and minimizing their impact on the environment.

A. PATHWAYS AND MECHANISMS OF ROOT UPTAKE

Root uptake of pesticides is similar to ion and nutrient uptake and the concepts developed from studies of ion uptake are usually applied to all compounds. The large surface area of the roots and the ability of plants to produce new root hairs at rapid rates provides plants with an extremely efficient uptake system.[91] Root hairs play the most important part in absorption of substances from the soil, although absorption by other regions of the root occurs. The external wall of the root hair, like the epidermis of the leaf, is covered with a layer of cuticle, although it is very thin.[33,173] Solutes and water readily pass through this barrier by mass flow and/or diffusion and enter the cell wall where they will be available for transport in the apoplast. Active uptake of herbicides and their transfer through the symplast is known to occur and is an essential step in the mode of action of herbicides.[55]

It has been reported for a number of pesticides that the time course of root uptake is characterized by an initial rapid uptake followed by a slower steady state phase.[105,154,155,189] The bi-phasic nature of root uptake indicates the involvement of two mechanisms.

The initial rapid phase is concentration dependent, has a Q_{10} of less than 2, and is not affected by metabolic inhibitors. Water soluble molecules diffuse freely into the apparent free space of the root during this phase. For some herbicides like 2,4-D and DNOC (4,6-dinitro-*O*-cresol), the initial rapid uptake rate was reported to fall off to zero and then become negative.[15,16,25,133] This was due to efflux of the molecules from the roots, the efflux being greater if the roots were in distilled water. For other herbicides different mechanisms might be involved in retaining the molecules inside the roots. These include electrical binding of dissociated cations to cell constituents, partitioning of lipid-soluble molecules into lipid constituents of the root, such as membranes, and degradation of the herbicide to polar products.[55,33] Moreover, a similar mechanism to the hypothesis of accumulation of weak acids (see Section III. B) might also occur in the root. Some of these mechanisms require metabolic energy and they can be considered as part of an active uptake process. It is difficult to separate the two phases of root uptake as they may work in tandem and be closely associated.

The second phase of root uptake is active, requires energy and oxygen, has a Q_{10} greater than 2, its rate does not show a linear relationship to the concentration, and it can be blocked by metabolic inhibitors.[33] This phase allows the molecules to cross the plasmalemma and enter the living cells, therefore it is an essential requirement for some pesticides, including all herbicides. The active uptake component is necessary for systemic soil-applied pesticides because of the structural anatomy of the root. Solutes have to pass through the endodermis, enter the stele, and be available for xylem transport. The endodermis cells are covered on their anticlinal walls by the Casparian strip which is made up of lipophilic substances that contain suberin.[57,58] The Casparian strip is impervious to water and solutes which have to by-pass it by crossing the plasmalemma membrane and entering the symplast of the endodermis.

The root system is the most efficient absorption organ, but other underground plant parts can absorb herbicides effectively. The hypocotyl, cotyledons, and epicotyl of dicotyledons, and coleoptile and mesocotyl of monocotyledons can absorb pesticides.[167] These are young and growing organs with relatively thin cuticles and an unsuberized epidermis that makes them more permeable to entry of pesticides. Modified stems such as rhizomes, stolons, tubers and bulbs can also absorb pesticides, and this can be an important step in the efficacy of soil-applied herbicides against certain weeds.[165]

B. TRANSPORT PATHWAYS IN THE XYLEM

Once inside the stele, pesticide molecules and water move toward xylem elements due to root pressure and transpiration pull, which together create a water potential difference in the xylem. This movement can be symplastic, i.e., the pesticide molecules can pass from cell to cell via plasmodesmata, or molecules may move back into the cell wall and move along an apoplastic route. Physicochemical properties determine which route is more important for a particular compound. For some pesticides, the total transport pathway is short and is referred to as short-distance transport. Germination inhibitors and growth inhibitor herbicides like EPTC [*S*-ethyl dipropyl carbamothioate] and trifluralin need to be carried only a few cell layers to reach their site of action.[22] The majority of soil-applied herbicides, however, need to reach the growing shoots and apical meristems above ground, to be effective. This long-distance transport is through xylem elements which are non-living and part of the apoplast. Water and solutes entering the xylem are rapidly carried upward in the transpiration stream. Transpiration pull is the major driving force, and it maintains a concentration gradient from the external medium to the apoplast, which helps root uptake.[33] However, transpiration pull does not dictate the uptake of all compounds since it has been shown that uptake curves over time were similar for isolated roots and intact plants.[25,133]

C. IMPORTANT FACTORS IN THE PERFORMANCE OF SOIL-APPLIED PESTICIDES

As a general rule, for pesticides to show their effect they must reach a specific target site. The main events involved in this process are application to the soil, availability of active ingredient for uptake by underground plant parts, and finally the amount of active ingredient absorbed and translocated to the target sites, normally above the ground.

1. Pesticide Application

The first key point is the effectiveness of application. Obviously, this should be performed in a way to maximize the likelihood of the active ingredient reaching the target sites and to minimize environmental contamination. The proportion of the active ingredient reaching the absorption sites is dependent upon plant, application, and environmental factors. For soil applied pesticides, additional physical factors operate which make the distribution of the active ingredient even more non-uniform and their availability to the intended target sites more difficult. The preparation of the soil prior to application is important, as a rough surface projects a larger surface area which tends to result in non-uniform distribution. Moreover, many weed seeds can be found inside soil clods and are thus protected from herbicides. If the herbicide is to be incorporated into the upper layer of the soil profile, vigorous cultivation usually helps to achieve a more uniform distribution. For many preemergence herbicides, a layer or a film of active ingredient on the soil surface should be maintained undisturbed for effective results. An important consideration in soil application is the dynamic nature of chemicals in the soil. The active ingredients are not static and can move to considerable distance from the site of application. The key elements of this movement are discussed in the following section.

2. Pesticide Availability

The availability of pesticides in the soil is greatly influenced by their sorption, transport and degradation. Compounds which are adsorbed very strongly to soil particles such as ethofumesate ((±)-2-ethoxy-2,3-dihydro-3,3-dimethyl-5-benzofuranyl methanesulfonate), may lose their activity, especially in dry soils.[126] Adsorption to soil is very strong from the vapor phase and involves direct contact of molecules with soil colloids. This is usually considered in application of herbicides to dry soils, where higher volume rates are recommended.

Mobility of active ingredients within the soil is the second step in determining the proportion of the applied pesticides reaching the intended targets. The preferred type of movement depends upon the physical properties of the molecules.[116,160] Compounds with high vapor pressure, such as EPTC and trifluralin, move rapidly by diffusion following a concentration gradient to fill the soil micropores. Pesticides with low vapor pressure, like triazines and urea compounds, are normally transported by mass flow with soil water and their rate of movement is subject to water velocity. As water movement under field conditions can be downward or upward, and because of sorption characteristics of different soils and solubility differences between active ingredients, the movement of pesticides by mass flow is very hard to predict, especially considering other variables such as decomposition, plant uptake, soil structure, soil moisture content, and climatic factors. General principles governing mobility of pesticides and appropriate measurement methods have been reviewed recently.[116,222]

As a general rule, compounds with high solubility have the potential of leaching out of the target soil layer. However, solubility is not the main determinant for distribution of pesticides in different soil profiles. Compounds of low to moderate solubility, like simazine

(6-chloro-*N,N*-diethyl-1,3,5-triazine-2,4-diamine) and atrazine (6-chloro-*N*-ethyl-*N*-(1-methylethyl)-1,3,5-triazine-2,4-diamine), are found frequently at low profile depths mainly due to their persistence.[89] This problem has received much attention in the past decade as a potential hazard for ground water contamination.[61,73,222] Rate adjustment or repeated application to compensate for the loss in active ingredient only adds to the contamination hazard and alternative ways should be pursued. Efforts to correlate formulation types such as emulsifiable concentrates, suspension concentrates, or dispersible granules to leaching or run-off of agrochemicals have been inconclusive and more work is required before any general model can be established.[217,218]

One of the attempts in reducing the risk of groundwater contamination is through manipulating the formulation of active ingredients. Controlled release technology has been an active area of research with many innovative formulations patented.[17,113] Cyclodextrins are polymers of D-glucose formed through enzymatic starch degradation.[49] They have a hydrophobic center and relatively hydrophilic outer surface allowing for inclusion complexes with many pesticides. Work with cyclodextrin complexes containing atrazine, simazine and metribuzin (4-amino-6-(1,1-dimethylethyl)-3-(methylthio)-1,2,4-triazin-5(4*H*)-one) has shown that despite the improvement in physical characteristics of the active ingredients, cyclodextrin formulations are inferior in herbicidal activity compared to the commercial formulations; this may be a result of slow-release properties of these complexes.[49] Another technique employed to reduce the contamination of the environment is through the use of slow-release herbicide tablets. Several herbicides were successfully formulated by this method and a combination of metolachlor (2-chloro-*N*-(2-ethyl-6-methylphenyl)-*N*-(2-methoxy-1-methylethyl)acetamide) and oxyfluorfen (2-chloro-1-(3-ethoxy-4-nitrophenoxy)-4-(trifluoromethyl)benzene) with the surfactant Triton X-100 was found to be very effective in broad-spectrum weed control.[186] Another formulation technique with a longer history is starch-encapsulation of pesticides.[178] Recent work shows that many pesticides, including herbicides and chemical and biological insecticides, can be encapsulated in starch granules by cross-linking reactions.[19,68,179,221] These formulations provide an improved release rate which can be controlled by varying the starch type and particle size. Microcapsules have been developed based on side chain crystallizable polymers with melting points in the 15-35°C range.[36] At the temperature of interest, the side chains melt to release the encapsulated active ingredient molecules. This allows for the accurate adjustment of the delivery time of pesticides to coincide with the emergence of pests. The system has been used successfully for controlled temperature-activated release of the herbicide trifluralin and the insecticide diazinon.[36]

In comparison to the amount of work on the effect of adjuvants on the performance of foliar applied pesticides, especially herbicides, relatively little work has been reported on soil-applied compounds. However, a review of literature indicates that adjuvants can affect the selectivity and behavior of certain herbicides in soil. The addition of soybean [*Glycine max* (L.) Merr.] oil contributed to improved selectivity of acifluorfen and oxyfluorfen in several crops without any effect on weed control.[1] A study on the movement of herbicides in columns containing fine sand showed that three humic substances and a synthetic polymer were only marginally effective in reducing leaching of dicamba (3,6-dichloro-2-methoxybenzoic acid) and bromacil (5-bromo-6-methyl-3-(1-methylpropyl)-2,4(1*H*,3*H*)pyrimidinedione).[5] Leaching of simazine, however, was significantly decreased in the presence of adjuvants. Another study reported an increase in solubility of petroleum hydrocarbons by both biosurfactants, which are produced by microorganisms, and commercial surfactants.[60]

3. Root Uptake and Translocation of Pesticides

Once in contact with plant roots, the pesticide molecules dissolved in the soil water are available for root uptake. The rate of uptake depends on factors such as temperature, pH, and concentration of pesticide in the soil solution as well as the physicochemical properties of active ingredient.[33] Higher temperatures, within the physiological range, normally enhance root uptake, the effect apparently being greater on the active phase of uptake rather than the passive phase.[155] The effect of pH on absorption is through the dissociation of molecules. Weak acids are dissociated at higher pH values and this reduces their uptake. At lower pH values the nonionized molecules are more lipophilic and penetrate the plasmalemma more readily.[13,33] Moreover, pH controls the availability of pesticides in the soil solution as their adsorption usually decreases and their mobility and persistence increases in more alkaline soils. This was reported both for acidic compounds such as 2,4-D, dicamba, chloramben (3-amino-2,5-dichlorobenzoic acid) and chlorsulfuron[24,40,132,180,206], and for basic compounds like *s*-triazines.[12,114,115,117,204] For most soil applied herbicides, the extent to which they can be taken up by roots and move via the xylem to the shoots determines their efficacy. Several extensive studies suggest that, at least for non-ionized compounds, the above two processes do not have similar requirements in terms of physicochemical characteristics of active ingredient.[20,182,183] Experiments with nine *O*-methylcarbamoyloxime insecticides and nine inactive analogues of substituted phenylurea herbicides showed that root uptake from nutrient solution rapidly reached equilibrium values that increased with increasing lipophilicity.[20] This was suggested to be a result of partitioning of lipophilic molecules into lipid components of root structure. The relationship between uptake and lipophilicity is more complicated in soil where adsorption to organic matter limits availability.[22] Root uptake of polar compounds follows a different route than the more lipophilic compounds with entry into the aqueous phase both in the free space and within the root cells. Translocation to the shoots, on the other hand, was found to be greatest for compounds of intermediate lipophilicity.[20] The challenge of finding an effective systemic active ingredient for soil application is complex. One must combine the desired physiological activity with physicochemical properties which provide for greater availability, optimum persistence, higher root uptake, and good translocation to the above-ground plant parts.

V. FUTURE DEVELOPMENTS

Formulation manipulation for the overall enhancement of pesticide efficacy has shown significant gains in recent years. The level of understanding of both the chemical and biological environments has improved and this, coupled with the introduction of new approaches to pesticide formulation and the introduction of new classes of adjuvants, has resulted in major gains in effectiveness. The extent of these developments is well chronicled in the proceedings from three International Symposia on Adjuvants for Agrochemicals, the first of which was held in 1986.

The significance of overall improvements in efficacy must be measured alongside the economic cost and environmental impact of all formulation components. Future opportunities are likely to be generated by the introduction of new chemistry and technology or by understanding and predicting the performance of present systems with greater accuracy.

A. NEW TECHNOLOGY

The introduction of new active ingredient chemistry is likely to have minimal impact if the physicochemical characteristics of new active ingredients are within the bounds of presently available materials. The exception may be chemical structures that can be linked to customized adjuvant systems that confer unique characteristics. The last 10 years have seen the introduction of a wider range of adjuvants including organosilicone surfactants and various complex adjuvant mixtures. For example, pyrrolidone-based composition adjuvants that are biologically efficacious may also offer major proprietary advantage to the developers.[151] The introduction of new adjuvant chemistry and the customization of products will occur against a background of increasing development costs and the need for these to be recouped. These new products will need to demonstrate cost effectiveness against existing formulations and be able to withstand rigorous scrutiny in terms of their environmental impact. Thus, the development of alkyl polyglycosides and sucroglycerides as biodegradable nonionic surfactants is typical of new technology.[4,59] At this stage there is no indication of the biological efficacy of such surfactants and the extent to which they can enhance the uptake and translocation of active ingredients.

A further area of new technology development will be techniques and research methods that permit a clearer understanding of the behavior of the components of formulations and their interaction with biological systems. A simple example is the current level of understanding of chemistry and structure in relation to transfer of active ingredient co-formulated with a surfactant. Exhaustive studies on isolated cuticles and on intact plants, using a wide range of formulation compounds, has yet to reveal a simple standardized understanding of the processes involved. There is a similar lack of information on the mechanisms regulating the phloem translocation of pesticides. Approaches that will unravel the underlying mechanisms would be hugely beneficial, particularly if it allowed the fate of individual formulation components to be determined. One of the major practical limitations in present studies is the inability to tag formulation components, other than the active ingredient, in a way that allows fate to be determined. Given that the overall processes are operating at the cellular and subcellular level, it is critical to understand the impact of formulation components on the system at these levels. At present there is only scant information, for example, on the co-penetration and co-translocation of surfactants and active ingredient. The present uncertainty about relationships at even this gross level is indicative of the strides that could be made by the introduction of new technological approaches.

B. MANIPULATION OF UPTAKE AND TRANSPORT PATHWAYS

The introduction of organosilicone surfactants has opened up a whole new opportunity for the rapid delivery of active ingredient to the internal tissues of plants.[62,63,169] These developments emphasize the importance of understanding the theoretically possible uptake pathways and matching them with formulations that meet the physical and chemical requirements imposed by the biological structure of stomata and associated structures.

The use of stomatal entry routes, especially in the development of organosilicone surfactants, have not been fully exploited. One of the obvious limitations to the present system is the degree of stomatal opening that determines the potential for inflow of solution. Closure of stomata prevents solution inflow.[32] The possibility of developing adjuvants that promote stomatal opening and hence consistent solution inflow is a clear opportunity for future development. Progress is likely to be made on the basis of a better understanding of the endogenous control mechanisms that regulate stomatal opening and closing.

While inflow of solutions through open stomata is an obvious opportunity for manipulation, there must also be possibilities for the development of co-penetrants for transcuticular movement.

Enhancement of translocation by adjuvants does not appear to be a profitable line for investigation. Overall improvements in the transfer of active ingredient are more likely to be accomplished by improvements in the uptake process rather than manipulation of translocation *per se*. The current evidence is that adjuvants rarely reach the phloem transport system, and there is certainly no universal view that adjuvants and active ingredients follow the same transport pathways. Future developments may see the production of an adjuvant that could promote sieve tube loading, but whether this could ever be made to be selective for a particular active ingredient seems a remote possibility. It is possible to consider a wider whole-plant view of the translocation process; to think in terms of source-sink relationships and the way that they may be manipulated to facilitate improved long-distance transport of pesticides. Several studies with synthetic plant growth regulators as adjuvants have demonstrated that this approach may have some major limitations.

C. MANAGEMENT OF THE WHOLE SYSTEM

One of the more obvious and implementable approaches to improving uptake and translocation of active ingredients will be the development of a better overall framework that identifies both what is known and what needs to be known about the system as a whole. At present there is insufficient coherence between the studies on the chemistry of active ingredients and adjuvants and on the biological systems in which they operate. The total system approach would see the development of more advanced predictive models that take into account the physicochemical characteristics of formulation components, the operating systems in plants, and the interaction between these two in a changing external environment of climatic and edaphic factors. The current attempts at predictive models have taken a major step forward, most notably the model developed by Holloway and co-workers.[97] This is a preferable approach and likely to have greater utility than the reductionist approach taken by other workers, who seem overly concerned with fragmenting the system. For example, work on isolated cuticles and cell suspension or tissue culture can yield valuable information, but only if it is placed in the context of a whole plant.

Further development is needed of mechanistic models that describe compartments within the biological system and their relationship to the transfer of active ingredient and other formulation components. The advantage of the mechanistic model approach is that it will identify the gaps in biological knowledge that have to be filled in order to understand the performance of xenobiotics. Approaches taken towards the development of mechanistic models are useful, but must be classed as very preliminary and of very limited use in predicting the behavior of specific formulations.[32,52,127]

REFERENCES

1. **Abada, M. and Aviram, H.,** Vegetable oil as adjuvant to improve herbicide effectivity, *Phytoparasitica*, 13, 240, 1985.
2. **Achhireddy, N. R., Kirkwood, R. C., and Fletcher, W. W.,** Foliar absorption and translocation of isoproturon, and its action of photosynthesis in wheat (*Triticum aestivum*) and slender foxtail (*Alopecurus myosuroides*), *Weed Sci.*, 33, 762, 1985.
3. **Akesson, N. B., Bayer, D. E., and Yates, W. E.,** Application effects of vegetable oil additives and carriers on agricultural sprays, in *Adjuvants and Agrochemicals*, Vol. II, Chow, P. N. P., Grant, C. A., Hinshalwood, A. M., and Simundsson, E., Eds., CRC Press, Boca Raton, FL, 1989, 121.
4. **Aleksejczyk, R. A.,** Alkyl polyglycosides: versatile, biodegradable surfactants for the agricultural industry, in *Pesticide Formulations and Application Systems*, Vol. 12, Devisetty, B. N., Chasin, D. G., and Berger, P. D., Eds., American Society for Testing and Materials, Philidelphia, 1993, 22.
5. **Alva, A. K. and Singh, M.,** Use of adjuvants to minimize leaching of herbicides in soil, *Environ. Manage.*, 15, 263, 1991.
6. **Ananthapadmanabhan, K. P., Goddard, E. D., and Chandar, P.,** A study of the solution, interfacial and wetting properties of silicone surfactants, *Colloids Surfaces*, 44, 281, 1990.
7. **Arnold, A. C. and Mumford, J. D.,** The development and use of vegetable oil adjuvants with pesticides in Western Europe, in *Adjuvants and Agrochemicals*, Vol. II, Chow, P. N. P., Grant, C. A., Hinshalwood, A. M., and Simundsson, E., Eds., CRC Press, Boca Raton, Fl, 1989, 25.
8. **Baker, E. A.,** Chemistry and morphology of plant epicuticular waxes, in *The Plant Cuticle*, Cutler, D. F., Alvin, K. L., and Price, C. E., Eds., Academic Press (for Linnean Society of London) London, 1980, 139.
9. **Baker, E. A. and Hunt, G. M.,** Factors affecting the uptake and translocation of pesticides, in *Pesticide Formulations - Innovations and Developments*, Cross, B. and Scher H. B., Eds., ACS Symp. Ser. 371, American Chemical Society, Washington, D.C., 1988, 8.

10. **Becher, P. and Becher, D.,** The effect of hydrophile-lipophile balance on contact angle of solutions of nonionic surface-active agents. Relation to adjuvant effects, *Advances in Chemistry Series. Pesticide Formulations Research,* 86, 15, 1969.
11. **Berndt, G. F.,** Efficiency of foliar sprays as influenced by the inclusion of surfactants, *Res. Dev.Agric.,* 4(3), 129, 1987.
12. **Best, J. A. and Weber, J. B.,** Disappearance of *s*-triazines as affected by soil pH using a balance-sheet approach, *Weed Sci.,* 22, 364, 1974.
13. **Beyer, E. M., Jr., Duffy, M. J., Hay, J. V., and Schlueter, D. D.,** Sulfonylureas, in *Herbicides: Chemistry, Degradation and Mode of Action,* Vol. 3, Kearney, P. C. and Kaufman, D. D., Eds., Marcel Dekker, New York, 1988, 117.
14. **Bishop, N. G. and Field, R. J.,** Improved performance of glyphosate in the full season control of perennial ryegrass, *Aspects Appl. Biol.,* 4, 363, 1983.
15. **Blackman, G. E. and Sargent, J. A.,** The uptake of growth substances, II. The absorption and accumulation of 2,3 5-Triiodobenzoic acid by the roots and frond of *Lemna minor, J. Exp. Bot.,* 10, 480, 1959.
16. **Blackman, G. E., Sen, G., Birch, W. R., and Powell, R. G.,** The uptake of growth substances, I. Factors controlling the uptake of phenoxyacetic acids by *Lemna minor, J. Exp. Bot.,* 10, 33, 1959.
17. **Bode, L. E. and Chasin, D. G.,** *Pesticide Formulations and Application Systems,* Vol. 11, American Society for Testing and Materials, Philadelphia, 1992.
18. **Bovey, R. W., Hein, H. J., and Meyer, R. E.,** Concentration of 2,4,5-T triclopyr, picloram, and clopyralid in honey mesquite (*Prosopis glandulosa*) stems, *Weed Sci.,* 34, 211, 1986.
19. **Boydston, R. A.,** Controlled release starch granule formulations reduce herbicide leaching in soil columns, *Weed Technol.,* 6, 317, 1992.
20. **Briggs, G. C., Bromilow, R. H., and Avis, A. E.,** Relationships between lipophilicity and root uptake and translocation of non-ionised chemicals by barley, *Pestic. Sci.,* 13, 495, 1982.
21. **Bromilow, R. H. and Chamberlain, K.,** Designing molecules for systemicity, in *Mechanisms and Regulation of Transport Processes,* Vol. 18, Atkin, R. K. and Clifford, D. R., Eds., British Plant Growth Regulator Group Monograph, London, U.K., 1989, 113.
22. **Bromilow, R. H. and Chamberlain, K.,** Pathways and mechanisms of transport of herbicides in plants, in *Target*

Sites for Herbicide Action, 2nd ed., Kirkwood, R. C., Ed., Plenum Press, New York, 1991, 245.

23. **Bromilow, R. H., Chamberlain, K., and Evans, A. A.,** Physicochemical aspects of phloem translocation of herbicides, *Weed Sci.*, 38, 305, 1990.
24. **Brown, H. M.,** Mode of action, crop selectivity, and soil relations of the sulfonylurea herbicides, *Pestic. Sci.*, 29, 263, 1990.
25. **Bruinsma, J.,** Uptake and translocation of 4,6-dinitro-O-cresol (DNOC) in young plants of winter rye (*Secale cereale* L.), *Acta Bot. Neerl.*, 16, 73, 1967.
26. **Buchan, G. D., Buick, R. D., and Field, R. J.,** The curious case of ... the changing wetness of droplets spreading on leaves, *WISPAS*, 55, 10, 1993.
27. **Buick, R. D.,** *Mode of Action of Organosilicone Surfactants in Enhancing the Performance of Triclopyr Herbicide*, Ph.D. thesis, Lincoln University, Canterbury, 1990, 350 pp.
28. **Buick, R. D., Buchan, G. D., and Field, R. J.,** The role of surface tension of spreading droplets in absorption of a herbicide formulation via leaf stomata, *Pestic. Sci.*, 38, 227, 1993.
29. **Buick, R. D. and Field, R. J.,** The mechanism of organosilicone surfactant-induced uptake of amine and ester formulations of triclopyr, in *Proc. First Int. Weed Control Congr.,* Vol. 2., Richardson, R. G., Ed., Monash University, Melbourne, Australia, 1992, 103.
30. **Buick, R. D., Field, R. J., and Robson, A. B.,** The effects of Silwet L-77® on triclopyr absorption, in *Proc. 43rd New Zealand Weed and Pest Control Conf.*, Popay, A. J., Ed., New Zealand Weed and Pest Control Society Inc, Palmerston North, New Zealand, 1990, 174.
31. **Buick, R. D., Field, R. J., Robson, A. B., and Buchan, G. D.,** A foliar uptake model of triclopyr, in *Adjuvants for Agrichemicals*, Foy, C. L., Ed., CRC Press, Boca Raton, FL, 1992, 87.
32. **Buick, R. D., Robson, B., and Field, R. J.,** A mechanistic model to describe organosilicon surfactant promotion of triclopyr uptake, *Pestic. Sci.*, 36, 127, 1992.
33. **Bukovac, M. J.,** Herbicide entry in plants, in *Herbicides, Physiology, Biochemistry, Ecology*, 2nd ed., Vol. 1., Audus, L. J., Ed., Academic Press, London, 1976, chap. 11.
34. **Bukovac, M. J., Rasmussen, H. P., and Shull, V. E.,** The cuticle: surface structure and function, *Scanning Electron Microscopyy*, 3, 213, 1981.

35. **Cantliffe, D. J. and Wilcox, G. E.,** Effect of surfactant on iron penetration through leaf wax and a wax model, *J. Am. Soc. Hort. Sci.*, 97, 360, 1972.
36. **Carter, D. H., Meyers, P. A and Lawrence Green, C.,** Temperature-activated release of trifluralin and diazinon, in *Pesticide Formulations and Application Systems*, Vol. 11th., Bode, L. E. and Chasin, D. G., Eds., American Society for Testing and Materials, Philadelphia, 1992, 57.
37. **Chaumet, E. and Chamel, A.,** Study of the cuticular retention and permeation of diuron using isolated cuticles of plants, *Plant Physiol. Biochem.*, 28, 719, 1990.
38. **Cook, G. T. and Duncan, H. J.,** Uptake of aminotriazole from humectant-surfactant combinations and the influence of humidity, *Pest. Sci.*, 9, 535, 1978.
39. **Cook, G. T. and Duncan, H. J.,** Foliar uptake enhancements by inorganic salts - an ion exchange approach, *Aspects Appl. Biol.*, 4, 371, 1983.
40. **Corbin, F. T., Upchurch, R. P., and Selman, F. L.,** Influence of pH on the phytotoxicity of herbicides in soil, *Weed Sci.*, 19, 233, 1971.
41. **Coupland, D.,** Factors affecting the phloem translocation of foliage-applied herbicides, in *Brit. Plant Growth Regul. Group, Monogr. 18*, Atkin, R. K. and Clifford, D. R., Eds., The British Plant Growth Regulator Group, London, 1988, 71.
42. **Coupland, D.,** Predicting and optimising the translocation of foliage-applied herbicides - a plant physiologist's perspective, *Proc. Brighton Crop Prot. Conf. - Weeds*, 2, 837, 1991.
43. **Coupland, D. and Caseley, J. C.,** Presence of ^{14}C activity in root exudates and guttation fluid from *Agropyron repens* treated with ^{14}C-labelled glyphosate, *New Phytologist*, 83, 17, 1979.
44. **Coupland, D. and Peel, A. J.,** The effect of temperature on the respiration of ^{14}C-labelled sugars in stems of willow, *Ann. Bot.*, 35, 9, 1971.
45. **Coupland, D. and Peel, A. J.,** Maleic hydrazide as an antimetabolite of uracil, *Planta*, 103, 249, 1972.
46. **Crisp, C. E.,** Pesticide Chemistry, in *Proc. 2nd IUPAC Congr.*, Vol. 1., Tahori, A. S., Ed., Gordon and Breach, New York, 1972, 211.
47. **Crisp, C. E. and Look, M.,** Advances in Pesticide Science, in *Proc. 4th IUPAC Congr.*, Vol. 3., Geissbühler, H., Ed., Pergamon Press, Oxford, 1979, 430.
48. **Currier, H. B. and Dybing, C. D.,** Foliar penetration of herbicides - review and present status, *Weeds*, 7, 195, 1959.
49. **Dailey, O. D., Dowler, C. C., and Glaze, N. C.,** Evaluation of cyclodextrin complexes of pesticides for use in minimization of groundwater contamination, in *Pesticide*

Formulation and Application Systems, Vol. 10, Bode, L. E., Hazen, J. L., and Chasin, D. G., Eds., American Society for Testing and Materials, Philadelphia, 1990, 26.

50. **Dastgheib, F.**, *Response of wheat cultivars to chlorsulforon and the effect of nitrogen availability*, Ph.D. thesis, Lincoln University, Canterbury, 1993.
51. **Dastgheib, F., Field, R. J., and Namjou, S.**, The mechanism of differential response of wheat cultivars to chlorsulfuron, *Weed Res.*, 34, 299, 1994.
52. **Devine, M. D.**, Phloem translocation of herbicides, *Rev. Weed Sci.*, 4, 191, 1989.
53. **Devine, M. D. and Bandeen, J. D.**, Fate of glyphosate in *Agropyron repens* (L.) Beauv. growing under low temperature conditions, *Weed Res.*, 23, 69, 1983.
54. **Devine, M. D. and Hall, L. M.**, Implications of sucrose translport mechanisms for the translocation of herbicides, *Weed Sci.* 38, 299, 1990.
55. **Devine, M. D. and Vanden Born, W. H.**, Absorption and transport in plants, in *Environmental Chemistry of Herbicides*, Vol. II., Grover, R. and Cessna, A. J., Eds., CRC Press, Boca Raton, FL., 1991, 119.
56. **Dunleavy, P. J., Cobb, A. H., Pallett, K. E., and Davies, L. G.**, The involvement of stomata in bentazone action in *Chenopodium album* L, *Proc. Brit. Crop Prot. Conf.*, 187, 1982.
57. **Epstein, E.**, Roots, *Sci. Am.*, 228 (5), 48, 1973.
58. **Esau, K.**, *Anatomy of Seed Plants*, 2nd ed., John Wiley, New York, 1977.
59. **Faird, J. F., Mercier, J. M., and Prevotat, M. L.**, Sucroglycerides: Novel biodegradable surfactants for plant protection formulations, in *Pesticide Formulations and Application Systems*, Vol. 12, Devisetty, B. N., Chasin, D. G., and Berger, P. D., Eds., American Society for Testing and Materials, Philidelphia, 1993, 33.
60. **Falatko, D. M. and Novak, J. T.**, Effect of biologically produced surfactants on the mobility and biodegradation of petroleum hydrocarbons, *Water Environ. Res.*, 64, 163, 1992.
61. **Felding, G.**, Leaching of atrazine into ground water, *Pestic. Sci.*, 35, 9, 1992.
62. **Field, R. J. and Bishop, N. G.**, The mechanism of action of Silwet L77 in improving the performance of glyphosate applied to perennial ryegrass, *Proc. 8th Austr. Weeds Conf.*, 411, 1987.
63. **Field, R. J. and Bishop, N. G.**, Promotion of stomatal infiltration of glyphosate by an organosilicone surfactant reduces the critical rainfall period, *Pestic. Sci.*, 24, 55, 1988.

64. **Field, R. J. and Peel, A. J.,** The movement of growth regulators and herbicides into the sieve elements of willow, *New Phytol.*, 70, 997, 1971.
65. **Field, R. J. and Peel, A. J.,** The longitudinal mobility of growth regulators and herbicides in sieve elements of willow, *New Phytol.*, 71, 249, 1972.
66. **Field, R. J., Dobson, N. N., and Tisdall, L. J.,** Species-specific sensitivity to organosilicone surfactant-enhancement of glyphosate uptake, in *Adjuvants for Agrichemicals*, Foy, C. L., Ed., CRC Press, Boca Raton, FL, 1992, chap. 40.
67. **Field, R. J. and Tisdall, L. J.,** The mechanism of organosilicone surfactant-induced antagonism of glyphosate uptake, in *Proc. 9th Austr. Weeds Conf.*, Heap, J. W., Ed., Crop Sci. Soc. South Australia (including Weed Science) Inc., Adelaide, South Australia, 1990, 332.
68. **Fleming, G. F., Wax, L. M., and Simmons, F. W.,** Leachability and efficacy of starch-encapsulated atrazine, *Weed Technol.*, 6, 297, 1992.
69. **Foy, C. L.,** Adjuvants: terminology, classification, and mode of action, in *Adjuvants and Agrochemicals*, Vol. I, Chow, P. N. P., Grant, C. A., Hinshalwood, A. M., and Simundsson, E., Eds., CRC Press, Boca Raton, FL, 1989, 1.
70. **Foy, C. L.,** Adjuvants for agrochemicals: introduction, historical overview, and future outlook, in *Adjuvants and Agrochemicals*, Vol. II, Chow, P. N. P., Grant, C. A., Hinshalwood, A. M., and Simundsson, E., Eds., CRC Press, Boca Raton, FL, 1989, 1.
71. **Foy, C. L., Whitworth, J. W., Muzik, T. J., and Currier, H. B.,** The penetration, absorption and translocation of herbicides, in *Environmental and Other Factors in the Response of Plants to Herbicides*, Oregon State University, Corvallis, 1967, 362.
72. **Franke, W.,** Ectodesmata and foliar absorption, *Am. J. Bot.*, 48(8), 683, 1961.
73. **Funari, E. and Bottoni, P.,** Groundwater contamination by herbicides - processes and evaluation criteria, in *Chemistry, Agriculture and the Environment*, Richardson, M. L., Ed., Birch Assessment Services for Information on Chemicals, London, 1991, 234.
74. **Gaskin, R. E. and Holloway, P. J.,** Some physicochemical factors influencing foliar uptake enhancement of glyphosate-mono(isopropylammonium) by polyoxyethylene surfactants, *Pestic. Sci.*, 34, 195, 1992.
75. **Gaskin, R. E. and Murray, B.,** Surfactant effects on activity of glyphosate and haloxyfop in pampas grass, *Proc. 41st New Zealand Weed Pest Control Conf.*, 153, 1988.

76. **Gaskin, R. E. and Stevens, P. J. G.,** Antagonism of the foliar uptake of glyphosate into grasses by organosilicone surfactants. 1. Effects of plant species, formulation, concentrations and timing of application, *Pestic. Sci.*, 38, 185, 1993.
77. **Gaskin, R. E. and Stevens, P. J. G.,** Antagonism of the foliar uptake of glyphosate into grasses by organosilicone surfactants. 2. Effects of surfactant structure and glycerol addition, *Pestic. Sci.*, 38, 193, 1993.
78. **Gaskin, R. E. and Zabkiewicz, J. A.,** Influence of plant development and adjuvant addition of glyphosate uptake and translocation in pampas grass, in *Proc. Eighth Austr. Weeds Conf.,* Lemerle, D. and Leys, A. R., Eds., Weed Society of New South Wales, Sydney, 1987, 416.
79. **Gaskin, R. E. and Zabkiewicz, J. A.,** Effect of Silwet L-77® on uptake and translocation of metsulfuron in gorse, *Proc. 41st New Zealand Weed Pest Control Conf.*, 128, 1988.
80. **Gaskin, R. E. and Zabkiewicz, J. A.,** The effect of surfactants on the uptake and translocation of glyphosate in Yorkshire fog, *Proc. 42nd New Zealand Weed Pest Control Conf.*, 42, 128, 1989.
81. **Geyer, V. and Schönherr, J.,** In vitro test for effects of surfactants and formulations on permeability of plant cuticles, in *Pesticide Formulations, Innovations and Developments,* Cross, B. and Scher, H. B., Eds., American Chemical Society, Washington, D.C., 1988, 22.
82. **Giaquinta, R. T.,** Phloem loading of sucrose, *Ann. Rev. Plant Physiol.*, 34, 347, 1983.
83. **Goddard, E. D. and Padmanabhan, P. A.,** A mechanistic study of the wetting, spreading and solution properties of organosilicone, in *Adjuvants for Agrichemicals,* Foy, C. L., Ed., CRC Press, Boca Raton, FL, 1992, 373.
84. **Graham-Bryce, I. J. and Hartley, G. S.,** The scope for improving pesticidal efficacy through formulations, in *Advances in Pesticidal Science Part 3.* IUPAC 4th Int. Congr. on Pest. Chem., Zurich, 1978, 718.
85. **Grayson, B. T. and Kleier, D. A.,** Phloem mobility of xenobiotics. IV. Modelling of pesticide movement in plants, *Pestic. Sci.*, 30, 299, 1990.
86. **Green, J. M., Brown, P. A., Berengut, D., and King, M. G.,** Nonionic surfactant property effects on thifensulfuron methyl performance in soybeans, in *Adjuvants for Agrichemicals,* Foy, C. L., Ed., CRC Press, Boca Raton, FL, 1992, 525.
87. **Haapala, E.,** The effect of a nonionic detergent on some plant cells, *Physiol. Plant.*, 23, 187, 1970.

88. **Hall, C., Edgington, L. V., and Siviter, C. M.,** Translocation of different 2,4-D, bentazon, diclofop or diclofop-methyl combinations in oat (*Avena sativa*) and soybean (*Glycine max*), *Weed Sci.*, 30, 767, 1982.

89. **Hall, J. K. and Hartwig, N. L.,** Triazine herbicide fate in a no-tillage corn (*Za mays* L.)-crownvetch (*Coronilla varia* L.) "living mulch" system, *Agric., Ecosyst. Environ.*, 30, 281, 1990.

90. **Hatzios, K. K. and Penner, D.,** Interaction of herbicides with other agrochemicals in higher plants. *Rev. Weed Sci.*, 1, 1, 1985.

91. **Hess, F. D.,** Herbicide absorption and translocation and their relationship to plant tolerance and susceptibility, in *Weed Physiology*, Vol. II, Duke, S. O., Ed., CRC Press, Boca Raton, 1985, 191.

92. **Higgins, J. M., Whitewell, T., Corbin, F. T., Carter, G. E., and Hill, H. S.,** Absorption, translocation and metabolism of acifluorfen and lactofen in pitted morningglory (*Ipomoea lacunosa*) and ivy leaf morningglory (*Ipomoea hederacea*), *Weed Sci.*, 36, 141, 1988.

93. **Ho, L. C.,** Metabolism and compartmentation of imported surgars in sink organs in relation to sink strength, *Ann. Rev. Plant Physiol. Plant Mol. Biol.*, 39, 355, 1988.

94. **Hodgson, R. H.,** Ed., *Adjuvants for Herbicides*, Monogr. 1, Weed Science Society of America, Champaign, IL, 1982.

95. **Holloway, P. J.,** Structure and histochemistry of plant cuticular membranes: an overview, in *The Plant Cuticle*, Cutler, D. F., Alvin, K. L., and Price, C. E., Eds., Academic Press (for Linnean Society of London), London, 1980, 1.

96. **Holloway, P. J. and Silcox, D.,** Behaviour of three nonionic surfactants following foliar application, *Br. Crop Prot. Conf. - Weeds*, 297, 1985.

97. **Holloway, P. J. and Stock, D.,** Factors affecting the activation of foliar uptake of agrochemicals by surfactants, *R. Soc. Chem.*, II, 303, 1989.

98. **Holloway, P. J., Wong, W. W.-C., Partridge, H. J., Seaman, D., and Perry, R. B.,** Effects of some nonionic polyoxyethylene surfactants on uptake of ethirimol and diclobutrazol from suspension formulations applied to wheat leaves, *Pestic. Sci.*, 34, 109, 1992.

99. **Holly, K.,** Selectivity in relation to formulation and application method, in *Herbicides: Physiology, Biochemistry and Ecology*, Vol. 2, Audus, L. J., Ed., Academic Press, London, 1976, 249.

100. **Horn, M. A., Heinstein, P. F., and Low, P. S.,** Receptor-mediated endocytosis in plant cells, *Plant Cell*, 1, 1003, 1989.

101. **Horn, M. A., Heinstein, P. F., and Low, P. S.,** Biotin-mediated delivery of exogenous macromolecules into soybean cells, *Plant Physiol.*, 76, 811, 1990.
102. **Hsu, F. C., Kleier, D. A., and Melander, W. R.,** Phloem mobility of xenobiotics. II. Bioassay testing of the unified mathematical model, *Plant Physiol.*, 86, 811, 1988.
103. **Humphreys, T. E.,** Phloem transport - with emphasis on loading and unloading, in *Mechanisms and Regulation of Transport Processes,* Vol. 18, Atkin, R. K. and Clifford, D. R., Eds., British Plant Growth Regulator Group Monograph, London, U.K., 1988, 305.
104. **Irons, S. M. and Burnside, O. C.,** Absorption, translocation and metabolism of bentazon in sunflower (*Helianthus annuus*), *Weed Sci.*, 30, 255, 1982.
105. **Isenee, A. R., Jones, G. E., and Turner, B. C.,** Root absorption and translocation of picloram by oats and soybeans, *Weed Sci.*, 19, 727, 1971.
106. **Jansen, L. L.,** Relation of structure of ethylene oxide ether-type nonionic surfactants to herbicidal activity of water-soluble herbicides, *J. Agric. Food Chem.*, 12, 223, 1964.
107. **Jermyn, J. M.,** *The Efficacy of the Organosilicone Surfactant, Silwet L77, and the Herbicide Glyphosate on a Range of Grass Species,* Honours dissertation, Lincoln University, New Zealand, 1993.
108. **Kanellopoulos, A. G.,** Additives in herbicide formulations, *Chem. Ind.*, 7, 951, 1974.
109. **Kirkwood, R. C.,** Pathways and mechanisms of uptake of foliage-applied herbicides with particular reference to the role of surfactants, in *Target Sites for Herbicide Action,* Kirkwood, R. C., Ed., Plenum Press, New York, 1991.
110. **Kirkwood, R. C.,** Use and mode of action of adjuvants for herbicides: a review of some current work, *Pestic. Sci.*, 38, 93, 1993.
111. **Kirkwood, R. C., McKay, I., and Livingstone, R.,** The use of model systems to study the cuticular penetration of ^{14}C-MCPA and ^{14}C-MCPB, in *The Plant Cuticle,* Cutler, D. F., Alvin, K. L., and Price, C. E., Eds., Academic Press, London, 1982, 253.
112. **Kleier, D. A.,** Phloem mobility of xenobiotics. I. Mathematical model unifying the weak acid and intermediate permeability theories, *Plant Physiol.*, 86, 803, 1988.
113. **Kydonieus, A. F.,** *Controlled Release Technologies: Methods, Theory and Applications,* Vol. 1, CRC Press, Boca Raton, FL, 1980.
114. **Ladlie, J. S., Meggitt, W. F., and Penner, D.,** Effect of pH on metribuzin activity in the soil, *Weed Sci.*, 24, 505, 1976.

115. **Lavy, T. L.,** Micromovement mechanisms of *s*-triazines in soil, *Soil Sci. Soc. Am. Proc.*, 32, 377, 1968.
116. **Leake, C. R.,** Fate of soil-applied herbicides: factors influencing delivery of active ingredients to target sites, in *Target Sites for Herbicide Action*, Kirkwood, R. C., Ed., Plenum Press, New York, 1991, 189.
117. **Leefe, J. S.,** Effect of soil pH on the phytotoxicity of simazine to strawberries, *Can. J. Plant Sci.*, 48, 424, 1968.
118. **Lindner, P.,** Effect of water in agricultural emulsions, in *Herbicides, Fungicides Formulation Chemistry*, Tahori, A. S., Ed., Gordon and Breach Science, London, 1972, 453.
119. **Linskens, H., Heinen, W., and Stoffers, A. L.,** Cuticula of leaves and the residue problem, *Res. Rev.*, 8, 136, 1965.
120. **Lord, W. G., Greene, D. W., and Emino, E. R.,** Absorption of silver nitrate and lead nitrate into leaves of McIntosh apples, *Can. J. Plant Sci.*, 59, 137, 1979.
121. **MacIsaac, S. A., Paul, R. N., and Devine, M. D.,** A scanning electron microscope study of glyphosate deposits in relation to foliar uptake, *Pestic. Sci.*, 31, 53, 1991.
122. **Manthey, F. A., Matysiak, R., and Nalewaja, J. D.,** Petroleum oil and emulsifier affect the phytotoxicity of imazethapyr, *Weed Technol.*, 6, 81, 1992.
123. **Manthey, F. A., Nalewaja, J. D., and Szelezniak, E. F.,** Esterified seed oils with herbicides, in *Adjuvants and Agrochemicals*, Vol. II, Chow, P. N. P., Grant, C. A., Hinshalwood, A. M., and Simundsson, E., Eds., CRC Press, Boca Raton, Fl, 1989, 139.
124. **Martin, J. T. and Juniper, B. E.,** *The Cuticles of Plants*, Edward Arnold, Edinburgh, 1970.
125. **Matsumoto, S., Suzuki, S., Tomita, H., and Shigematsu, T.,** Effect of humectants on pesticides uptake through plant leaf surfaces, in *Adjuvants and Agrochemicals*, Vol. II, Chow, P. N. P., Grant, C. A., Hinshalwood, A. M., and Simundsson, E., Eds., CRC Press, Boca Raton, FL, 1989, 261.
126. **McAuliffe, D. and Appleby, A. P.,** Activity loss of ethofumesate in dry soil by chemical degradation and adsorption, *Weed Sci.*, 32, 468, 1984.
127. **McCall, P. J., Stafford, L. E., and Gavit, P. D.,** Compartmental model describing the foliar behavior of tridiphane on giant foxtail, *J. Agric. Food Chem.*, 34, 229, 1986.
128. **McCann, A. W. and Whitehouse, P.,** More reliable herbicide performance: improvement through formulation, *Span*, 28(3), 98, 1985.
129. **McWhorter, C. G.,** The use of adjuvants, in *Adjuvants for Herbicides*, Hodgson, R.H., Ed., Monogr. 1, Weed Science Society of America, Champaign, IL, 1982, 10.

130. **McWhorter, C. G.,** The physiological effects of adjuvants on plants, *Weed Physiology*, Vol II, Duke, S. O., Ed., CRC Press, Boca Raton, FL, 1985, 141.

131. **McWhorter, C. G. and Barrentine, W. L.,** Spread of paraffinic oil on leaf surfaces of Johnsongrass (*Sorghum halepense*), *Weed Sci.*, 36, 111, 1988.

132. **Mersie, W. and Foy, C. L.,** Phytotoxicity and adsorption of chlorsulfuron as affected by soil properties, *Weed Sci.*, 33, 564, 1985.

133. **Moody, K., Kust, C. A., and Buchholtz, K. P.,** Uptake of herbicides by soybean roots in culture solutions, *Weed Sci.*, 18, 642, 1970.

134. **Nalewaja, J. D. and Matysiak, R.,** Species differ in response to adjuvants with glyphosate, *Weed Technol.*, 6, 561, 1992.

135. **Nalewaja, J. D. and Skrzypczak, G. A.,** Absorption and translocation of fluazifop with additives, *Weed Sci.*, 34, 572, 1986.

136. **Nalewaja, J. D. and Skrzypczak, G. A.,** Absorption and translocation of sethoxydim with additives, *Weed Sci.*, 34, 657, 1986.

137. **Nalewaja, J. D. and Matysiak, R.,** Salt antagonism of glyphosate, *Weed Sci.*, 39, 622, 1991.

138. **Nalewaja, J. D., Matysiak, R., and Freeman, T. P.,** Spray droplet residual of glyphosate in various carriers. *Weed Sci.*, 40, 576, 1992.

139. **Nelson, P. V. and Garlich, H. H.,** Relationship of chemical classification and hydrophile-liphophile balance of surfactants to enhancement of foliar uptake of iron, *J. Agric. Food Chem.*, 17, 148, 1969.

140. **Neumann, P. M. and Prinz, R.,** Evaluation of surfactants for use in the spray treatment of iron chloris in citrus trees, *J. Sci. Food Agric.*, 25, 221, 1974.

141. **Neumann, P. M. and Prinz, R.,** The effect of organosilicone surfactants in foliar nutrient sprays on increased adsorption of phosphate and iron salts through stomatal infiltration, *Isr. J. Agric. Res.*, 23(3-4), 123, 1974.

142. **Neumann, S. and Jacob, F.,** Aufnahme von Amino-buttersaure durch die Blätter von *Vicia faba*, *Naturwissenschaffen*, 55, 89, 1968.

143. **Nobel, P. S.,** *Biophysical Plant Physiology and Ecology*, W. H., Freeman,, San Francisco, 1983.

144. **Noga, G. J., Knoche, M., Wolter, M., and Barthlott, W.,** Changes in leaf micromorphology induced by surfactant application, *Angew. Bot.*, 61(5-6), 521, 1987.

145. **Norris, L. A. and Freed, V. H.,** The absorption and translocation characteristics of several phenoxyalkyl acid herbicides in bigleaf maple, *Weed Res.*, 6, 203, 1966.

146. **Norris, L. A. and Freed, V. H.,** The metabolism of a series of chlorophen-oxyalkyl acid herbicides in bigleaf maple, *Acer macrophyllum* Pursh., *Weed Res.*, 6, 212, 1966.

147. **Norris, L. A. and Freed, V. H.,** The absorption, translocation and metabolism characteristics of 4-(2,4-dichlorophenoxyl)butyric acid in bigleaf maple, *Weed Res.*, 6, 283, 1966.

148. **Norris, R. F.,** Surfactants and cuticular penetration, *Weed Sci. Soc. Am., Abstr.*, 11, 1971.

149. **Norris, R. F.,** Action and fate of adjuvants in plants, in *Adjuvants for Herbicides,* Hodgson, R. D., Ed., Monogr. 1, Weed Science Society of America, Champaign, IL, 1982, 68.

150. **Norris, R. F. and Bukovac, M. J.,** Structure of the pear leaf cuticle with special reference to cuticular penetration, *Am. J. Bot.*, 55, 975, 1968.

151. **Parker, T.,** Multipurpose Agrimax adjuvant system, *Proc. 10th Austr. and 14th Asian-Pacific Weed Conf.*, Weed Society of Queensland, Ed., Queensland, Australia, 1993, 73.

152. **Parr, J. F.,** Toxicology of adjuvants, in *Adjuvants for Herbicides,* Hodgson, R.D., Ed., Monogr.1, Weed Science Society of America, Champaign, IL, 1982, 93.

153. **Parr, J. F. and Norman, A. G.,** Considerations in the use of surfactants in plant systems: a review, *Bot. Gaz.*, 126(2), 86, 1965.

154. **Prasad, R. and Blackman, G. E.,** Studies in the physiological action of 2,2-dichloropropionic acid, II. The effect of light and temperature on the factors responsible for the inhibition of growth, *J. Exp. Bot.*, 16, 86, 1965.

155. **Prasad, R. and Blackman, G. E.,** Studies in the physiological action of 2,2-dichloropropionic acid, III. Factors affecting the level of accumulation and mode of action, *J, Exp. Bot.*, 16, 545, 1965.

156. **Price, C. E.,** Penetration and movements within plants of pesticides and other solutes: uptake mechanisms, *Reports of the Progress of Applied Biology, Agriculture, Food and General Microbiology,* 310, 1975.

157. **Price, C. E.,** A review of the factors influencing the penetration of pesticides through plant leaves, in *The Plant Cuticle,* Cutler, D. F., Alvin, K. L., and Price, C. E., Eds., Academic Press, London, 1982, 237.

158. **Riederer, M. and Schönherr, J.,** Effect of surfactants on water permeability of isolated plant cuticles and on the

composition of their cuticular waxes, *Pestic. Sci.*, 29, 85, 1990.

159. **Rigitano, R. L. O., Bromilow, R. H., Briggs, G. G., and Chamberlain, K.**, Phloem translocation of weak acids in *Ricinus communis*, *Pestic. Sci.*, 19, 113, 1987.
160. **Riley, D. and Morrod, R. S.**, Relative importance of factors influencing the activity of herbicides in soil, *Proc. 1976 Br. Crop Prot. Conf.*, 971, 1976.
161. **Ripley, B. D. and Edginton, L. V.**, Internal and external plant residues and relationships to activity of pesticides, *Proc. 10th Int. Congr. Plant Prot.*, 2, 545, 1983.
162. **Ripp, K. G., Viitanen, P. V., Hitz, W. D., and Franceschi, V. R.**, Identification of a membrane protein associated with sucrose transport into developing cells of developing soybean cotyledons, *Plant Physiol.*, 88, 1435, 1988.
163. **Robards, A. W. and Lucas, W. J.**, Plasmodesmata, *Ann. Rev. Plant Physiol. Plant Mol. Biol.*, 41, 369, 1990.
164. **Røyneberg, T., Balke, N. E., and Lund-Høe, K.**, Cycloxydim absorption by suspension-cultured velvetleaf (*Abutilon theophrasti* Medic.) cells, *Weed Res.*, 34, 1, 1994.
165. **Sagar, G. R., Caseley, J. C., Kirkwood, R. C., and Parker, C.**, Herbicides in plants, in *Weed Control Handbook: Principles*, 7th ed., Roberts, H. A., Ed., Blackwell Scientific, Oxford, 1982, chap. 3.
166. **Sargent, J. A. and Blackman, G. E.**, Studies on foliar penetration. I. Factors controlling the entry of 2,4-dichlorophenoxyacetic acid, *J. Exp. Bot.* 16, 24, 1962.
167. **Schmidt, R. R. and Pestemer, W.**, Plant availability and uptake of herbicides from the soil, in *Interaction Between Herbicides and the Soil*, Hance, R. J., Ed., Academic Press, London, 1980, chap. 7.
168. **Schönherr, J. and Bukovac, M. J.**, Preferential pathways in the cuticle and their relationship to ectodesmata, *Planta*, 92(3), 189, 1970.
169. **Schönherr, J. and Bukovac, M. J.**, Preferential pathways in the cuticle and their relationship to ectodesmata, *Plant Physiol.*, (Lancaster), 49, 813, 1972.
170. **Schönherr, J. and Riederer, M.**, Desorption of chemicals from plant cuticles: evidence for asymmetry, *Arch. Environ. Contam. Toxicol.*, 17, 13, 1988.
171. **Schönherr, J., Riederer, M., Schreiber, L., and Bauer, H.**, Foliar uptake of pesticides and its activation by adjuvants: Theories and methods for optimization, in *Pesticide Chemistry: Advances In International Research Development and Legislation, Proc. 7th Int. Cong. Pestic. Chem. (IUPAC),*

Frehse, H., Ed., VCH Verlagsgesellschaft mbH, Weinheim, Germany, 1991, 237.

172. **Schwarz, E. G. and Reid, W. G.,** Surface active agents - their behaviour and industrial use, *Ind. Eng. Chem.*, 56, 26, 1964.

173. **Scott, F. M., Bystrom, B. G., and Bowler, E.,** Root hairs, cuticle and pits, *Science*, 140, 63, 1963.

174. **Shafer, W. E. and Bukovac, M. J.,** Studies on octylphenoxy surfactants, III. Sorption of Triton X-100 by isolated tomato fruit cuticles, *Plant Physiol.*, 85, 965, 1987.

175. **Shafer, W. E. and Bukovac, M. J.,** Studies on octylphenoxy surfactants. 7. Effects of Triton X-100 on sorption of 21(1-Naphthy) acetic acid by tomato fruit cuticles, *Agric. Food Chem.*, 37, 486, 1989.

176. **Shafer, W. E., Bukovac, M. J., and Fader, R. G.,** Adjuvants and agrochemicals, in *Recent Developments, Application and Bibliography of Agro-adjuvants*, Vol. II., Chow, P. N. P., Grant, C. A., Hinshalwood, A. M., and Simundsson, E., Eds., CRC Press, Boca Raton, FL., 1989, 39.

177. **Sharma, M. P., Chang, F. Y., and Van den Born, W. H.,** Penetration and translocation of picloram in Canada thistle, *Weed Sci.*, 64, 519, 1971.

178. **Shasha, B. S., Doane, W. M., and Russell, C. R.,** Starch-encapsulated pesticides for slow release, *J. Polym. Sci., Polym. Lett. Ed.*, 14, 417, 1976.

179. **Shasha, B. S. and McGuire, M. R.,** Starch matrices for slow release of pesticides, in *Pesticide Formulations and Application Systems*, Vol. 11, Bode, L.E., and Chasin, D.G., Eds., American Society for Testing and Materials, Philadelphia, 1992.

180. **Shea, P.,** Chlorsulfuron dissociation and absorption on selected adsorbents and soils, *Weed Sci.*, 34, 474, 1986.

181. **Sherrick, S. L., Holt, H. A., and Hess, D. F.,** Absorption and translocation of MON 0818 adjuvant in field bindweed (*Convolvulus arvensis*), *Weed Sci.*, 34, 817, 1986.

182. **Shone, M. G. T., Bartlett, B. O., and Wood, A. V.,** Relationship between uptake by roots and translocation to shoots, *J. Exp. Bot.*, 25, 401, 1974.

183. **Shone, M. G. T. and Wood, A. V.,** Factors affecting absorption and translocation of simazinc by barlcy, *J. Exp. Bot.*, 23, 141, 1972.

184. **Silcox, D.,** *Studies on the Uptake, Translocation and Metabolism of Three Non-Ionic Ethoxylate Surfactants Following Foliar Application*, Ph.D. thesis, University of Bristol, 1988.

185. **Silcox, D. and Holloway, P. J.,** Foliar absorption of some nonionic surfactants from aqueous solutions in the absence

and presence of pesticidal active ingredients, in *Adjuvants and Agrochemicals*, Vol. I, Chow, P. N. P., Grant, C. A., Hinshalwood, A. M., and Simundsson, E., Eds., CRC Press, Boca Raton, FL, 1989, 115.

186. **Smith, E. M. and Treaster, S. A.,** Improving slow-release herbicide tablets for container nursery stock, *Combined Proc., Int. Plant Propagators Soc.*, 39, 214, 1989.
187. **Smith, L. W. and Foy, C. L.,** Herbicide activator, penetration and distribution studies in bean, cotton and barley from foliar and root application of Tween 20-C^{14}, fatty acid and oxyethylene labelled, *J. Agric. Food Chem.*, 14, 117, 1966.
188. **Smith, L. W., Foy, C. L., and Bayer, D. E.,** Structure-activity relationships of alkylphenol ethylene dioxide ether nonionic surfactants and three water-soluble herbicides, *Weed Res.*, 6, 233, 1966.
189. **Smith, J. W. and Sheets, T. J.,** Uptake, distribution and metabolism of monuron and diuron by several plants, *J. Agric. Food Chem.*, 15, 577, 1967.
190. **St. John, J. B. and Hilton, J. L.,** Surfactant effects on isolated plant cells, *Weed Sci.*, 22, 233, 1974.
191. **Steurbaut, W., Melkebeke, G., and Dejonckheere, W.,** The influence of nonionic surfactants on the penetration and transport of systemic fungicides in plants, in *Adjuvants and Agrochemicals*, Vol. I, Chow, P. N. P., Grant, C. A., Hinshalwood, A. M., and Simundsson, E., Eds., CRC Press, Boca Raton, FL, 1989, 93.
192. **Stevens, P. J. G.,** Organosilicone surfactants as adjuvants for agrochemicals, *Pestic. Sci.*, 38, 103, 1993.
193. **Stevens, P. J. G. and Baker, E. A.,** Factors affecting the foliar absorption and redistribution of pesticides. 1. Properties of leaf surfaces and their interactions with spray droplets, *Pestic. Sci.*, 19, 265, 1987.
194. **Stevens, P. J. G. and Bukovac, M. J.,** Studies on octylphenoxy surfactants. 2. Effect on foliar uptake and translocation, *Pestic. Sci.*, 20, 37, 1987.
195. **Stevens, P. J. G., Gaskin, R. E., Hong, Sung-O., and Zabkiewicz, J. A.,** Contributions of stomatal infiltration and cuticular penetration to enhancements of foliar uptake by surfactants, *Pestic. Sci.*, 33, 371, 1991.
196. **Stevens, P. J. G., Gaskin, R. E., and Zabkiewicz, J. A.,** Silwet L-77®: a new development in spray adjuvants, *Proc. 41st New Zealand Weed Pest Control Conf.*, 141, 1988.
197. **Stevens, P. J. G., Kimberley, M. O., Murphy, D.S., and Policello, G.A.** Adhesion of spray droplets to foliage: the role of dynamic surface tension and advantages of organosilicone surfactants, *Pestic. Sci.*, 38, 237, 1993.

198. **Stock, D., Edgerton, B. M., Gaskin, R. E., and Holloway, P. J.,** Surfactant-enhanced foliar uptake of some organic compounds: interactions with two model polyoxyethylene aliphatic alcohols, *Pestic. Sci.*, 34, 233, 1992.
199. **Stock, D. and Holloway, P. J.,** Possible mechanisms for surfactant-induced foliar uptake of agrochemicals, *Pestic. Sci.*, 38, 165, 1993.
200. **Stock, D., Holloway, P. J., Grayson, B. T., and Whitehouse, P.,** Development of a predictive uptake model to rationalise selection of polyoxyethylene surfactant adjuvants for foliage-applied agrochemicals, *Pestic. Sci.*, 37, 233, 1993.
201. **Stolzenberg, G. E.,** The analysis of surfactants and some of their plant metabolites, in *Adjuvants and Agrochemicals,* Vol. I, Chow, P. N. P., Grant, C. A., Hinshalwood, A. M., and Simundsson, E., Eds., CRC Press, Boca Raton, FL, 1989, 17.
202. **Stolzenberg, G. E., Olson, P. A., Zaylskie, R. G., and Mansager, E. R.,** Behaviour and fate of ethoxylated alkylphenol nonionic surfactant in barley plants, *J. Agric. Food Chem.*, 30, 637, 1982.
203. **Tadros, T. F.,** Interactions at interfaces and effects on transfer and performance, *Aspects Appl. Biol.*, 14, 1, 1987.
204. **Talbert, R. E. and Fletchall, O. H.,** The absorption of *s*-triazines in soils, *Weeds*, 13, 46, 1965.
205. **Taylor, F. E., Davies, L. G., and Cobb, A. H.,** Analysis of the epicuticular wax of *Chenopodium album* leaves in relation to environmental change, leaf wettability and the penetration of the herbicide, bentazone, *Ann. Bot.*, 98, 471, 1981.
206. **Thirunarayan, K., Zimbdahl, R. L., and Smike, D. E.,** Chlorsulfuron absorption and degradation in soil, *Weed Sci.*, 33, 558, 1985.
207. **Towne, C. A., Bartels, P. G., and Hilton, J. L.,** Interaction of surfactant and herbicide treatments on single cells of leaves, *Weed Sci.*, 26, 182, 1978.
208. **Turner, D. J.,** Effects of glyphosate performance of formulation, additives and mixing with other herbicides, in *The Herbicide Glyphosate*, Grossbard, E. and Atkinson, D., Eds., Butterworths, London, 1985, 221.
209. **Turner, D. J. and Loader, E.,** Effect of ammonium sulphate and other additives upon the phytotoxicity of glyphosate to *Agropyron repens L.* Beauv., *Weed Res.*, 20, 139, 1980.
210. **Tyree, M. T., Peterson, C. A., and Edgington, L. V.,** A simple theory regarding ambimobility of xenobiots with special reference to the nematicide oxamyl, *Plant Physiol.*, 63, 367, 1979.
211. **Urvoy, C. and Gauvrit, C.,** Seed oils as adjuvants: penetration of glycerol trioleate, methanol oleate diclofop-

methyl in maize leaves, *Brighton Crop Prot. Conf.* Weeds, 1, 337, 1991.

212. **Urvoy, C., Pollacsek, M., and Gauvrit, C.,** Seed oils as additives: penetration of trilein, methyloleate and diclofop-methyl in maize leaves, *Weed Res.* 32, 375, 1992.
213. **Van Bel, A. J. E.,** Strategies of phloem loading, *Ann. Rev. Plant Physiol. Plant Mol. Biol.*, 44, 253, 1993.
214. **Van Overbeek, J.,** Absorption and translocation of plant regulators, *Annu. Rev. Plant Physiol.*, 7, 355, 1956.
215. **Van Valkenburg, J. W.,** Terminology, classification and chemistry, in *Adjuvants for Herbicides*, Hodgson, R. D., Ed., Monogr. 1, Weed Science Society of America, Champaign, IL, 1982, 1.
216. **Warmbrodt, R. D., Buckhout, T. J., and Hitz, W. D.,** Localization of a protein, immunologically similar to a sucrose-binding protein from developing soybean cotyledons, *Planta*, 180, 105, 1989.
217. **Wauchope, R. D.,** The pesticide content of water draining from agricultural fields - a review, *J. Environ. Qual.*, 7, 459, 1978.
218. **Wauchope, R. D., Williams, R. G., and Marti, L. R.,** Runoff of sulfometuron-methyl and cyanazine from small plots: effects of formulation and grass cover, *J. Environ. Qual.*, 19, 119, 1990.
219. **Whitehouse, P., Holloway, P. J., and Caseley, J.,** The epicuticular wax of wild oats in relation to foliar entry of the herbicides diclofop-methyl and difenzoquat, in *The Plant Cuticle*, Cutler, D. F., Alvin, K. L., and Price, C. E., Eds., Academic Press, London, 1982, 315.
220. **Wills, G. D. and McWhorter, C. G.,** Absorption and translocation of herbicides: effect of environment, adjuvants, and inorganic salts, in *Pesticide Formulations: Innovations and Developments,* Cross, B. and Scher, H. B., Eds., ACS Symp. Ser. 371, American Chemical Society, New Orleans, LA, 1988, 90.
221. **Wing, R. E., Carr, M. E., Doane, W. M., and Schrieber, M. M.,** Starch encapsulated herbicide formulations: scale-up and laboratory evaluations, *Pesticide Formulations and Application Systems*, Vol. 11, Bode L.E. and Chasin, D.G., Eds., American Society for Testing and Materials, Philadelphia, 1992, 40.
222. **Yaron, B.,** General principles of pesticide movement to groundwater, *Agric. Ecosyst. Environ.*, 26, 275, 1989.
223. **Zabkiewicz, J. A., Coupland, D., and Ede, F.,** Effects of surfactants on droplet spreading and drying rates in relation to foliar uptake, in *Pesticide Formulations: Inovations and*

Developments, Cross, B. and Scher, H. B., Eds., ACS Symp. Ser. 371, American Chemical Society, New Orleans, 1988, 1.

224. **Zabkiewicz, J. A. and Gaskin, R. E.,** Effect of adjuvants on uptake and translocation of glyphosate in gorse (*Ulex europaeus* L.), in *Adjuvants and Agrochemicals,* Vol. I, Chow, P. N. P., Grant, C. A., Hinshalwood, A. M., and Simundsson, E., Eds., CRC Press, Boca Raton, FL, 1989, 141.
225. **Zabkiewicz, J. A., Gaskin, R. E., and Balneaves, J. M.,** Effect of additives on foliar wetting and uptake of glyphosate into gorse (*Ulex europaeus*), *Br. Crop Prot. Counc. Monogr.*, 28, 127, 1985.
226. **Zabkiewicz, J. A., Stevens, P. J. G., Forster, W. A., and Steele, K. D.**, Foliar uptake of organosilicone surfactant oligomers into bean leaf in the presence and absence of glyphosate, *Pestic. Sci.*, 38, 135, 1993.

Chapter 14

Herbicide Leaching and Its Prevention

Megh Singh and Siyuan Tan

CONTENTS

1-8493-7678-5/96/$0.00+$.50

ABSTRACT

Groundwater contamination resulting from leaching of pesticides is a growing concern in major agricultural regions of the world. Leaching of soil applied herbicides can be affected by edaphic factors, climatic conditions, physical-chemical properties of the active ingredient and formulations, crop management practices, and herbicide application techniques. Principal methods of studying pesticide leaching include leaching columns, soil thin-layer chromatography, lysimeter, residue monitoring, and computer modeling. For leaching prevention, it is usually difficult to modify climatic conditions and soil related factors in a given agricultural region. Therefore, focuses should be on selecting proper herbicides, improving formulations, adapting appropriate crop management practices, and modifying application techniques. Leaching characteristics of many herbicides have been well documented. Studies about impacts of crop managements on leaching are often inclusive. Limited information is available on leaching minimization by improving formulations and using adjuvants. However, studies in this area have increased considerably over the last decade. Both naturally occurring and synthetic additives or adjuvants have been tested to reduce herbicide leaching. Active charcoal, pine kraft lignins, and starch granulation have been shown to decrease leaching of herbicides. Synthetic polymers and surfactants have also shown potential for reducing leaching. The additives in formulations or adjuvants may act not only as controlled release agents that release herbicide in small amounts over prolonged period but also function as agents to inhibit movement of herbicides in soil. Another approach to reduce herbicide leaching is enhancing herbicide efficacy thus decreasing herbicide input and amount of herbicide available for leaching. Additional measures such as use of postemergence herbicides, application timing in relation to rainfall, and improvement of application technology may also contribute toward minimized leaching.

I. INTRODUCTION

There is growing public concern about effects of pesticide residues in groundwater.[7,104] Continued news media coverage of pesticides escalates the concern.[7] Groundwater contamination resulting from application of pesticides is a general and increasing problem in the major agricultural regions of the world.[27,28,35,53,83,97,117] The loss of pesticides to ground or surface water accounts for less than 5% of the amount applied.[53] However, pesticide residues in groundwater were increasing similarly to that of nitrate.[52] Leaching of pesticide metabolites and formulation impurities also raises concerns.[7]

Koterba et al.[70] reported that the most commonly detected residues were atrazine, cyanazine, simazine, alachlor, metolachlor, and dicamba. Concentrations were low with few exceeding 3 μg L^{-1}. Most detections correlate with the intensive use of these herbicides in widely distributed and commonly rotated crops, particularly when grown in well-drained soils. Detections often occurred in samples collected from shallow wells.

The most frequently detected herbicides in groundwater in western Europe were triazines, such as atrazine and simazine, and some of their transformation products, in concentrations ranging from 0.01 to 1 μg L^{-1}.[73] Schreiber et al. stated that herbicide atrazine was one of the most commonly detected contaminants in groundwater.[101] The highest levels of atrazine found appeared to be associated with heavy spring rains following its application.

A regional assessment of non-point-source contamination of pesticide residues in groundwater was made of the San Joaquin Valley.[32] About 10% of the total pesticide use in the U.S.is in the San Joaquin Valley. Pesticides detected included atrazine, bromacil, 2,4-DP, diazinon, dibromochloropropane, 1,2-dibromoethane, dicamba, 1,2-dichloropropane, diuron, prometon, prometryn, propazine, and simazine, all soil-applied except for diazinon.

Herbicide leaching not only can cause groundwater contamination but also can reduce efficacy by decreasing the amount of herbicide in surface soil layers where weed roots or seeds are located. For instance, propyzamide, alachlor, and carbetamide lost their effectiveness on dodder (*Cuscuta campestris* Yuncker) when applied during the rainy season due to intensive leaching.[46] Deeply leached pesticides tend to degrade more slowly and have greater potential to contaminate the groundwater than pesticides which remain in the surface soil. Pothuluri et al. reported that under aerobic conditions, the half-life of alachlor in surface soil was less than that in the vadose zone and aquifer samples.[89] The lower degradation rates in vadose zone and aquifer materials may be due to less microbial activity.

II. FACTORS AFFECTING LEACHING

Movement of herbicides in soils is a complex process affected by many factors.[19,121] The risk of groundwater contamination with herbicides depends on soil properties, agricultural practices, climatic influences, and properties of the herbicides themselves.[75,85,120,123]

A. HERBICIDES

1. Type of Herbicides

In a study involving four herbicides and two types of soils, it was found that the magnitude of leaching was dependent on the properties of herbicides and soil characteristics.[1] It has also been observed that the relative mobility of atrazine on soil thin-layer chromatography plates was greater than that of norflurazon.[59] In studies of several herbicides commonly used in citrus, it was found the leaching potential of the herbicides was in the order: bromacil > atrazine > simazine > terbumeton > terbuthylazine > terbutryn = diuron > trifluralin.[44] Metsulfuron-methyl was shown to be much more mobile than triasulfuron in two soils with different sorption capacities.[49]

Glufosinate ammonium was found to be slightly more mobile than amitrole and less mobile than picloram in Fox sandy loam and Guelph loam soils.[42] Two glufosinate ammonium metabolites, 3-(methyl-phosphinyl) propionic acid (MPPA-3) and 2-(methylphosphinyl) acetic acid (MPAA-2), were more mobile than the parent herbicide in the Fox sandy loam. In Guelph loam, the MPAA-3 metabolite was similarly more mobile than the parent herbicide, but the MPAA-2 metabolite was significantly less mobile than the parent herbicide in Guelph loam.

The leaching potential of atrazine, alachlor, cyanazine, and metribuzin to shallow groundwater was reported to depend on the chemical properties of herbicides (solubility, persistence, degradation, adsorption, uptake, mineralization, denitrification, and volatilization).[65]

2. Herbicide Water Solubility

The amount of herbicide that leached significantly following irrigation increased with the water solubility of each herbicide.[81] Beckie and McKercher used intact field soil cores to compare the mobility of DPX-A7881, a new sulfonylurea herbicide, and chlorsulfuron in four soils.[11] DPX-A7881 was generally less mobile than chlorsulfuron in soil. This was explained on the basis of the reduced water solubility of DPX-A7881 relative to chlorsulfuron (1.7 and 300 ppm, respectively, at pH 5.0). Mobility of five thiocarbamate herbicides was found to be directly correlated with water solubility.[47,69] Similar relationship between solubility and herbicide leaching has been observed.[6]

Rao et al. studied the leaching potential of 41 pesticides based on their relative travel time needed to migrate through the vadose zone and on the mass emission from the vadose zone.[94] Pesticides with solubilities exceeding 10 mg L^{-1} were founded to have the highest leaching potential. Examples were bromacil, terbacil, simazine, and cyanazine, which had considerable potential to contaminate groundwater. Nicholls believed that lipophilicity of pesticides was the most important physical-chemical property influencing the movement of un-ionized pesticides in the soil.[85] Anions and weak acids can be weakly adsorbed and hence might be subject to rapid leaching.[85]

3. Herbicide Half Life

Herbicide leaching is dependent on the half-life of the herbicide in the soil.[32] Truman and Leonard indicated that changes in herbicide half-life (persistence) in surface and subsurface horizons of different soils influenced potential herbicide leaching from the root zone.[116] The transport of dicamba in soils was greatly reduced if sufficient dicamba was degraded before irrigation or precipitation.[29] In their study on 41 pesticides, Rao et al. concluded that pesticides with half-lives longer than 50 d seemed to have the highest leaching potential.[94] Boesten and Linden observed that leaching of herbicides was very sensitive to their degradation rate: altering the degradation rate by a factor of two changed the magnitude of leaching typically by about a factor of ten.[20] Estimates of pesticide degradation rates in subsoils was essential for models predicting pesticide movement to groundwater.[89]

4. Application Rate and Pattern

Kotoula Syka et al.[71] found that leaching of chlorsulfuron, tribenuron-methyl, triasulfuron, and metsulfuron-methyl in sandy loam, sandy clay loam, and silty clay loam soils increased with increasing rate of application except for tribenuron-methyl in the sandy clay loams. They also observed that leaching of chlorsulfuron, metsulfuron, and triasulfuron in sandy loam, sandy clay loam, and silty clay loams increased with increasing concentration of application.[72] Horowitz and Elmore examined the leaching of oxyfluorfen in container media and found that raising the dose of oxyfluorfen from 20 to 200 ppm increased the depth of leaching and concentration of herbicide in the leachate.[58] Leaching was also dependent on use patterns of herbicides.[32]

B. SOIL PROPERTIES

1. Soil Types

Angemar et al. observed that leachability of bromacil in four soils was in the order: Sharon > Sa'ad > Newe Ya'ar > Hula.[5] Radder et al. reported that leaching of carbofuran and fenitrothion followed the order: red soil > lateritic soil > black soil > saline alkali soil.[91] The low rate of leaching of these pesticides in saline alkali soil was probably due to its low hydraulic conductivity (0.25 cm hr^{-1}) and its sorption. Truman and Leonard[116] noticed that potential pesticide leaching was greatest for Lakeland sand and least for Greenville sandy clay loam. Mueller and Banks observed that 75% applied ^{14}C-flurtamone remained in the 0- to 4-cm soil depth in Greenville sandy clay loam, with less than 5% moving to a depth > 4 cm.[79] However, in the Cecil loam and the Dothan loamy sand, flurtamone moved to a depth of 16 and 12 cm, respectively.

2. Organic Matter and Clay Content

Organic matter content is an important property of the soil for unionized pesticides.[85] Soil organic matter content influenced adsorption, phytotoxicity, and leachability of bromacil in various soils.[5] Biljon et al.[18] observed that the leaching depth of metolachlor was better correlated with carbon content than with clay and cation exchange capacity of the soil. Norflurazon mobility decreased and adsorption increased as soil organic matter and clay content increased.[59] Beckie and McKercher reported that enhanced mobility of DPX-A7881 and chlorsulfuron was related to low organic matter content of the soil.[11]

Organic carbon distribution in soil determined solute transport in the soil.[21] Immobile organic carbon provides a hydrophobic environment that could enhance sorption of neutral organic compounds from solution, resulting in decreased mobility of such compounds. Conversely, a mobile organic carbon phase, such as colloidal organic carbon dispersed in soil water, could act as a carrier of hydrophobic, neutral organic compounds thereby increasing their mobility.

In studies on the interactions of water-soluble soil organic matter with bromacil, metribuzin, alachlor, diquat, and paraquat, it was concluded that binding of these herbicides to water-soluble soil organic matter was not significant in increasing their mobility, thus their groundwater contamination potential.[88] However, gravitational displacement of soil colloids containing the adsorbed pendimethalin was attributed to the herbicide in the leachate.[108] Vink and Robert studied alachlor leaching in undisturbed soil columns and showed that alachlor behaved differently in topsoil and subsoil layers.[118] This may be primarily due to the difference in organic matter content between the soil layers.

3. Soil Texture

Pesticide leaching is dependent on soil texture.[32] Radder et al.[91] observed that leaching, movement, and distribution of carbofuran and fenitrothion in soil columns were greater in coarse-textured red and laterite soils than in fine-textured black and saline alkali soils. Leaching of imazaquin was found to be greater in sand than in clay.[66]

The fine-textured soil in the western San Joaquin Valley was found to inhibit pesticide leaching because of either low vertical permeability or high surface area, both enhancing adsorption onto solid phases.[32] Soils in the eastern part of the valley are coarse grained with low total organic carbon. Most pesticide leaching occurred in these alluvial soils, particularly in areas where depth to groundwater was less than 30 m. The lowest soil adsorption and the greatest movement of ethylmetribuzin and metribuzin were in coarser textured soils.[86] Atrazine was also found to move rapidly through sandy soil.[107] The effect of soil type on adsorption, mobility, and efficacy of imazaquin and imazethapyr was studied and the pesticides were found to be more efficacious and mobile in the more coarse-textured soils.[111]

The leaching depth of oxyfluorfen in container media was redwood bark and sand (3:1) mix > Yolo fine sandy loam > Stockton clay > peat and sand (1:1).[58] Depth of leaching was not related to organic matter content. Equilibration experiments showed that peat adsorbed four to five times more oxyfluorfen than did redwood bark, thus less leaching was observed in potting mixtures containing peat rather than redwood bark.

4. Sorption

Sorption characteristics of soils can affect the leaching of herbicides and their potential to pollute groundwater.[64] Peek and Appleby observed that metribuzin was absorbed less and moved more than ethyl-metribuzin in all soils.[86] Soil with low sorptive capacity showed little retention of metsulfuron-methyl and triasulfuron in the columns, while soil with high sorptive capacity retained triasulfuron for a long period, even under high flow conditions.[49]

Weber and Miller[121] reviewed the processes involved in the transport of organic chemicals in the environment. They concluded that chemicals with strong sorption by soils were less mobile than those with weak sorption of soils. Andrew et al.[4] also reported a reversal relationship between herbicide adsorption and mobility. They found that the order of adsorption of three herbicides by five Alabama soils was atrazine > metribuzin > chlorimuron, while the order of mobility was in the reverse order. Pesticide leaching was reported to be very sensitive to sorption (characterized by the organic-matter/water distribution coefficient): increasing the sorption by a factor of two decreased the magnitude of the leaching typically by about a factor of ten.[20]

Stougaard et al.[111] studied the effect of soil type on adsorption, mobility, and efficacy of imazaquin and imazethapyr and found that adsorption was greatest in the silty clay loam and least in the sandy loam soil. Greater adsorption was observed for flurtamone than atrazine in three soils.[79] The order of adsorption to soil for flurtamone and atrazine was Greenville sandy clay loam > Cecil loam > Dothan loamy sand. Greater adsorption of each herbicide corresponded to soils with greater organic matter and clay content.

5. Soil Moisture

Vink and Robert studied alachlor leaching in soil columns and found that leaching under saturated conditions was more rapid than leaching under unsaturated conditions, even though the same amount of water passed through the column.[118] Chemicals that are mobile can move both downward (leaching) and upward (capillary flow) in soils.[121] Capillary water caused considerable upward movement of norflurazon and atrazine in subirrigated columns containing herbicide-treated Hebert silt loam.[59]

6. Soil pH

The mobility of weak acids depends on soil pH.[85] Leaching of imazaquin increased with increasing pH, but was less than that of picloram under similar conditions.[66] The mobility of several sulfonylurea herbicides in four soils was found to increase at higher soil pH.[11]

In studies on the effect of pH on adsorption, mobility, and efficacy of imazaquin and imazethapyr, it was found that both herbicides were more strongly adsorbed, less mobile, and less efficacious at a lower pH.[111] These results were attributed to ionic bonding resulting from protonation of the herbicide's basic functional groups as pH decreased. In contrast, the greatest leaching of hexazinone and tebuthiuron was found to occur on sites with the lowest pH and base saturation.[110]

7. Macropore and Preferential Flow

Preferential (uneven) flow has been shown to be an important mechanism affecting water and solute movement in some soils. Alachlor, cyanazine, and pendimethalin were detected only in drainage from columns with a continuous macropore.[30] The leaching of herbicides dichlorprop and bentazon was generally greater in clay monoliths than in sand monoliths and this was explained in terms of macropore flow.[16] Strong evidence was found for preferential flow through the soil, with the chemicals by-passing much of the soil-matrix under recently plowed soils as well as no-till soils.[74]

Rice et al.[96] discovered that preferential flow phenomena resulted in solute and herbicide moving velocities 1.6 to 2.5 times faster than calculated by traditional water balance methods and piston flow models. Little preferential flow was observed under continuously flooded conditions on a loam soil. Generally, preferential flow occurred in coarse-textured soils, cracked soils, or in macropores such as root or worm holes.

Edwards et al.[34] investigated factors affecting preferential water and chemical transport in burrows formed by earthworms in a field subjected to simulated rainfall. Atrazine transport was affected by the factors influencing the amount of preferential flow, e.g., its movement was reduced by low-intensity rain prior to high-intensity, percolation-producing rain. Stehouwer et al. reported that earthworm burrows functioned as preferential flow conduits and enhanced downward transport of surface-applied pesticides.[109] Earthworm burrow walls have altered soil characteristics that might affect pesticide sorption and transport.

Zins et al. studied the movement of atrazine and alachlor through soil as affected by alfalfa (*Medicago sativa* L.) roots in three stages of decay and found greater preferential movement in columns with roots than in columns without roots.[128] Higher levels of atrazine and alachlor were bound to soil at lower depths in the presence of roots than without roots. Although only small amounts of the applied herbicide leached through the columns, preferential flow of herbicides through root-mediated soil macropores and cracks could be a mechanism of herbicide transport through soil under appropriate conditions.

C. PRECIPITATION

1. Time of Precipitation

Rainfall, irrigation, and uniformity of water flow in soil affected herbicide leaching.[19] Rainfall timing relative to herbicide application was critically important to herbicide leaching. Anderson[3] reported that movement of diclofop in soil was reduced if rainfall was delayed 4 d after application to a sand soil, indicating the need for precipitation within a few days after application. Edwards et al.[34] observed that atrazine transport in soils was affected by the time of storms relative to the time of herbicide application. Atrazine movement in earthworm burrows was greatest when high-intensity rainfall occurred shortly after application. Atrazine transport was reduced by a delay in rainfall. Rainfall timing also affected simulated pesticide loss by percolation, especially for nonpersistent pesticides.[116] For short pesticide half-life, excessive rainfall events within one half-life were largely responsible for simulated pesticide loss by percolation.

2. Amount of Precipitation

Herbicide soil movement is governed largely by the amount and frequency of the water applied and by the evapotranspiration potential of the soil.[121] However, Horowitz and Elmore[58] examined the leaching of oxyfluorfen in container media and found that a tenfold increase in water volume had only a limited effect on the leaching.

D. MANAGEMENT PRACTICES

1. Tillage and Cover Crops

Kanwar et al. found that the leaching potential of atrazine, alachlor, cyanazine, and metribuzin into shallow groundwater depended on the agricultural production systems (tillage systems and crop rotations).[65] Atrazine had consistently greater leaching under conventional-till than no-till field conditions.[60] Levanon et al. studied the impact of two contrasting tillage systems, plow-tillage and no-tillage, on the movement of agrichemicals in soil.[74] The agrichemicals used in the study were NH_4NO_3, atrazine, carbofuran, diazinon, and metolachlor. The results showed there were greater leaching losses of surface-applied agrichemicals to groundwater under plow-tillage than under no-tillage.

This is not consistent with the results of Clay et al.[24] who observed that twice as much alachlor leached from surface no-till than from surface conventional tillage columns. The differences in leaching patterns from the surface soil were attributed to the effect of tillage on soil physical and chemical properties. Similar results were obtained by Troiano et al. on atrazine and cyanazine.[115] Hall et al.[51] also reported that leaching of several herbicides was greater under no-tillage than under conventional tillage, while the reverse was true for runoff.

Taylor and Weber[113] investigated the effect of two tillage systems (conventional and reduced), three surface cover types (bermudagrass, soybean, and fallow), and two water input treatments (normal and double average rainfall) on the distribution of metolachlor. They found that reduced tillage + soybean + normal irrigation resulted in the greatest amount of metolachlor in the leachate (2.24%); by contrast, conventional tillage + soybean + double irrigation resulted in the least amount (0.01%).

Jones et al.[63] found that alachlor movement was greater in tilled plots compared to no-till plots. Metribuzin movement was greater in no-till plots. Straw cover had little effect on the movement of alachlor, but the presence of 2800 kg ha^{-1} of straw on the soil surface increased the downward movement of metribuzin compared to soil with no straw cover.

2. Irrigation

The greatest leaching of hexazinone and tebuthiuron were found to occur after high applications of water.[110] The accelerated leaching of solutes and a herbicide under intermittent flood and sprinkler irrigation has been reported,[96] and leaching has been shown to be enhanced by flood-irrigation except where the pesticide was foliar applied such as diazinon.[32]

Magnitude of atrazine leaching differed between the irrigation methods and increased in the order: sprinkler $<$ basin $<$ furrow. The total amount of applied water was similar, but sprinkler irrigations were more frequent, resulting in more evaporation and less water available for deep percolation.[115] Both irrigation amount and method were important in affecting herbicide movement.

III. EXPERIMENTAL METHODS FOR LEACHING STUDIES

A. LEACHING COLUMNS

Different approaches have been employed to study pesticide leaching.[54] Herbicide leaching through soil columns is believed to approximate field conditions more closely than most other methods.[123] Methods of studying herbicide leaching with soil columns have been described in detail by Weber et al.[122] In soil column studies, methods of water application could affect the extent of herbicide leaching.[123]

Czapar et al.[30] suggested that herbicides were not detected in column drainage unless a continuous macropore was present. Leaching studies using packed unstructured soil columns may significantly underestimate the extent of herbicide movement through a structured soil since they do not allow for preferential or macropore movement. Herbicide leaching experiments using soil columns with an artificial macropore may provide estimates that are more representative of field behavior. Others believed that soil column-leaching studies tended to show leaching to greater depths than that would occur under field conditions.[54]

B. SOIL THIN-LAYER CHROMATOGRAPHY

Soil thin-layer chromatography is also used to study herbicide leaching in soils.[42,48,59,62,88] It is easy to be standardized and is an ideal method to study herbicide upward movement in capillary water.

C. LYSIMETERS

Lysimeters have been used to estimate leaching of herbicides.[14,15] Outdoor lysimeter experiments take in account the comprehensive influence of environmental variables on the mobility and fate of a chemical, and give valid information on its potential for groundwater contamination. The interpretation of the studies has to consider that (a) migration behavior under environmental conditions does not correspond to 'ideal chromatographic behavior', and (b) lysimeter studies include the variables of field experiments and are not fully standardized.[67]

D. HERBICIDE RESIDUE MONITORING

Monitoring herbicide levels is another effective way to understand herbicide leaching characteristics. Data from groundwater monitoring studies are used both to determine the likelihood that a herbicide will leach, and to detect the presence of a herbicide in groundwater after years of use.[13] The residual level of a herbicide reflects the possibility of the herbicide leaching into the groundwater resulted from its using pattern and characteristics of downward movement in soils.

E. MODELING

Many pesticide simulation models are available and new ones are being developed to study pesticide behavior including leaching in soils.[87] Examples are Chemical Movement in Layered Soils (CMLS), Method of Underground Solute Evaluation (MOUSE), and Pesticide Root Zone Model (PRZM). The Pesticide Root Zone Model has been evaluated to predict pesticide movement in the soil.[100] Clemente et al.[25] described the development and verification of a new contaminant transport model, a one-dimensional transient mathematical model, and reported that there was a close agreement between the model's predictions and the analytical results. The model simulated simultaneous movement of water and solutes in the soil environment, as affected by runoff, leaching, dispersion, uptake, macropore flow, sorption/desorption, volatilization, heat flow, and chemical and microbial degradation.

IV. LEACHING PREVENTION MEASURES

As the concern about groundwater contamination grows, researchers are examining various approaches to reduce herbicide leaching. Schweizer[103] listed several possible approaches, both direct and indirect, to reduce herbicide leaching in soils. Indirect ways include two approaches. One is the development of new chemicals that are either less mobile in soil or need low application rates. The other is to use alternatives to chemicals such as microbial herbicides, allelochemicals, weed/crop modeling, herbicide resistant crops, and integrated pest management. Chemicals most likely to succeed as alternative herbicides are those that have a high adsorption capacity, low water solubility, and a short persistence in soil. For example, sulfonylureas are applied at very low rates and have less chance to reach groundwater.[103] Several naturally occurring products such as allelochemicals have the potential to become alternative herbicides.[23,33,90,103] The indirect approaches have limitations in terms of availability in the near future, efficacy, and cost. Direct approaches include the selection of less-leachable herbicides, formulation and adjuvant technologies, and improved herbicide application and farm management practices.

A. HERBICIDE SELECTION

The Soil Conservation Service has developed a formula to rank pesticides and soils based on their leaching potential.[45] The intent was to predict soil-pesticide interactions that field personnel could use to determine if a particular pesticide on a given soil could result in leaching. They presented a table in which pesticides were assigned to one of four categories of leaching potential. Overall leaching potential can thus be determined by the interaction of the leaching potential of the soil and that of the pesticide. The U.S. Environmental Protection Agency uses environmental transport and fate models to screen pesticides during the registration process.[12] Hornsby and Augustijin-Beckers developed practical guides and classification systems to characterize potentials for chemical mobility.[56,57]

Franklin et al.[41] used a simple computer code for microcomputers to predict persistence and migration of 17 herbicides through a hypothetical, coarse-textured soil. They concluded that data relating to soil and herbicide characteristics could not be used to override cost effectiveness and efficacy for weed control. They stated that steps currently being taken to ensure judicious pesticide use, including combining chemical and cultural controls to reduce the need for chemical inputs, are the most effective precautions a farmer can take to protect the environment. Franklin et al.[41] believed that farmers base their decisions largely on efficacy and cost when deciding which pesticide to use. They assume that if pesticides are labeled and available, they are safe to use. A decision to use a high-cost alternative based on the likelihood of less leaching to groundwater may not be justified, given the uncertainties of predicting the fate of pesticides in the environment.

B. HERBICIDE FORMULATIONS

Controlled release formulations may have promise for reducing the leaching of herbicides in soils.[102] Meyers et al.[77] reported that capsule suspensions, a controlled-release formulation that releases active ingredients in response to small changes in temperature, significantly decreased groundwater leaching potential of sulfonylurea and acetanilide herbicides. Dailey et al.[31] observed that cyclodextrin complexes of herbicides significantly decreased atrazine leaching but also reduced its efficacy.

Extensive research has been done in granulating herbicides with starch to reduce leaching. Considerable progress has been made in the last decade.[22,37,38,105,125] Fleming et al. found that starch encapsulation was more effective than polymer additives in retarding atrazine movement in soils even though polymer treatments reduced atrazine movement from the soil surface by 9 to 21% compared to atrazine without the additives.[37] Starch encapsulation retained 99% atrazine in the top 5 cm of the column compared to only 18 and 13% of the dry flowable formulation when 0.44 and 0.88-pore volumes of water were applied, respectively.

Larger starch granules had less atrazine leached from the surface 5 cm of soil than did the smaller starch granules.[37] However, unlike large granules, the small granules provided equivalent efficacy to the dry flowables formulation.[38] They concluded that finer starch granules should result in faster atrazine release and improved weed control while reducing the leaching potential of atrazine. The release rate of some herbicides and the amount of herbicide encapsulation can also be modified by changing the ratio of amylose to amylopectin of the starch.[127] Others have also reported the release rate was increased by decreasing granule size.[101,114,125]

The depth of atrazine leaching was reported to be considerably less when applied in a controlled release starch granule than in conventional dry granules.[102,119] Schreiber et al.[101] reported that starch-encapsulated granular formulations of atrazine significantly reduced the initial mobility of atrazine without appreciable loss of weed control efficacy. Leaching studies in the laboratory indicate that starch-encapsulated formulations reduced movement of atrazine by 70% compared to commercial formulations in both silty clay loam and sandy soils following a 75-mm hr^{-1} simulated rainfall. Starch-encapsulated formulations gave good to excellent control of most weeds at ten sites in six Midwest states and resulted in corn yields equal to or better than with commercial formulations. Schreiber et al.[101] concluded that starch-encapsulated formulations offered an excellent potential technology for significant reduction of atrazine and other pesticide contaminants of groundwater with minimal loss of efficacy.

Boydston[22] found that controlled release starch granules reduced the leaching of norflurazon and simazine. Barley (*Hordeum vulgare* L.) bioassays indicated norflurazon and simazine when leached with 6 cm water remained in the surface 0-2.5 cm of soil when applied as controlled release starch granule formulations, but moved to a depth of 15 cm when applied as commercial dry formulations.

Corn starch was used experimentally as a matrix in controlled release herbicide formulations[26,114,119,124,125,126,127] because it is inexpensive, nontoxic, and compatible with many herbicides, and the procedures for formulating herbicides in starch are simple and rapid.[126]

C. ADJUVANTS

Many adjuvants have been found to affect herbicide leaching. Both natural and synthetic adjuvants have been tested to reduce herbicide leaching.

1. Naturally Occurring Adjuvants

Certain pine kraft lignins were found to decrease leaching losses of herbicides.[98,99] A pine kraft lignin controlled the release of chloramben, metribuzin, and alachlor as measured by water leaching in soil columns. As more pine kraft lignin was used, more of the herbicides were retained in the top portion of the soil columns.[98]

Activated charcoal also could reduce herbicide leaching in soils.[2] Stone et al.[110] investigated the effects of litter-humus treatments and precipitation on mobility of herbicides hexazinone and tebuthiuron, and found that leaching of both herbicides was affected significantly by litter-humus treatment and the amount of applied rainwater.

Mullins et al. tested various types of biologically-based materials to use for removal of pesticides from aqueous solutions as a matrix.[80] They found that relatively high concentrations (5000 mg L^{-1}) of formulated chlorpyrifos and metolachlor could be removed by using those materials.

2. Synthetic Adjuvants

Disadvantages to the use of naturally occurring additives, such as starch, are the lack of reproducible product specifications and bio-erosion of the matrix in the soil, often resulting in inconsistent results.[61,82] In contrast, better specifications can be achieved easily with synthetic adjuvants.[61,82] Further, there are more synthetic adjuvant types available than naturally occurring ones.[82] It is also possible to synthesize an adjuvant to fit into a specific situation.

a. Polymers

Synthetic polymers were reported to reduce the leaching of some herbicides.[37,61,82,95] Narayanan et al.[82] reported some polymers could reduce leaching of certain herbicides in sandy soil up to 25%. They suggested that polymers inhibiting herbicide leaching all have an anchoring group, which can bind to the negatively charged silicates in the soil, and a balanced hydrophobic-hydrophilic group which will interact with herbicides. This possible binder function of polymers was also suggested for cationic surfactants as discussed in the following paragraphs.[21,106]

b. Surfactants

Surfactants were also found to have an impact on herbicide leaching. Koren[68] reported that three nonionic or mixed surfactants increased both depth of water penetration and herbicide movement into soil. There was no direct relationship between the ability of surfactants to improve water penetration and their effect on leaching of the tested herbicides. A leaching increase of 2,4-D by surfactants was observed by Helling.[55] Two nonionic surfactants were observed to have no influence on herbicide mobility at low concentrations but increased the mobility of some herbicides at higher concentrations. Three tertiary amine adjuvants were reported to slightly reduce movement of metribuzin in thin layer soil chromatography.[112] In a study of herbicide mobility as affected by surfactants, Foy[39] found that surfactant at high rates markedly enhanced herbicide mobility. Anionic, nonionic, and cationic surfactants caused variable effects on water movement and herbicide movement, depending on the herbicide, surfactant and its concentration, soil type, and preleaching conditions. Foy and Takeno[40] studied the impact of surfactants on methazole leaching in the soil, and found surfactants slightly retarded or had no effect on leaching.

Certain cationic surfactants were observed to reduce or prevent the leaching of substituted urea herbicides in soils, while other cationic surfactants had no effect or increased leaching.[9,10] Most of the tested nonionic surfactants increased the leaching regardless of surfactant concentrations and amount of water applied.[9,10] The reduction of herbicide leaching by cationic surfactants was attributed to an increase in adsorption of the herbicide onto the soil; nonionic surfactants did not affect adsorption.[106] An increased adsorption of picloram by soil was observed from cationic surfactant solutions than that from aqueous and anionic or nonionic surfactant solutions.[43] It has been suggested that cationic surfactants can replace metals on the mineral surface exchange complex due to ionic attraction of surfactants by soil particles.[21,106] The mineral surface of the soil is transformed from hydrophilic to hydrophobic by the presence of the hydrocarbon moiety of the sorbed surfactant cations. The result can be a great increase in mineral surface sorptivity for neutral organic compounds and a decrease in herbicide mobility. Cationic surfactants may be a promising solution to the problem of herbicide leaching.

D. LEACHING REDUCTION THROUGH IMPROVING MANAGEMENT PRACTICES

1. Irrigation

Ranjha et al.[92] compared impacts of alternative irrigation system designs, water, and other managements on potential groundwater contamination due to pesticides at several sites in Utah. They concluded that pesticide contamination of groundwater could be reduced by careful selection of pesticides, proper design of irrigation systems, and improved water management practices. They also reported a simulation of effects of furrow irrigation designs, water management practices, and pesticide parameters on pesticide leaching, and believed that potential groundwater contamination by pesticides could be reduced by the use of the best management practices.[93]

2. Application Technology

Improved application technology can reduce the amount of herbicide available for leaching. Use of air curtain sprayers, spinning-disc sprayers, electrostatic sprayers, and rope-wick applicators helped to prevent over-application.[76] Control applications, spraying only areas with target weeds rather than the complete field by using either weed detector or navigation technology,[8,36,78] provides the possibility of reducing total herbicide input and decreasing herbicide leaching.

3. Herbicide Handling

Neary[84] pointed out that the greatest hazards to groundwater and surface water quality arise from possible mishandling of pesticide products during transportation, storage, mixing-loading, equipment cleaning, and container disposal phases of the pesticide use.

4. Herbicide Sorbents

Guo et al.[50] determined the effect of waste-activated carbon, digested municipal sewage sludge, and animal manure on the sorption and leaching of alachlor. The digested municipal sewage sludge was found to be most effective in reducing the mobility of alachlor; waste-activated carbon was superior to animal manure. Amounts of alachlor recovered in the leachates depended upon the carbon loading rate of the wastes and the source of carbon-containing species. The dissipation of alachlor from soil was greater when soil was amended with digested municipal sewage sludge or animal manure than when amended with waste-activated carbon. Sorption of alachlor was generally inversely related to its leaching in the amended soil, suggesting that sorption by soil organic matter controlled the mobility of alachlor. Application of carbon-rich wastes to sandy coarse-textured soils may be useful for reducing herbicide leaching to groundwater.

5. Other Methods

In preventing leaching of herbicide waste, Berry et al.[17] reported that leachability of atrazine and carbofuran was dramatically reduced by a process called solid-state fermentation, a disposal technique for pesticides. The amount of leachable pesticide remaining in peat sorbent following bioreactor start-up was dependent on the initial pesticide loading rate and the length of bioreactor operation time. Based on reviews of factors affecting leaching of herbicides earlier in this chapter, we can take other possible measures to reduce the leaching. Examples are using postemergence herbicides, utilizing synergism of herbicides to reduce application rate, and changing cultivation methods. A successful strategy of leaching prevention should be the one to utilize potentials of all possible measures and to integrate them into the farm operation.

In summary, groundwater contamination resulting from herbicide leaching is a growing concern in agriculture. Herbicide leaching in soils is affected by properties of the herbicides, soil characteristics, climatic conditions, crop management practices, and methods of herbicide handling and application. Primary methods of studying herbicide leaching include leaching columns, soil thin-layer chromatography, lysimeter, residue monitoring, and computer modeling. Herbicide leaching can be reduced by choosing less-mobile herbicides, modifying formulations, improving farm management and herbicide-application practices, and using spray adjuvants. Most of the current approaches to decrease herbicide leaching have been emphasized on reduction of herbicide downward movement in soils. Little attention has been given to the approach of decreasing herbicide leaching through reducing herbicide input or available amounts of herbicides in soils. The reduction of available herbicides for leaching in soils can be achieved by enhancing efficacy of the herbicides or by using controlled-release formulations. Adjuvants have been demonstrated to have potentials for not only inhibiting herbicide downward movement but also acting as controlled-release agents for herbicides. Based on the evidences of adjuvant enhancing efficacy of foliar-applied herbicides, we may also expect adjuvants to improve efficacy of soil-applied herbicides.

V. APPENDIX

Table 1. Common and Chemical Names of Herbicides and Other Pesticides Mentioned in the Text.

Common name or designation	Chemical name
Alachlor	2-Chloro-*N*-(2,6-diethylphenyl)-*N*-(methoxymethyl) acetamide
Amitrole	1*H*-1,2,4-Triazol-3-amine
Atrazine	6-Chloro-*N*-ethyl-*N'*-(1-methylethyl)-1,3,5-triazine-2,4-diamine
Bentazon	3-(1-Methylethyl)-(l*H*)-2,1,3-benzothiadiazin-4(3*H*)-one-2,2-dioxide
Bromacil	5-Bromo-6-methyl-3-(1-methylpropyl)-2,4-(1*H*,3*H*) pyrimidinedione
Carbetamide	*N*-Ethy-2[[(phenylamino)carbonyl]oxy]propanamide
Carbofuran	2,3-Dihydro-2,2-dimethyl-7-benzofuranylmethylcarbamate
Chloramben	3-Amino-2,5-dichlorobenzoic acid
Chlorimuron	2-[[[[(4-Chloro-6-methoxy-2-pyrimidinyl)amino]carbonyl] amino]sulfonyl]benzoicacid
Chlorpyrifos	*O*,*O*-Diethyl-*O*-(3,5,6-trichloro-2-pyridyl) phosphorothioate
Chlorsulfuron	2-Chloro-*N*-[[(4-methoxy-6-methyl-1,3,5-triazin-2-yl)amino]carbonyl]benzenesulfonamide
Cyanazine	2-[[4-Chloro-6-(ethylamino)-1,3,5-triazin-2-yl]amino] -2-methylpropanenitrile
2,4-D	(2,4-Dichlorophenoxy)acetic acid
2,4-DP	2-(2,4-Dichlorophenoxy)propionic acid
DPX-A7881	2-[[[[[4-Ethoxy-6-(methylamino)-1,3,5-triazin-2-yl]amino]carbonyl]amino]sulfonyl]benzoate
Diazinon	0,0-Diethyl-*O*-(6-methyl-2(1-methethyl)-4-pyramidinyl phosphorothioate
Dibromochloropropane	1,2-Dibromo-3-chloropropane

Table 1. Continued.

Common name or designation	Chemical name
Dicamba	3,6-Dichloro-2-methoxybenzoic acid
Dichlorprop	(±)-2-(2,4-Dichlorophenoxy)propanoic acid
Diclofop	(±)-2-[4-(2,4-Dichlorophenoxy)phenoxy]propanoic acid
Diquat	6,7-Dihydrodipyrido[1,2-∝:2',1'-*c*]pyrazinediium ion
Diuron	*N*'-(3,4-Dichlorophenyl)-*N*,*N*-dimethylurea
Fenitrothion	*O*,*O*-Dimethyl-*O*-4-nitro-m-tolylphosphorothioate
Flurtamone	(±)-5-(Methylamino)-2-phenyl-4-[3-(trifluoromethyl) phenyl]-3(2*H*)-furanone
Glufosinate	2-Amino-4-(hydroxymethylphosphinyl)butanoic acid
Hexazinone	3-Cyclohexyl-6-(dimethylamino)-1-methyl-1,3,5-triazine-2,4(1*H*,3*H*)-dione
Imazaquin	2-[4,5-Dihydro-4-methyl-4-(1-methylethyl)-5-oxo-1*H*-imidazol-2-yl]-3-quinolinecarboxylic acid
Imazethapyr	2-[4,5-Dihydro-4-methyl-4-(1-methylethyl)-5-oxo-1*H*-imidazol-2-yl]-5-ethyl-3-pyridinecarboxylic acid
Methazole	2-(3,4-Dichlorophenyl)-4-methyl-1,2,4-oxadiazolidine-3,5-dione
Metolachlor	2-Chloro-*N*-(2-ethyl-6-methylphenyl)-*N*-(2-metboxy -1-methylethyl)acetamide
Metribuzin	4-Amino-6-(1,1-dimethylethyl)-3-(methylthio)-1,2,4-triazin-5(4*H*)-one
Metsulfuron	2-[[[[(4-Methoxy-6-methyl-1,3,5-triazin-2-yl)amino]carbonyl]amino]sulfonyl]benzoic acid
Norflurazon	4-Chloro-5-(metbylamino)-2-(3-(trifluoromethyl) phenyl)-3(2*H*)-pyridazinone
Oxyfluorfen	2-Chloro-1-(3-ethoxy-4-nitrophenoxy)-4-(trifluoromethyl) benzene
Paraquat	1,1'-Dimethyl-4,4'-bipyriduniumion
Pendimethalin	*N*-(1-Ethylpropyl}-3,4-dimethyl-2,6-dinitrobenzenamine
Picloram	4-Amino-3,5,6-trichloro-2-pyridinecarboxylic acid
Prometon	6-Methoxy-*N*,*N*'-bis(1-methylethyl)-1,3,5-triazine-2,4-diamine
Prometryn	*N*,*N*'-Bis(1-methylethyl)-6-(methylthio)-1,3,5-triazine-2,4-diamine
Propazine	6-Chloro-*N*,*N*'-bis(1-methylethyl)-1,3,5-triazine-2,4-diamine
Propyzamide	3,5-Dichloro-*N*-(1-dimethyl-2-propynyl)benzamide
Simazine	6-Chloro-*N*,*N*'-diethyl-1,3,5-triazine-2,4-diamine
Tebuthiuron	*N*-[5-(1,1-Dimethylethyl)-1,3,4-thiadiazol-2-yl] -*N*,*N*'-dimethylurea
Terbacil	5-Chloro-3-(1,1-dimethylethyl)-6-methyl-2,4-(1*H*,3*H*)-pyrimidinedione
Terbumeton	2-Tert-butylamino-4-ethylamino-6-methoxy-s-triazine
Terbuthylazine	4-Tert-butylamino-2-chloro-6-ethylamino-s-triazine
Terbutryn	*N*-(1,1-Dimethylethyl)-*N*'-ethyl-6-(methylthio)-1,3,5-triazine-2,4-diamine

Table 1. Continued.

Common name or designation	Chemical name
Triasulfuron	2-[[[[[4-(Dimethylamino)-6-(2,2,2-trifluoroethoxy)-1,3,5-triazin-2-yl]amino]carbonyl]amino]sulfonyl]-3-methylbenzoic acid
Tribenuron	2-[[[[(4-Methoxy-6-methyl-1,3,5-triazin-2-yl) methylamino]carbonyl]amino]sulfonyl]benzoic acid
Trifluralin	2,6-Dinitro-*N*,*N*-dipropyl-4-(trifluoromethyl)benzenamine

AKNOWLEDGMENT

Florida Agricultural Experiment Station Journal Series No. R-04343.

REFERENCES

1. **Alva, A. K. and Singh, M.,** Sorption of bromacil, diuron, norflurazon, and simazine at various horizons in two soils, *Bull. Environ. Contam. Toxicol.*, 45, 365, 1990.
2. **Alva, A. K. and Singh, M.,** Use of adjuvants to minimize leaching of herbicides in soil, *Environ. Manage.*, 15, 263, 1991.
3. **Anderson, R. L.,** Factors affecting preemergence bioactivity of diclofop: rainfall, straw retention, and plant growth stage, *Agron. J.*, 80, 952, 1988.
4. **Andrew, J. G., Walker, R. H., Glenn, W., and Hajek, B. F.,** Sorption and mobility of chlorimuron in Alabama soils, *Weed Sci.*, 37, 428, 1989.
5. **Angemar, Y., Rebhun, M., and Horowitz, M.,** Adsorption, phytotoxicity, and leaching of bromacil in some Israeli soils, *J. Environ. Qual.*, 13, 321, 1984.
6. **Baker, H. R., Talbert, R. E., and Frans, R. E.,** The leaching characteristics and field evaluation of various derivatives of amiben for weed control in soybeans, *Proc. South. Weed Sci. Soc.*, 19, 117, 1966.
7. **Barrett, M. R., Williams, W. M., and Wells, D.,** Use of ground water monitoring data for pesticide regulation, *Weed Technol.*, 7, 238, 1993.
8. **Barton, K.,** A new age of weed control, *Citrus & Vegetable Magazine,* 20, 1993.
9. **Bayer, D. E.,** Effect of surfactants on leaching of substituted urea herbicides in soil, *Weeds*, 15, 246, 1967.
10. **Bayer, D. E. and Foy, C. L.,** Action and fate of adjuvants in soils, in *Adjuvants for Herbicides,* Hodgson, R.H., Ed., Monograph I, Weed Sci. Soc. Amer., Champaign, IL, 1982, 84 pp.
11. **Beckie, H. J. and McKercher, R. B.,** Mobility of two sulfonylurea herbicides in soil, *J. Agric. Food Chem.*, 38, 310, 1990.
12. **Behl, E.,** Computer models for rate assesement during the registration process: Data needs, *Weed Technol.*, 6, 696, 1992.
13. **Behl, E. and Eiden, C. A.,** Field-scale monitoring studies to evaluate mobility of pesticides in soils and groundwater, *Am. Chem. Soc. Symp. Ser. 465*, 27, 1991.
14. **Bergstrom, L.,** Use of lysimeters to estimate leaching of pesticides in agricultural soils, *Environ. Pollut.*, 67, 325, 1990a.
15. **Bergstrom, L. F.,** Leaching of chlorsulfuron and metsulfuron-methyl in three Swedish soils measured in field lysimeters, *J. Environ. Qual.*, 19, 701, 1990b.

16. **Bergstrom, L. F. and Jarvis, N. J.,** Leaching of dichlorprop, bentazon, and ^{36}Cl in undisturbed field lysimeters of different agricultural soils, *Weed Sci.*, 41, 251, 1993.
17. **Berry, D. F., Tomkinson, R. A., Hetzel, G. H., Mullins, D. E., and Young, R. W.,** Evaluation of solid-state fermentation techniques to dispose of atrazine and carbofuran, *J. Environ. Qual.*, 22, 366, 1993.
18. **Biljon van, J. J., Nel, P. C., and Groeneveld, H. T.,** Leaching of metolachlor in some South African soils, *Proc. 1st Int. Weed Control Congr.*, 2, 86, 1992.
19. **Boesten, J. J. T. I.,** Leaching of herbicides to ground water: A review of important factors and of available measurements, *1987 Br. Crop Prot. Conf. Weeds*, 559, 1987.
20. **Boesten, J. J. T. I. and Linden, A. M. A.,** Modeling the influence of sorption and transformation on pesticide leaching and persistence, *J. Environ. Qual.*, 20, 425, 1991.
21. **Bouchard, D. C., Enfield, C. G., and Piwoni, M. D.,** Transport processes involving organic chemicals, in *Reactions and Movement of Organic Chemicals in Soils*, Sawhney, B. L. and Brown, K., Eds., Soil Sci. Soc. Am., Inc., Madison, WI, 1989, 329 pp.
22. **Boydston, R. A.,** Controlled release starch granule formulations reduce herbicide leaching in soil columns, *Weed Technol.*, 6, 317, 1992.
23. **Brandt, A., Bresler, E., Diner, N., Ben-Asher, I., Heller, J., and Goldberg, D.,** Infiltration from a trickle source. I. Matematical models, *Soil Sci. Soc. Am. Proc.*, 35, 675, 1971.
24. **Clay, S. A., Koskinen, W. C., and Carlson, P.,** Alachlor movement through intact soil columns taken from two tillage systems, *Weed Technol.*, 5, 485, 1991.
25. **Clemente, R. S., Prasher, S. O., and Barrington, S. F.,** PESTFADE, a new pesticide fate and transport model: model development and verification, *Trans. Am. Soc. Agric. Eng.*, 36, 357, 1993.
26. **Coffman, C. B. and Gentner, W. A.,** Persistence of several controlled release formulations of trifuralin in greenhouse and field, *Weed Sci.*, 28, 21, 1980.
27. **Cohen, S. Z., Carsel, R. F., Creeger, S. M., and Enfield, G. G.,** Potential for pesticide contamination of groundwater resulting from agricultural uses, in *Treatment and Dispersal of Pesticide Wastes,* Krueger, F. F. and Seiker, J. N., Eds., Am. Chem. Soc., 1984, 297 pp.
28. **Cohen, S. Z., Eiden, C., and Lorber, M. N.,** Monitoring ground water for pesticides, in *Evaluation of Pesticides in Groundwater,* Garner, W. Y., Honeycut, R. C., and Nigg, H. N. Eds., ACS Symp. Ser. No. 315, Am. Chem. Soc., Washington, D.C., 1986, 170 pp.
29. **Comfort, S. D., Inskeep, W. P., and Macur, R. E,** Degradation and transport of dicamba in a clay soil, *J. Environ. Qual.*, 21, 653, 1994.
30. **Czapar, G. F., Horton, R., and Fawcett, R. S.,** Herbicide and tracer movement in soil columns containing an artificial macropore, *J. Environ. Qual.*, 21, 110, 1992.
31. **Dailey, O. D., Jr., Dowler, C. C., and Glaze, N. C.,** Evaluation of cyclodextrin complexes of pesticides for use in minimization of groundwater contamination, in *Pesticide Formulations and Application Systems: 10th Vol. STP 1078*, ASTM, Philadelphia, PA, 1990, 26 pp.
32. **Domagalski, J. L. and Dubrovsky, N. M.,** Pesticide residues in ground water of the San Joaquin Valley, California, *J. Hydrol.*, 130, 299, 1992.
33. **Duke, S. O.,** Naturally occurring chemical compounds as herbicides, *Rev. Weed Sci.*, 2, 15, 1986.
34. **Edwards, W. M., Shipitalo, M. J., Owens, L. B., and Dick, W. A.,** Factors affecting preferential flow of water and atrazine through earthworm burrows under continuous no-till corn, *J. Environ. Qual.*, 22, 453, 1993.
35. **Felding, G.,** Leaching of atrazine into ground water, *Pestic. Sci.*, 35, 39, 1992.

36. **Felton, W. C., McClory, K. R., Doss, A. F., and Burger, A. E.**, Evaluation of a weed detector, *Proc. 8th Austr. Weeds Conf.*, 8, 80, 1987.
37. **Fleming, G. F., Simmons, F. W., Wax, L. M., Wing, R. E., and Carr, M. E.**, Attrazine movement in soil columns as influenced by starch-encapsulation and acrylic polymer additives, *Weed Sci.*, 40, 465, 1992a.
38. **Fleming, G. F., Wax, L. M., Simmons, F. W., and Felsot, A. S.**, Movement of alachlor and metribuzin from controlled release formulations in a sandy soil, *Weed Sci.*, 40, 606, 1992b.
39. **Foy, C. L.**, Influence of certain surfactants on the mobility of selected herbicides in soil, in *Adjuvants for Agrichemicals,* Foy, C. L., Ed., CRC Press, Boca Raton, FL, 1992, 349 pp.
40. **Foy, C. L. and Takeno, T.**, Effect of polysorbate surfactants with various hydrophylic-lipophilic balance (HLB) values on leaf surface ultrastructure and mobility of methazole in plants and soil, in *Adjuvants for Agrichemicals,* Foy, C. L., Ed., CRC Press, Boca Raton, FL, 1992, 169 pp.
41. **Franklin, R. E., Quisenberry, V. L, Gossett, B. J., and Murdock, E. C.**, Selection of herbicide alternatives based on probable leaching to groundwater, *Weed Technol.*, 8, 6, 1994.
42. **Gallina, M. A. and Stephenson, G. R.**, Dissipation of ^{14}C glufosinate ammonium in two Ontario soils, *J. Agric. Food. Chem.*, 40, 165, 1992.
43. **Gaynor, J. D. and Volk, V. V.**, Surfactant effects on picloram adsorption by soils, *Weed Sci.*, 24, 549, 1976.
44. **Gomez-de-Barreda, D., Gamon, M., Lorenzo, E., and Saez, A.**, Residual herbicide movement in soil columns, *Sci. Total Environ.*, 132, 155, 1993.
45. **Goss, D. W.**, Screening procedures for soils and pesticides for potenial water quality impacts, *Weed Technol.*, 6, 701, 1992.
46. **Graph, S., Herzlinger, G., Kleifeld, Y., Bargutti, A., Retig, B., and Lehrer, W.**, Control of dodder in chickpeas, tomatoes and pumpkins, *Phytoparasitica*, 13, 243, 1985 (Abstr.).
47. **Gray, R. A. and Weirich, A., J.**, Leaching of five thiocarbamate herbicides in soils, *Weed Sci.*, 16, 77, 1968.
48. **Gruber, V. F., Halley B. A., Hwang, S. C., and Ku, C. C.**, Mobility of avermectin B1a in soil, *J. Agric. Food Chem.*, 38, 886, 1990.
49. **Gunther, P., Pestemer, W., Rahman, A., and Nordmeyer, H.**, A bioassay technique to study the leaching behaviour of sulfonylurea herbicides in different soils, *Weed Res.*, 33, 177, 1993.
50. **Guo, L., Bicki, T. J., Felsot, A. S., and Hinesly, T. D.**, Sorption and movement of alachlor in soil modified by carbon-rich wastes, *J. Environ. Qual.*, 22, 186, 1993.
51. **Hall, J. K., Mumma, R. O., and Watts, D. W.**, Leaching and runoff losses of herbicides in a tilled and untilled field, *Agric. Ecosys. Environ.*, 37, 303, 1991.
52. **Hallberg, G. R.**, Groundwater quality and agricultural chemicals: a perspective from Iowa, *Proc. North Cent. Weed Control Conf.*, 40, 130, 1985.
53. **Hallberg, G. R.**, Agricultural chemicals in ground water: extent and implications, *Am. J. Alternative Agric.*, 2, 3, 1988.
54. **Hamaker, J. W.**, The interpretation of soil leaching experiments, in *Environmental Dynamics of Pesticides*, Haque, R. and Freed, V. H., Eds., Plenum Press. NY, 1976, 115 pp.
55. **Helling, C. S.**, Pesticide mobility in soils. 2. applications of soil thin-layer chromatography, *Soil Sci. Soc. Am. Proc.*, 35, 732, 1971.
56. **Hornsby, A.**, Site-specific pesticide recommendations: The final step in environmental impact prevention, *Weed Technol.*, 6, 736, 1992.

57. **Hornsby, A. G. and Augustijin-Beckers, P. W.**, *Handbook, Managing Pesticides for Crop Production and Water Quality,* Florida Coop. Ext. Ser., Inst. Food Agric. Sci., Univ. Florida, Gainesville, FL, 1991.

58. **Horowitz, M. and Elmore, C. L.**, Leaching of oxyfluorfen in container media, *Weed Technol.*, 5, 175, 1991.

59. **Hubbs, C. W. and Lavy, T. L.**, Dissipation of norflurazon and other persistent herbicides in soil, *Weed Sci.*, 38, 81, 1990.

60. **Isensee, A. R. and Sadeghi, A. M.**, Laboratory apparatus for studying pesticide leaching in intact soil cores, *Chemosphere*, 25, 581, 1992.

61. **Jain, R. and Singh, M.**, Effect of a synthetic polymer on adsorption and leaching of herbicides in soil, in *Adjuvants for Agrichemicals,* Foy, C. L., Ed., CRC Press, Boca Raton, FL, 1992, 329 pp.

62. **Jamet, P. and Thoisy Dur, J. C.**, Pesticide mobility in soils: assessment of the movement of isoxaben by soil thin-layer chromatography, *Bull. Environ. Contam. Toxicol.*, 41, 135, 1988.

63. **Jones, R. E., Jr., Banks, P. A., and Radcliffe, D. E.**, Alachlor and metribuzin movement and dissipation in a soil profile as influenced by soil surface condition, *Weed Sci.*, 38, 589, 1990.

64. **Jury, W. A, Focht, D. C., and Farmer, W. J.**, Evaluation of pesticide groundwater pollution potential from standard indices of soil chemical adsorption and degradation, *J. Environ. Qual.*, 16, 422, 1987.

65. **Kanwar, R. S., Baker, D. G., Czapar, G. F., Ross, K. W., Shannon, D., and Honeyman, M.**, A field monitoring system to evaluate the impact of agricultural production practices on surface and ground water quality, *Proc. Sixth Int. Drainage Symp.*, St. Joseph, MI, 361, 1992.

66. **Ketchersid, M. L. and Merkle, M. G.**, Effect of soil type and pH on persistence and phytotoxicity of imazaquin (Scepter), *Proc. South. Weed Sci. Soc.*, 39, 420, 1986.

67. **Kordel, W., Herrchen, M., and Klein, W.**, Experimental assessment of pesticide leaching using undisturbed lysimeters, *Pestic. Sci.*, 31, 337, 1991.

68. **Koren, E.**, Leaching of trifluralin and oryzalin in soil with three surfactants, *Weed Sci.*, 20, 230, 1972.

69. **Koren, E., Foy, C. L., and Ashton, F. M.**, Adsorption, leaching and lateral diffusion of four thiocarbamate herbicide in soils, *Weed Sci. Soc. Am. Abstr.*, Champaign, IL, 72, 1967.

70. **Koterba, M. T., Banks, W. S. L., and Shedlock, R. J.**, Pesticides in shallow groundwater in the Delmarva Peninsula, *J. Environ. Qual.*, 22, 500, 1993.

71. **Kotoula Syka, E., Eleftherohorinos, I. G., Gagianas, A. A., and Sficas, A. G.**, Phytotoxicity and persistence of chlorsulfuron, metsulfuron-methyl, triasulfuron and tribenuron-methyl in three soils, *Weed Res.*, 33, 355, 1993a.

72. **Kotoula Syka, E., Eleftherohorinos, I. G., Gagianas, A. A., and Sficas, A. G.**, Persistence of preemergence applications of chlorsulfuron, metsulfuron, triasulfuron, and tribenuron in three soils in Greece, *Weed Sci.*, 41, 246, 1993b.

73. **Leistra, M.**, Behavior and significance of pesticide residues in ground water, *Aspects Appl. Biol.*, 17, 223, 1988.

74. **Levanon, D., Codling, E. E., Meisinger, J. J., and Starr, J. L.**, Mobility of agrochemicals through soil from two tillage systems, *J. Environ. Qual.*, 22, 155, 1993.

75. **Matthies, M. and Behrendt, H.**, Pesticide transport modelling in soil for risk assessment of groundwater contamination, *Toxicol. Environ. Chem.*, 31/32, 357, 1991.

76. **McWhorter, C. G., Shaw, W. C., and Schweizer, E. E.**, Present status and future needs in weed control, in *Technology: Public Policy, and the Changing Structure of Ameican Agriculture,* Office of Technology Assessment, Washington, DC, 2, 19, 1986.

77. **Meyers, P. A., Greene, C. L., and Springer, J. T.**, Use of Intelimer microcapsules to control the release of agricultural products and reduce leaching, in *Pesticide Formulations and Application Systems: 13th Vol. STP 1183*, ASTM, Philadelphia, PA, 1993, 335 pp.
78. **Miller, P. C. H. and Stafford, J. V.**, Herbicide application to targeted patches, *Proc. Brighton Crop Prot. Conf. - Weeds*, 1249, 1991.
79. **Mueller, T. C. and Banks, P. A.**, Flurtamone adsorption and mobility in three Georgia soils, *Weed Sci.*, 39, 275, 1991.
80. **Mullins, D. E., Young, R. W., Berry, D. F., Gu, J. D., and Hetzel, G. H.**, Biologically based sorbents and their potential use in pesticide waste disposal during composting, *Am. Chem. Soc. Symp. Ser.*, 522, 113, 1993.
81. **Nakamura, K.**, Behaviour of pesticides in soil and other environments, *J. Pestic. Sci.*, 15, 271, 1990.
82. **Narayanan, K. S., Singh, M., and Chaudhuri, R. K.**, The reduction of herbicide leaching using vinyl pyrrolidone copolymers and methyl vinyl ether maleic acid ester copolymers, in *Pesticide Formulations and Application Systems: 13th Vol. STP 1183*, ASTM, Philadelphia, PA, 1993, 57 pp.
83. **National Research Council,** *Pesticides and Groundwater Quality: Issues and Problems in Four States*, Natl. Acad. Press, Washington, DC, 1986.
84. **Neary, D. G.**, Fate of pesticide in Florida's forests: An overview of potential impacts on water quality, *Proc. Soil Crop Sci. Soc. Fla.*, 44, 18, 1985.
85. **Nicholls, P. H.**, Factors influencing entry of pesticides into soil water, *Pestic. Sci*,, 22, 123, 1988.
86. **Peek, D. C. and Appleby, A. P.**, Phytotoxicity, adsorption, and mobility of metribuzin and its ethylthio analog as influenced by soil properties, *Weed Sci.*, 37, 419, 1989.
87. **Pennell, K. D., Hornsby, A. G., Jessup, R. E., and Rao, P. S. C.**, Evaluation of five simulation models for predicting aldicarb and bromide behavior under field conditions, *Water Resour. Res.*, 26, 2679, 1990.
88. **Pennington, K. L., Harper, S. S., and Koskinen, W. C.**, Interaction of herbicides with water-soluble soil organic matter, *Weed Sci.*, 39, 667, 1991.
89. **Pothuluri, J. V., Moorman, T. B., Obenhuber, D. C., and Wauchope, R. D.**, Aerobic and anaerobic degradation of alachlor in samples from a surface-to-groundwater profile, *J. Environ. Qual.*, 19, 525, 1990.
90. **Putnam, A. R. and Tang, C. S.**, Allelopathy: state of the science, in *The Science of Allelopathy,* Putnam, A. R. and Tang, C. S., Eds., John Wiley & Sons, New York, 1986, 1 pp.
91. **Radder, B. M., Parama, V. R. R., and Siddaramappa, R.**, Leaching loss, movement and distribution of carbofuran and fenitrothion in different soil columns, *Pestic. Res. J.*, 1, 59, 1989.
92. **Ranjha, A. Y., Peralta, R. C., Hill, R. W., Requena, A. M., Allen, L. N., Deer, H. M., and Ehteshami, M.**, Sprinkler irrigation-pesticide best management systems, *Appl. Eng. Agric.*, 8, 347, 1992a.
93. **Ranjha, A. Y., Peralta, R. C., Requena, A. M., Deer, H. M., Ehteshami, M., Hill, R. W., and Walker, W. R.**, Best management of pesticide--furrow irrigation systems, *Irrig. Sci.*, 13, 9, 1992b.
94. **Rao, P. S., Hornsby, A. G., and Jessup, R. E.**, Indices for ranking the potential for pesticide contamination of groundwater, *Proc. Soil Crop Sci. Soc. Fla.*, 44, 1, 1985.
95. **Reddy, K. N. and Singh, M.**, Effect of acrylic polymer adjuvants on leaching of bromacil, diuron, norflurazon, and simazine in soil columns, *Bull. Environ. Contam. Toxicol.*, 50, 449, 1993.

96. **Rice, R. C., Jaynes, D. B., and Bowman, R. S.,** Preferential flow of solutes and herbicide under irrigated fields, *Trans. Am. Soc. Agric. Eng.*, 34, 914, 1991.
97. **Richard, J. J., Junk, G. A., Avery, M. J., Fritz, J. S., and Svec, H. J.,** Analysis of various Iowa waters for selected pesticides: atrazine, DDE, and dieldrin, *Pestic. Monit. J.*, 9, 117, 1975.
98. **Riggle, B. D. and Penner, D.,** Evaluation of pine kraft lignins for controlled release of alachor and metribuzin, *Weed Sci.*, 35, 243, 1987.
99. **Riggle, B. D. and Penner, D.,** Controlled release of three herbicides with kraft lignan PC940C, *Weed Sci.*, 36, 131, 1988.
100. **Sauer, T. J., Fermanich, K. J., and Daniel, T. C.,** Comparison of the pesticide root zone model simulated and measured pesticide mobility under two tillage systems, *J. Environ. Qual.*, 19, 727, 1990.
101. **Schreiber, M. M., Hickman, M. V., and Vail, G. D.,** Starch-encapsulated atrazine efficacy and transport, *J. Environ. Qual.*, 22, 443, 1993.
102. **Schreiber, M. M., Shasha, B. S., Trimnell, D., and White, M. D.,** Controlled release herbicides, in *Methods of Applying Herbicides,* McWhorter, C. G. and Gebhards, M. R., Eds., Weed Sci. Soc. Am., 177, 1987.
103. **Schweizer, E. E.,** New technological developments to reduce groundwater contamination by herbicides, *Weed Technol.*, 2, 223, 1988.
104. **Shankland, D. L.,** Environmental toxicants and public risk in Florida, *Proc. Soil Crop Sci. Soc. Fla.*, 44, 14, 1985.
105. **Shasha, B. S., Trimnell, D. T., and Otey, F. H.,** Encapsulation of pesticides in a starch-calcium adduct, *J. Polym. Sci.*, 19, 1891, 1981.
106. **Smith, L. W. and Bayer, D. E.,** Soil adsorption of diuron as influenced by surfactants, *Soil Sci.*, 103, 328, 1967.
107. **Smith, M. C., Thomas, D. L., Bottcher, A. B., and Campbell, K. L.,** Measurement of pesticide transport to shallow ground water, *Trans. Am. Soc. Agric. Eng.*, 33, 1573, 1990.
108. **Stahnke, G. K., Shea, P. J., Tupy, D. R. Stougaard, R. N., and Shearman, R. C.,** Pendimethalin dissipation in Kentucky bluegrass turf, *Weed Sci.*, 39, 97, 1991.
109. **Stehouwer, R. C., Dick, W. A., and Traina, S. J.,** Characteristics of earthworm burrow lining affecting atrazine sorption, *J. Environ. Qual.*, 22, 181, 1993.
110. **Stone, D. M., Harris, A. R., and Koskinen, W. C.,** Leaching of soil-active herbicides in acid, low base saturated sands: worst-case conditions, *Environ. Toxicol. Chem.*, 12, 399, 1993.
111. **Stougaard, R. N., Shea, P. J., and Martin, A. R.,** Effect of soil type and pH on adsorption, mobility, and efficacy of imazaquin and imazethapyr, *Weed Sci.*, 38, 67, 1990.
112. **Street, J. E., Wehtje, G., Walker, R. H., and Patterson, M. G.,** Effects of adjuvants on behavior of metribuzin in soil and soybean injury, *Weed Sci.*, 35, 422, 1987.
113. **Taylor, K. A., and Weber, J. B.,** Distribution of metolachlor in a Dothan soil as influenced by soil surface management under two irrigation regimes, *Proc. South. Weed Sci. Soc.*, 46, 339, 1993.
114. **Trimmnell, D. and Shasha, B. S.,** Controlled release formulations of atrazine in starch for potential reduction of groundwater pollution, *J. Controlled Release*, 12, 256, 1990.
115. **Troiano, J., Garretson, C., Krauter, C., Brownell, J., and Huston, J.,** Influence of amount and method of irrigation water application on leaching of atrazine, *J. Environ. Qual.*, 22, 290, 1993.
116. **Truman, C. C. and Leonard, R. A.,** Effects of pesticide, soil, and rainfall characteristics on potential pesticide loss by percolation--a GLEAMS simulation, *Trans. Am. Soc. Agric. Eng.*, 34, 2461, 1991.

117. **USEPA,** *Agricultural Chemicals in Ground Water: Proposed Pesticide Strategy,* Office of Pesticides and Toxic Substances, Washington, DC, 1987, 150 pp.
118. **Vink, J. P. M. and Robert, P. C.,** Adsorption and leaching behaivour of the herbicide alachlor (2-chloro-2',6' diethyl-N-(methoxymethyl) acetanilide) in a soil specific management, *Soil Use Manage.*, 8, 26, 1992.
119. **Wauchope, R. D., Glaze, N. C., and Dowler, C. C.,** Mobility and efficacy of controlled release formulations of atrazine, *Weed Sci. Soc. Am. Abstr.*, 30, 216, 1990.
120. **Weber, J. B.,** Interaction of organic pesticides with particulate matter in aquatic and soil systems, in *Fate of Organic Pesticies in the Aquatic Environment,* Gould, R. E., Ed., Am. Chem. Soc., Washington, DC, 1972, 55 pp.
121. **Weber, J. B. and Miller, C. T.,** Organic chemical movement over and through soil, in *Reactions and Movement of Organic Chemicals in Soils,* Sawhney, B. L. and Brown, K., Eds., Soil Sci. Soc. Am., Inc. Madison, WI, 1989, 305 pp.
122. **Weber, J. B., Swain, L. R., Strek, H. J., and Sartori, J. L.,** Herbicide mobility in soil leaching columns, in *Research Methods in Weed Science,* Camper, N. D., Ed., South. Weed Sci. Soc., 1986, 189 pp.
123. **Weber, J. B. and Whitacre, D. M.,** Mobility of herbicides in soil columns under saturated- and unsaturated-flow conditions, *Weed Sci.*, 30, 579, 1982.
124. **White, M. D. and Schreiber, M. M.,** Herbicidal activity of starch encapsulated trifluralin, *Weed Sci.*, 32, 387, 1984.
125. **Wing, R. E., Doane, W. M., and Schreiber, M. M.,** Starch-encapsulated herbicides approach to reduce groundwater contamination, in *Pesticide Formulations and Application Systems: Tenth Symp. ASTM Pub. No. 1078*, ASTM, Philadelphia, PA, 1990, 45 pp.
126. **Wing, R. E., Maiti, S., and Doane, W. M.,** Effectiveness of jetcooked pearl cornstarch as a controlled release matrix, *Starch/Starke*, 39, 422, 1987.
127. **Wing, R. E., Maiti, S., and Doane, W. M.,** Amylose content of starch controls the release of encapsulated bioactive agents, *J. Controlled Release*, 7, 33, 1988.
128. **Zins, A. B., Wyse, D. L., and Koskinen, W. C.,** Effect of alfalfa (*Medicago sativa*) roots on movement of atrazine and alachlor through soil, *Weed Sci.*, 39, 262, 1991.

Chapter 15

Adjuvants - Current Trends and Technology

Chester L. Foy

CONTENTS

ABSTRACT

Three international symposia on adjuvants and agrichemicals have been held to date (1986, 1989, and 1992). The publications based on the symposia are excellent sources of information. The majority of non-proprietary, published research on adjuvant use with pesticides has focused on herbicides and many

1-8493-7678-5/96/$0.00+$.50

articles on herbicides and adjuvants have appeared in the literature during 1992-1994. Some topics addressed included: organosilicone surfactants; fertilizer additives; pyrrolidones; effects of various other adjuvants on efficacy, antagonism, uptake, penetration, mobility, and translocation; effects on droplet size, drift and spreading; adjuvant evaporation; bioherbicides; safeners; and guidelines for adjuvant use. The effects of ethylene oxide (EO) content of surfactants and oxyethylene chain lengths were studied. Approximately 36 herbicides, 24 weed species, 18 crops, and numerous adjuvants were involved.

Favorable results were generally achieved with the organosilicone surfactant, Silwet L-77®. Fertilizer additives enhanced the activity of several herbicides and overcame antagonism caused by calcium chloride, sodium bicarbonate, and some herbicides. Pyrrolidones exhibited excellent wetting, low surface tension and low contact angle, and enhanced the performance of herbicides. Positive results were obtained with various other adjuvants in most cases. Nonionic surfactants and crop oil concentrates are the adjuvants most widely used with herbicides.

Agrichemical manufacturers are moving toward recommending specific adjuvants that research has identified as maximizing the efficacy of their products. The use of standardized laboratory procedures for evaluating adjuvants is vital in order to ensure that results are both reproducible and valid.

I. INTRODUCTION

The world's major source of food is plants. Crop losses are severe in many areas of the world, particularly in developing countries and where environmental conditions allow pests to survive throughout the year. From planting to consumption, an estimated 50% or more of the world's food production is lost to weeds, insects, diseases, rodents, and other factors.[22] Losses from these harmful pests are estimated at $100 billion annually.[75]

Pesticides help farmers compete with at least 10,000 species of insects, 1,500 plant diseases, and 1,800 kinds of weeds to produce the world's food. Pesticides represent a $25 billion industry worldwide and sales total nearly $7 billion annually in the U.S. More than 1,400 trade names of pesticides are provided in the *Farm Chemicals Handbook* 1993.[45]

Current trends relating to pesticides in the U.S. include: a stable or declining market; lower dosage pesticides; more complex/fragile active ingredients; formulation changes; more premixes and tank mixes; more application changes; different types and lower volumes of carriers; environmental concerns; biotechnology; more postemergence use; regulations and registrations; consolidations/mergers of pesticide manufacturers; and fewer products.

Adjuvants [ingredients in pesticidal or other agricultural chemical prescriptions which aid or modify the action of the primary ingredient(s)] are important to the production, marketing, application and effective use of pesticide products.[20] Adjuvants are of two general types: (a) formulation adjuvants --- additives already present in the container when purchased by the dealer or grower and (b) formulation adjuvants --- additives added along with the formulated or proprietary product to the diluent just before spray application in the field. The adjuvant market is increasing and adjuvants now number in the thousands. As pesticide formulations change and environmental concerns become more of an issue, the need for different adjuvants is increased. New developments in spray adjuvant technology include organosilicone surfactants, methylated seed oils, saponified seed oils, polymers, carrier conditioning agents, fertilizer based adjuvants, pyrollidones, clathrates, sucroglycerides, "designer" adjuvants, and hybrids. A partial listing of functions of adjuvants that are required or suggested by U.S. pesticide manufacturer's Environmental Protection Agency (EPA)-approved labels are shown in Table 1. A partial listing of types of spray adjuvants required or recommended by EPA-registered pesticide labels are shown in Table 2.

Table 1
Partial Listing of Functions of Adjuvants Required or Suggested on EPA-Registered Pesticide Labels[a]

Absorption	Enhanced biological activity	Soil infiltration
Anti-evaporation	Line cleaning	Soil wetting
Antifoaming	Marking	Spray thickening
Buffering	Non-phytotoxic	Spreading
Compatibility	Penetration	Sticking
Control/retard drift	Preventing hydrolysis	Suspension
Coverage	Rainfastness	Tank cleaning
Defoaming	Reduced water volume	UV degradation/ inhibition
Deposition	Runoff	
Dew suppression	Sinking Agent	Wetting

[a]Source: Helena Chemical Company, Memphis, TN with permission.

Table 2
Partial Listing of Types of Spray Adjuvants Required/Recommended on EPA-Registered Pesticide Labels[a]

Surfactants/Adjuvants/Utility/Etc.
Adjuvant
Agriculturally approved drift retarding agent
Agriculturally approved surfactant
Antifoam agents
Buffer
Colorant
Compatibility agent
Confinement agent
Defoaming agent
Drift control agent
Drift retardant
Dye
Emulsifier
EPA approved surfactant
EPA exempt ingredients
Foam marker
Foam reducing agent
Invert emulsion
Marking agent
Nonionic surfactant
Organosilicone surfactant
Organosilicone based surfactant
Silicone based surfactant
Sinking agent
Spray adjuvant
Spray pattern indicator
Spreader
Spreader-sticker
Sticker
Surfactant
Tank cleaner
Thickening agent
Tolerance exempt surfactant
Wetting agent
Oil or Oil Based
Cottonseed oil
Crop derived oil concentrate
Crop oil
Crop oil concentrate
Diesel fuel
Diesel oil
Emulsifiable oil
EPA approved petroleum based crop oil concentrate
Fuel oil
Herbicidal oil
Horticultural oil
Methylated oil
Methylated seed oil
Methylated seed oil based crop oil concentrate
Non-phytotoxic oil concentrate
Non-volatile vegetable oil
Oil concentrate
Once refined soybean oil concentrate
Once refined vegetable oil
Paraffin based herbicidal oil
Penetrating oil
Petroleum based crop oil concentrate
Petroleum derived emulsifiable oil
Petroleum derived oil concentrate
Petroleum oil
Refined cottonseed oil
Refined soybean oil
Refined vegetable oil
Spray oil
Summer oil
Vegetable oil
Vegetable oil concentrate
Fertilizers
10-34-0
21-0-0
28-0-0
28% N
N-28
28% urea ammonium nitrate
30% N
30% urea ammonium nitrate
32-0-0
32% N
32% urea ammonium
Ammonium nitrogen fertilizer
Ammonium sulfate
Ammonium sulfate solution
AMS
Aqueous ammonium polyphosphate
Diammonium phosphate
Liquid nitrogen fertilizer solution
Nitrogen solution
Spray grade ammonium sulphate
Sprayable fluid fertilizer
UAN
Urea ammonium nitrate
Urea ammonium nitrate solution

[a]Source: Helena Chemical Company, Memphis, TN with permission.

Megatrends impacting agriculture that have a direct effect on both pesticides and adjuvants include: internationalization of agriculture, restructuring of farms and firms, consumer focus vs. producer focus, new technology, development and adoption, reductions in production costs, diversity in the farm sector, capitalization of agriculture, strategic planning, and environmental and agricultural policies. Four trends (consumers, technology, regulatory environment, and producers) will drive change.

Three international symposia on adjuvants for agrochemicals have been held to date. The first symposium was held in Brandon, Manitoba, Canada in 1986; the second in Blacksburg, Virginia in 1989; and the third in Cambridge, United Kingdom in 1992. The publications [3-6, 12, 21] based on the three symposia are excellent sources of information on adjuvants and their uses with agrichemicals. Topics include: a bibliographic survey of research and development of agro-adjuvants; regulation, registration, importance, and environmental fate of adjuvants; rationale for adjuvant use; concerns within the pesticide industry relating to adjuvants; a review of the methods employed in laboratory evaluation of adjuvants; model systems for evaluating adjuvant action; results of current research on various adjuvants with herbicides, fungicides, insecticides, or growth regulators; and other agrichemicals; properties and mode of action of surfactant-based adjuvants, oil-based adjuvants, and organosilicone-based adjuvants, spray-modifier adjuvants, and field, greenhouse and laboratory methods for evaluating adjuvants. Bibliographies of adjuvant literature are provided. Other books related to this subject have also been published recently.[15,42]

A large portion of the recent work on adjuvants has focused on herbicides. This chapter presents many of those research findings. The results represent a snapshot view under sometimes limited testing conditions. Therefore, these results should not be used to make specific recommendations for use of the products mentioned. Always refer to the pesticide label, the pesticide manufacturer's recommendations, or extension service recommendations for use and application instructions regarding the use of spray adjuvants. Trade names and manufacturers/distributors of adjuvants mentioned in the text are shown in Table 3. Common and chemical names of herbicides, bioherbicides, defoliants, insecticides, and safeners are presented in Table 4. Common and scientific names of weed species and crops are listed in Tables 5 and 6, respectively.

II. IMPORTANCE OF ADJUVANT USE WITH HERBICIDES

Estimated annual costs and losses attributed to weeds were calculated to be over $20 billion in the U.S. in 1993.[10] Herbicides account for about 65% of the total dollar value of sales of all pesticides and plant growth regulators in the U.S.[20] Worldwide, herbicides account for about 44% of the pesticide market. Most herbicides require adjuvants, either in the formulated product, as spray tank additives, or both. The reasons for using adjuvants with herbicides are (a) to improve or otherwise facilitate the physical handling characteristics of herbicides;

(b) to improve performance effectiveness and consistency; and (c) to comply with legal requirements.[54] Spray adjuvants are used more extensively with herbicides than with other classes of pesticides.[22] The ideal adjuvant will enhance weed control, but not cause crop injury. Nonionic surfactants and crop oil concentrates are generally recommended with herbicides. Nonionic surfactants are the most widely used adjuvants for foliarly applied treatments where water (the primary diluent for all pesticides) is used as the carrier for herbicide spray solutions.

III. ORGANOSILICONE SURFACTANTS

The first significant report on the use of organosilicones as adjuvants for herbicides appeared in the literature in 1973.[32] Special properties of organosilicones include their extreme spreading, and by virtue of extremely low static surface tensions, their ability to induce infiltration of spray formulations into foliage via stomata.[67] Positive reports on enhancement form an unusually high proportion of the literature on organosilicones, although numerous examples have been cited where they were not beneficial. However, where they are advantageous, their effects are often outstanding.[67] In 1985, Silwet L-77® was the first organosilicone adjuvant to be commercialized, and various other silicone-based surfactants have subsequently been introduced.

Approximately two-thirds of the research with organosilicones has been directed at herbicides.[67] Potential applications of organosilicones are far wider and more numerous than have been explored to date. A 1993 report by Stevens[67], which included organosilicone uses with herbicides, foliar nutrients, growth regulators, insecticides and fungicides, cited 160 total references on adjuvants with agrochemicals, and almost 90% of them involve organosilicones.

Additional studies have been conducted and results have been reported. Examples of these research findings on the use of organosilicone surfactants with herbicides follow.

Silwet L-77® significantly reduced the time to reach half the maximum uptake of triclopyr in field bean by providing an alternate route via stomatal pores for herbicide entry into the plant.[9] The use of abscisic acid pretreatment confirmed this route.

Table 3
Trade Names or Code Numbers, Chemical Composition, and Manufacturers/Distributors of Adjuvants Mentioned in the Text

Trade name or code number	Chemical composition	Manufacturer/Distributor
Activate Plus	Alkylaryl polyoxyethylene glycols + free fatty acids + isopropanol	Terra/Riverside, Sioux City, IA
Activator 90	Alkyl polyoxyethylene ether and free fatty acids	Loveland Industries, Greeley, CO
Adwet[a]		
Agral 90	90% nonylphenoxy polyethoxy ethanol	ICI Chipman, Stoney Creek, Ontario
Agri-Dex®	Paraffinic mineral oil + polyoxyethylene sorbitan fatty acid esters	Helena Chemical Co., Memphis, TN
Agrimax™ 3	Microemulsified water-insoluble polymer concentrate	International Specialty Products, Wayne, NJ
Agrimer™ VA6	60:40 copolymer of vinyl pyrrolidone and vinyl acetate	International Specialty Products, Wayne, NJ
Agrimer™ AL 10	Graft copolymer of vinyl pyrrolidone (90%) and butene (10%)	International Specialty Products, Wayne, NJ
Agro Flow[b]		BoCo Chemicals Int., Winnie, TX
Agsol EX® 1	*N*-methylpyrrolidone	International Specialty Products, Wayne, NJ
Agsol EX® 8	*N*-octylpyrrolidone	International Specialty Products, Wayne, NJ
Agsol EX® 12	*N*-dodecylpyrrolidone	International Specialty Products, Wayne, NJ
Armoblen T/25	Polyoxyethylene (15) tallow amine	Akzo Chemicals, Amersfoort, The Netherlands
Ban Drift	Homopolymer of acrylamide	Allied Colloids Ltd., Bradford, U.K.
BCH 815	A component of Dash®	BASF Corp., Parsippany, NJ
Bond®	Synthetic latex with emulsifier	Loveland Industries, Greeley, CO
Booster Plus Crop Oil Concentrate	Mixture of nonphytotoxic petroleum oils and surfactant	Agway, Inc., Syracuse, NY
Camplus 411	Paraffin base petroleum oil (83%) and 17% polyol fatty acid esters and polyethoxylated derivatives thereof	Atkemix Inc., Brantford, Ontario
CC 16255	Not available	Rhone-Poulenc Canada, Inc. Mississauga, Ontario
Charge[a]		
Clean Crop®	Paraffin base petroleum oil	Loveland Industries, Greeley, CO

C12 E16	Hexadecaethylene glycol monododecyl ether	Diiachi Pure Chemical Co., Tokoyo, Japan
C13 E6	Hexaethylene glycol monotridecyl ether	Shell Development Corp., Houston, TX
Dash® HC	Mixture of petroleum hydrocarbons, naphthalene and oleic acid	BASF Corp., Parsippany, NJ
Direct	Drift controlling polymeric adjuvant	Precision Laboratories, Northbrook, IL
Driftgard	Drift reduction agent, polyacrylamide polymer	Custom Chemicides, Fresno, CA
Enhance	Tallow fatty acid amine ethoxylate (64%) and 14% nonylphenoxy polyethoxy ethanol	Elanco, Scarborough, Ontario
Formula 358	Drift control agent	Exacto Chemical Co., Solon Mills, IL
41A	Polyacrylamide, polysaccharide	Sanitek Products, Los Angeles, CA
Herbimax®	Petroleum hydrocarbons (light paraffinic distillate, odorless alphatic petroleum solvent) plus surfactant (mono and diesters of omega hydroxypolyoxythylene)	Loveland Industries, Greeley, CO
High Trees Mixture B	Nonyl phenol ethylene oxide condensate (50%) and 50% alcohol ethylene oxide condenstate	Service Chemicals Ltd.
Intac®	Blend of polymers and copolymers	Loveland Industries, Greeley, CO
Kinetic®	Blend of polyalkyleneoxide-modified poly-dimethylsiloxane and nonionic organosilicone surfactant	Helena Chemical Co., Memphis, TN
K1000, K2000, K3000[a]	Materials derived from soybeans	
Li 700®	80% phosphatidylcholine and methylacetic acid	Loveland Industries, Greeley, CO
Marlipal 34	Alcohol polyoxyethylene surfactant	Hiils, Germany
Maximizer 420	Paraffinic oil, emulsifier	Loveland Industries, Greeley, CO
Merge	50% surfactant blend and 50% petroleum hydrocarbon solvents	BASF Canada Inc., Toronto, Ontario
Meth Oil	Multi-component, methylated seed oil	Terra/Riverside, Sioux City, IA
Mor-Act	Petroleum oil adjuvant	Wilbur-Eillis, Fresno, CA
Nalcotrol®	Polyvinyl polymer	Nalco Chemical, Naperville, IL
1412-60	Linear alcohol additive	Vista Chemical Co., Saddle Brook, NJ
1412-70	Linear alcohol additive	Vista Chemical Co., Saddle Brook, NJ
Nu-Film P®	Di-1-p-menthene	Miller Chemical and Fertilizer Corp., Hanover, PA
Orchex 796	Paraffinic hydrocarbons, C_{23}	Exxon Corp., Baytoum, TX
Paratac®	Oil-soluble polyisobutylene polymer	
Penetrator®	Paraffin base petroleum oil, polyol fatty acid esters and polyethoxylated derivatives thereof	Helena Chemical Co., Memphis, TN
Plex		Terra/Riverside, Sioux City, IA
Preference[a]		

Table 3
Trade Names or Code Numbers, Chemical Composition, and Manufacturers/Distributors of Adjuvants Mentioned in the Text

Trade name or code number	Chemical composition	Manufacturer/Distributor
Prime Oil	Petroleum-based paraffinic oil and nonionic surfactants	Terra/Riverside, Sioux City, IA
R-11®	Octyl phenoxy polyethoxy ethanol, isopropanol, and compounded silicone	Wilbur-Ellis, Fresno, CA
Savol	83% paraffin base petroleum oil, 6% alkyloxy-polyoxyethylene, and 6% alklyphenoxy-polyoxyethylene	Uniroyal Chemical Ltd., Elmira, Ontario
Scoil	Methylated vegetable oil and emulsifier	Agsco, Inc., Grand Forks, ND
Silwet L-77®	Silicone polyalkylene oxide-modified dimethyl polysiloxane	Union Carbide Co., Tarrytown, NY
Spray Booster S[b]	Nonionic surfactant	Cenex Ind., St. Paul, MN
Spray Fuse 90	Alkyarylpolyoxyethylene glycols + free fatty acids + isopropanol	Cornbelt Chemical Co., McCook, NE
Sta-Put®	Polyvinyl polymer	Nalco Chemical Co., Naperville, IL
Sterox NJ	Nonylphenol ethoxylate (9.5 ETO)	Monsanto Co., St. Louis, MO
Sun-It II	Modified vegetable oil plus surfactant	Agsco Inc., Grand Forks, ND
Surfix®	Beta pinene polymer	Helena Chemical Co., Memphis, TN
SurpHtac	Monocarbamide dihydrogensulfate surfactant solution	Unocal, Los Angeles, CA
Sylgard 309® (DCX2-5309)	2-(3-hydroxypropyl-heptanethyltrisiloxane, ethoxylated and acetate	Dow Corning, Midland, MI
Target	Polymers of acrylamide, acrylate and saccharides	Loveland Industries, Greeley, CO
38F™	Polyacrylamide polymers	Loveland Industries, Greeley, CO
Turbocharge™[a]		
Windfall	Drift reduction agent - discontinued in 1992	Terra/Riverside, Sioux City, IA
X-77®	Mixture of alkylpolyoxethylene glycols, free fatty acids, and isopropanol	Loveland Industries, Greeley, CO (formerly by Valent USA)
XE 1167	Not available	Valent U.S.A. Corp., Walnut Creek, CA

[a]Names of manufacturers/distributor were not provided. These adjuvants were mentioned only in abstracts.
[b]Chemical composition was not given in the abstract or article.

Table 4
Common and Chemical Names of Herbicides, Bioherbicides, Defoliants, Insecticides, and Safeners Mentioned in the Text

Common name or code number	Chemical name
Herbicides	
Acifluorfen	5-[2-Chloro-4-(trifluoromethyl)phenoxy]-2-nitrobenzoic acid
Alachlor	2-Chloro-*N*-(2,6-diethylphenyl)-*N*-(methoxymethyl)acetamide
Atrazine	6-Chloro-*N*-ethyl-*N'*-(1-methylethyl)-1,3,5-triazine-2,4-diamine
Bentazon	3-(1-Methylethyl)-(1*H*)-2,1,3-benzothiadiazin-4(3*H*)-one 2,2-dioxide
Bromoxynil	3,5-Dibromo-4-hydroxybenzonitrile
Clopyralid	3,6-Dichloro-2-pyridinecarboxylic acid
Cyanazine	2-[[4-Chloro-6-(ethylamino)-1,3,5-triazin-2-yl]amino]-2-methylpropanenitrile
Cycloxydim	2-[1-(Ethoxyimino)-butyl]-3-hydroxy-5-(2H-tetrahydrothiopyran-3-yl)-2-cyclohexen-1-one
Dicamba	3,6-Dichloro-2-methoxybenzoic acid
Dichlorprop	(±)-2-(2,4-Dichlorophenoxy)propanoic acid
Diclofop	(±)-2-[4-(2,4-Dichlorophenoxy)phenoxy]propanoic acid
Difenzoquat	1,2-Dimethyl-3,5-diphenyl-1*H*-pyrazolium
DPX-PE350	Sodium 2-chloro-(4,5-dimethoxypyrimidin-2-ylthio)benzoate
Fluroxypyr	[(4-Amino-3,5-dichloro-6-fluoro-2-pyridinyl)oxy]acetic acid
Glufosinate	2-Amino-4-(hydroxymethylphosphinyl)butanoic acid
Glyphosate	*N*-(Phosphonomethyl)glycine
Imazamethabenz	(±)-2-[4,5-Dihydro-4-methyl-4-(1-methylethyl)-5-oxo-1*H*-imidazol-2-yl]-4(and 5)-methylbenzoic acid (3:2)
Imazapyr	(±)-2-[4,5-Dihydro-4-methyl-4-(1-methylethyl)-5-oxo-1*H*-imidazol-2-yl]-3-pyridinecarboxylic acid
Imazethapyr	2-[4,5-Dihydro-4-methyl-4-(1-methylethyl)-5-oxo-1*H*-imidazol-2-yl]-5-ethyl-3-pyridinecarboxylic acid
MCPA	(4-Chloro-2-methylphenoxy)acetic acid
Metolachlor	2-Chloro-*N*-(2-ethyl-6-methylphenyl)-*N*-(2-methoxy-1-methylethyl)acetamide
Metsulfuron	2-[[[[(4-Methoxy-6-methyl-1,3,5-triazin-2-yl)amino]carbonyl] amino]sulfonyl]benzoic acid
Nicosulfuron	2-[[[[(4,6-Dimethoxy-2-pyrimidinyl)amino]carbonyl]amino] sulfonyl]-*N,N*-dimethyl-3-pyridinecarboxamide
Primisulfuron	2-[[[[(4,6-Bis(difluoromethoxy)-2-pyrimidinyl]amino]carbonyl] amino]sulfonyl]benzoic acid
Pyridate	O-(6-Chloro-3-phenyl-4-pyridazin-yl)5-octylcarbonothioate
Quizalofop	(±)-2-[4-[(6-Chloro-2-quinoxalinyl)oxy]phenoxy]propanoic acid
Rimsulfuron	*N*-[[4,6-Dimethoxy-2-pyrimidinyl)amino]carbonyl]-3-(ethylsulfonyl)-2-pyridinesulfonamide
Sethoxydim	2-[1-(Ethoxyimino)butyl]-5-[2-(ethylthio)propyl]-3-hydroxy-2-cyclohexen-1-one
Terbacil	5-Chloro-3-(1,1-dimethylethyl)-6-methyl-2,4(1H,3H)-pyrimidimedione
Thifensulfuron	3-[[[[(4-Methoxy-6-methyl-1,3,5-triazin-2-yl)amino]carbonyl] amino]sulfonyl]-2-thiophenecarboxylic acid
Tralkoxydim	2-[1-(Ethoxyimino)propyl]-3-hydroxy-5-(2,4,6-trimethylphenyl) cyclohex-2-enone
Tribenuron	2-[[[[4-Methoxy-6-methyl-1,3,5-triazin-2-yl)methylamino] carbonyl]amino]sulfonyl]benzoic acid

Triclopyr	[(3,5,6-Trichloro-2-pyridinyl)oxy]acetic acid
WL 110547	1-(3-Trifluoromethylphenyl)-5-phenoxy-1,2,3,4-tetrazole
Bioherbicides	
Colletotrichum truncatum	
Schlerotina sclerotiorum	
Defoliants	
Thidiazuron	*N*-Phenyl-*N*'-1,2,3-thidiazol-5-ylurea
Insecticides	
Terbufos	S-[[1,1-Dimethylethyl)thio]methyl]0,0-diethyl phosphorodithioate
Safeners	
BAS 145138	1,Dichloro-acetyl-hexahydrio-3,3,8-trimethylpyrrolo-(1,2-α)-pyrimidin-6-(2*H*)-one
CGA 133205	0-(1,3-Dioxolan-2-yl-methyl)-2,2,2-trifluoro-4'-chloro-acetophenone oxime
Dichlormid	2,2-Dichloro-*N*,*N*-di-2-propenylacetamide
Flurazole	Phenylmethyl 2-chloro-4-(trifluoromethyl)-5-thiazolecarboxylate
Naphthalic anhydride	1*H*,3*H*-Naphtho[1,8-cd]pyran-1,3-dione
Oxabetrinil	*N*-(,3-Dioxolan-2-yl-methoxy)-iminobenzene-acetonitrile
R-29148	3-(Dichloroacetyl)-2,2,5-trimethyl-1,3-oxazolidine

Table 5
Common and Scientific Names of Weed Species Mentioned in the Text

Common name	**Scientific name**
Barnyardgrass	*Echinochloa crus-galli* (L.) Beauv.
Canada thistle	*Cirsium arvense* (L.) Scop.
Common cocklebur	*Xanthium strumarium* L.
Common lambsquarters	*Chenopodium album* L.
Common ragweed	*Ambrosia artemisiifolia* L.
Entireleaf morningglory	*Ipomoea hederacea* var. *integriuscula* Gray
Giant foxtail	*Setaria faberi* Herrm.
Green foxtail	*Setaria viridis* (L.) Beauv.
Hemp sesbania	*Sesbania exaltata* (Raf.) Rydb. ex. A. W. Hill
Johnsongrass	*Sorghum halepense* (L.) Pers.
Kochia	*Kochia scoparia* (L.) Schrad.
Largeleaf lantana	*Lantana camara* L.
Nightflowering catchfly	*Silene noctiflora* L.
Prickly sida	*Sida spinosa* L.
Quackgrass	*Elytrigia repens* (L.) Beauv.
Sicklepod	*Cassia obtusifolia* L.
Velvetleaf	*Abutilon theophrasti* (L.) Medicus
Venice mallow	*Hibiscus trionum* L.
Virginia buttonweed	*Diodia virginiana* L.
Wild buckwheat	*Polygonum convolvulus* L.
Wild oat	*Avena fatua* L.
Wild proso millet	*Panicum milaceum* L. ssp. *ruderale*
Yellow foxtail	*Setaria glauca* (L.) Beauv.
----	*Rhododendron ponticum*

Table 6
Common and Scientific Names of Crop Species Mentioned in the Text

Common name	Scientific name
Alfalfa	*Medicago sativa* L.
Apple	*Malus pumila* Mill.
Barley	*Hordeum vulgare* L.
Bean	*Phaseolus vulgaris* L.
Bitter orange	*Citrus aurantium* L.
Canola	*Brassica campestris* L.
Coco, novo	*Erythroxylum* spp.
Corn (maize)	*Zea mays* L.
Cotton	*Gossypium hirsutum* L.
Field bean	*Vicia faba* L. minor
Kohlrabi	*Brassica oleracea* var. *gongylodes* L.
Oat	*Avena sativa* L.
Sorghum	*Sorghum bicolor* (L.) Moench.
Soybean	*Glycine max* (L.) Merr.
Spearmint	*Mentha spicata* L.
Sugarbeet	*Beta vulgaris* L.
Sunflower	*Helianthus annuus* L.
Tomato	*Lycopersicon esculentum* Mill.
Wheat	*Triticum aestivum* L.

Silwet L-77® increased the efficacy of metsulfuron against largeleaf lantana in pastures more than did a conventional surfactant (nonylphenoxy-polyethoxyethanol).[46] Further, a single application of metsulfuron plus Silwet L-77® was as effective as three and four split applications and more effective than two split applications. Silwet L-77® was among the most effective adjuvants used with nicosulfuron to control foxtail spp. and wild proso millet at one location and was intermediate at another.[51] Kinetic® was intermediate at one location and among the least effective adjuvants at another. Sylgard 309® was less effective than a linear alcohol additive in reducing shoot growth of quackgrass by primisulfuron.[24] Several adjuvants were evaluated with nicosulfuron, primisulfuron, and thifensulfuron on several weed species.[38] The efficacy of these adjuvants was herbicide and weed specific.

Water mixtures of Silwet L-77® spread much better on water-sensitive paper and on johnsongrass leaves than did water with conventional adjuvants.[43] Water droplets containing Silwet L-77® spread better than did those containing Agri-Dex® or Sterox NJ on leaf surfaces of *Ethroxylum* sp.[44]

Silwet L-77® consistently increased the activity of imazapyr and metsulfuron-methyl on *Rhododendron ponticum*.[39]

DCX2-5309 (Sylgard 309®) increased efficacy of bentazon, pyridate, and terbacil on kochia in the greenhouse, and also decreased the rain-free period needed after application of bentazon and bromoxynil.[2] In field studies, bentazon and bromoxynil rates required to control kochia were reduced by one-half with DCX2-5309 compared to Mor Act. DCX2-5309 added to spray solution with bentazon did not injure spearmint and controlled kochia better than did bentazon plus Mor Act. DCX2-5309 increased uptake of ^{14}C-bentazon and ^{14}C-bromoxynil by kochia compared to Mor Act.

IV. FERTILIZER ADDITIVES

A noteworthy development in the last several years has been the evaluation and development of various fertilizer additives for use with herbicides.[76] These herbicides include imidazolinones, sulfonylureas, cyclohexanediones, diphenyl ethers, glyphosate, and others. Fertilizer based spray adjuvants enhance herbicide activity by improving penetration/absorption and increasing translocation. Fertilizer additives may be used to replace conventional adjuvants, supplement conventional adjuvants, enhance activity on specific weeds, improve action under stress conditions, and overcome water quality problems. The most common fertilizer additives are 28% nitrogen [a mixture of urea and ammonium nitrate (UAN)] and ammonium sulfate. The recent trend has been to evaluate and use a combination of adjuvants together in a tank mix with one or more herbicides.[76] Results of several recent studies involving fertilizer additives with herbicides are summarzied below.

A. EFFICACY WITH HERBICIDES

Ammonium sulfate applied with sethoxydim was effective against green foxtail, wheat, wild oat, and barley seeded in a canola crop.[30] Sethoxydim activity on barley was increased when the herbicide was applied with an oil concentrate plus ammonium sulfate.[71]

Nicosulfuron and primisulfuron applied with a petroleum oil adjuvant (Herbimax®) plus UAN provided better control of quackgrass than with the petroleum oil alone in greenhouse studies.[7] However, few differences among adjuvants were observed in the field. Addition of UAN to a petroleum oil adjuvant increased control of giant foxtail by nicosulfuron.[35] K2000, a soybean derived adjuvant, alone or with UAN, and X-77® with UAN provided the maximum enhancement of giant foxtail control with nicosulfuron. Enhancement of nicosulfuron, primisulfuron, and thifensulfuron activity on several weed species by UAN was herbicide, nonionic surfactant, and weed specific.[38]

Addition of ammonium sulfate to imazethapyr plus nonionic surfactant (X-77®) increased quackgrass control, especially at low rates of imazethapyr in the field.[27] Ammonium sulfate slightly improved control of sicklepod, prickly sida, and barnyardgrass with glyphosate.[34]

B. ANTAGONISM

Diammonium sulfate overcame calcium chloride antagonism of diethanolamine 2,4-D and sodium 2,4-D, dimethylamine MCPA, sodium bentazon, dimethylamine dicamba and sodium dicamba, sodium acifluorfen, and imazamethabenz, but not glyphosate or ammonium imazethapyr to kochia.[48]

Further, it or ammonium nitrate overcame both calcium chloride and sodium bicarbonate antagonism of dicamba to kochia and enhanced the performance of the sodium salt of dicamba to nearly that of dimethylamine salt of dicamba.

BCH 815 plus ammonium sulfate added to sethoxydim plus bentazon spray mixture overcame the antagonism on ^{14}C-sethoxydim absorption in quackgrass.[74] Ammonium phosphate, ammonium nitrate, or ammonium sulfate were equally effective in overcoming antagonism betwee Na-bentazon plus Na-acifluorfen.

Scanning electron micrographs revealed that spray droplets of glyphosate applied with nonantagonistic diammonium sulfate contained distinct crystals and evenly spread residue beneath the crystals when applied to wheat.[50] Glyphosate applied with antagonistic calcium chloride salt formed spray deposits that were amorphous, thick, and without crystals. Glyphosate spray droplet residue contact with wheat, sunflower, and kochia leaf surfaces correlated with observed differences in glyphosate toxicity to these species. Nalewaja and Matysiak[49] have derived an equation to determine the amount of diammonium sulfate required to overcome glyphosate antagonism based upon the sodium, potassium, calcium, and magnesium cations in spray carriers.

C. ABSORPTION, UPTAKE

Absorption of ^{14}C-bentazon by common cocklebur and velvetleaf were increased more with 28% UAN than with Prime Oil.[40]

Cycloxydim absorption by suspension-cultured velvetleaf cells was increased by ammonium sulfate;[60] glyphosate absorption was not affected.[61]

The UAN plus a nonionic surfactant (X-77®) or petroleum oil adjuvant (Herbimax®) significantly increased ^{14}C-nicosulfuron and^{14} C-primisulfuron uptake by quackgrass.[7] Absorption was greatest with the following adjuvants: petroleum oil adjuvant + UAN > nonionic surfactant + UAN = methylated seed oil (Scoil).

Approximately twice as much imazethapyr was absorbed by quackgrass leaves when the herbicide was applied with a nonionic surfactant (X-77®) plus ammonium sulfate than when it was applied with surfactant alone.[27] Addition of ammonium sulfate to the external medium of Black Mexican Sweet corn (maize) cells enhanced both the rate of uptake and medium acidification.

V. *N*-ALKYL PYRROLIDONES

Pyrrolidones, like *N*-methylpyrrolidone and higher alkyl pyrrolidones, *N*-octyl- and *N*-dodecylpyrrolidone have common features: low vapor pressure, high flash point, chemical and thermal stability, low toxicity, biodegradability and exceptional solvency; however, they differ in water solubility and surfactancy.[52] Higher alkyl pyrrolidones possess surfactant properties and limited water solubility; whereas, lower alkyl pyrrolidones do not have surfactant properties and are readily soluble in water. Mixtures of lower and higher alkyl pyrrolidones, often including co-surfactants and co-solvents, can, by varying the component ratios allow the production of emulsifiable concentrates of the most difficult active ingredients.[52]

The coupling of (1) microemulsion technology with (2) the synergy of pyrrolidones with anionic surfactants, and (3) the ability of certain pyrrolidones to enhance the penetration and translocation of some active ingredients, produces adjuvant systems that significantly increase the efficacy of a number of pesticides when added as adjuvants to the diluted, ready-to-spray commercial pesticide formulations.[52] Many different compositions were prepared and evaluated (several in the field). Compositions were the spreader-activator type, and film-forming polymers were incorporated into others of the spreader-sticker type.

All formulations exhibited excellent wetting, low surface tension, and low contact angle. Spreader-activator adjuvants enhanced the performance of glufosinate on both grass and broadleaf weeds. The commercial formulation of glyphosate was less than 50% as herbicidally effective as when 0.25% of the spreader-sticker adjuvants were included. The components evaluated were as follows: AgsolEx® 1, *N*-methylpyrrolidone; AgsolEx® 8 (*N*-octylpyrrolidone); AgsolEx® 12 (*N*-dodecylpyrrolidone); Agrimer™ VA 6, a 60:40 copolymer of vinylpyrrolidone and vinyl acetate; and Agrimer™ AL10, a graft copolymer of vinylpyrrolidone (90%) and butene (10%).[52]

Agrimax™ 3, combines the advantages of spreaders, stickers, and penetrants in a single adjuvant to enhance efficacy by increasing uptake and reducing wash-off by rain or irrigation water.[1] The adjuvant is a microemulsified water-insoluble polymer concentrate. At typical concentrations contacting the leaf, Agrimax™ 3 causes the pesticide spray to wet the leaf surface instantaneously. The low volatility of the adjuvant keeps the active ingredient in solubilized form and enhances cuticular penetration. Agrimax™ 3 forms a droplet structure which enables it to form a rainfast blanket of the polymer over the active ingredient.

VI. VARIOUS OTHER ADJUVANTS WITH HERBICIDES

A. EFFICACY

Adjuvants varied considerably in their ability to enhance sethoxydim activity on green foxtail, wheat, wild oat, and barley.[30] CC 16255 was the most effective, followed by ammonium sulfate and Merge. Enhance, Savol and XE 1167 were moderately effective. Canplus plus Merge was usually not beneficial, but Canplus plus Enhance often increased sethoxydim activity compared with sethoxydim and Enhance alone. Agral 90 and Li-700® were of little or no value as adjuvants with sethoxydim.

Seaweed extract or calcium ammonium salt of alginic acid applied as adjuvants with sethoxydim to barley resulted in a significant increase in sethoxydim activity.[71] Sodium salts of alginic acid were much less effective. Quackgrass shoot growth was reduced more when primisulfuron was applied with the linear alcohol additive 1412-70 compared with Scoil, Sylgard® 309, Agri-Dex®, or X-77®.[24]

Efficacy of K1000, K2000, and K3000 (three materials derived from soybean) compared favorably with some other adjuvants when applied with the postemergence herbicides, nicosulfuron and primisulfuron, in corn and soybean.[59] Maximum efficacy was obtained when they were applied at 1% concentration. Giant foxtail control was lowest with nicosulfuron when Herbimax® and X-77® were used.[35] Greatest control was obtained with Scoil or K2000 as the adjuvant. Similar results were observed with primisulfuron.

Piperonyl butoxide enhanced nicosulfuron, primisulfuron, and thifensulfuron activity on several weed species.[38] The effect of piperonyl butoxide on sulfonylurea herbicide metabolism as observed by enhanced weed control was evident only when herbicide absorption had not been enhanced by the inclusion of nonionic surfactants and 28% UAN to reach its maximum potential.

Control of johnsongrass and quackgrass with primisulfuron was greatest with primisulfuron plus 1412-60 and 1412-70 --- products having chain lengths of 14 and 12 carbon atoms (60:40 ratio) and 60 or 70% ethoxylation.[17] In this case, control was equal to or greater than control obtained with primisulfuron plus the commercial adjuvants Scoil, Sylgard® 309, Agri-Dex®, or X-77®.

Surfactants with a wide range of 14 chemical, physical, and surface properties were evaluated for their effect on rimsulfuron applied to giant foxtail, velvetleaf, and corn.[26] The most effective surfactant type and concentration increased rimsulfuron's activity tenfold. The most effective were nonionic surfactants with HLB from 12 to 17 and that had more than 12 CH_2 and 6 ethylene oxide units which formed a moist, gel-like spray. A surfactant concentration of 0.1% (w/w) was optimal.

Relative effectiveness of commercial oil adjuvants at 3.5 L/ha with nicosulfuron at 17 g/ha to control foxtail spp. and wild proso millet generally was: Scoil > MSO = Meth Oil = Sun-IT II > Clean Crop® = Mor-Act > Herbimax®.[51] Oils at 3.5 L/ha were more effective than surfactants at 0.25% (v/v). Relative effectiveness of surfactants was R11® > X-77® = Spray Booster S = Preference.

Adjuvant enhancement of nicosulfuron phytotoxicity was influenced by oil type, type and percent emulsifier, and type of surfactant.[47] Surfactant enhancement of nicosulfuron phytotoxicity was dependent upon specific chemistry and could not be explained by HLB or chemical groupings, indicating that adjuvants efficacy cannot generally be predetermined by type of oil, emulsifier type or percent, or surfactant HLB, but will require comprehensive plant testing on a variety of weed species under a wide variety of environmental factors.

Control of entireleaf morningglory with DPX-PE350 was enhanced by both nonionic surfactant or crop oil concentrate, the crop oil being superior to the nonionic surfactant.[33]

Relative adjuvant enhancement of imazethapyr for control of green foxtail, yellow foxtail, barnyardgrass, wild oat, common lambsquarters, common ragweed, common cocklebur, wild buckwheat, Venice mallow, and nightflowering catchfly in soybean and dry edible beans was methylated seed oils > petroleum oils > nonionic surfactants.[78]

Oat control with glyphosate at 70 g/ha depended on the surfactant.[51] R-11®, Adwet, and Activator 90 performed in the top one-third and Spray Fuse 90, Maximizer 420, and Penetrator® were in the lower-performing third of the surfactants evaluated. Surfactants having an intermediate effect included Spray Booster S, Preference, Activate Plus, X-77®, and Li 700®.

Glyphosate, imazapyr, and triclopyr alone and metsulfuron-methyl with added surfactant were all phytotoxic to *Rhododendron ponticum*.[39] Thifensulfuron-methyl and tribenuron-methyl with High Trees Mixture B were ineffective.

The addition of adjuvants to 2,4-DB amine increased winter annual broadleaf weed control by 50 to 97% in seedling alfalfa.[16] Adjuvants X-77®, Booster Plus Crop Oil Concentrate, and Dash® HC were superior to SURpHTAC.

B. ABSORPTION, UPTAKE, PENETRATION, MOBILITY, TRANSLOCATION

Quizalofop absorption and phytotoxicity to oat were greater when applied with sunflower oil, sunflower oil free fatty acids, and sunflower oil fatty acid methyl esters than with corresponding linseed oil derivatives.[41] Emulsifier addition generally reduced the differences between the linseed and sunflower oil derivatives.

Ban Drift increased cycloxydim and glyphosate absorption by suspension-cultured velvetleaf cells at all times, whereas Armoblen T/25 increased absorption at only a short incubation time (30 min).[60,61] Armoblen T/25 was toxic to the cells.

Radiolabeled diclofop-methyl penetration into maize (corn) leaves was dramatically increased by the seed oil additives, triolen and methyloleate.[73] However, penetration into leaves of glossy hybrids was low and not affected by triolen while methyloleate increased it sevenfold; this again shows the often seen species selectivity.

K2000 and K3000 enhanced ^{14}C-nicosulfuron absorption by seedling johnsongrass and giant foxtail over that observed with X-77® or Scoil.[59] The ethoxylated alcohol adjuvant 1412-70 provided equal or greater uptake and translocation of ^{14}C-primisulfuron in johnsongrass and quackgrass compared to Scoil, Sylgard 309®, Agri-Dex®, or X-77®.[17]

The enhancement of herbicide uptake by wheat and field bean by surfactants were most pronounced at 0.5 g/L glyphosate with surfactants of high EO content (15-20).[23] Surfactants of 5 to 10 EO units frequently reduced or failed to improve glyphosate absorption. Aliphatic amines at optimal EO content caused greatest enhancement in wheat. Aliphatic alcohols gave the best performance on field bean; nonylphenols were the least efficient on both species. EO content on glyphosate at 5 and 10 g/L uptake was less marked in both species, especially with aliphatic amines.

Interactions occurring during the surfactant-enhanced foliar uptake of model organic compounds [glyphosate, 2,4-D, difenzoquat, cyanazine, and WL110547 (Shell Research Ltd., Sittingbourne, U.K.)] were examined using two homogeneous surfactants, hexaethylene glycol monotridecyl ether (C13E6) and hexadecaethylene glycol monododecyl ether (C12E16).[68] Surfactant-compound interaction was found to vary according to the physicochemical properties of both the compound, surfactant type and its concentration, and by the target plant species. Penetration of C13E6 into the leaf appeared to be essential in order to activate the uptake of the compound while pre-penetration of C12E16 was not required to activate transport.

The effect of the oxyethylene chain length of three homologous series of nonionic surfactants (allinol, nonoxynol, octoxynol) on glyphosate uptake was markedly affected by the leaf surface structure of sugarbeet and kohlrabi.[37] Uptake without surfactant averaged 4% for sugarbeet without surfactant, the droplets not being retained by kohlrabi leaves in the absence of surfactant. Glyphosate absorption with octoxynol (9 to 10 oxyethylene units) was rapid initially and leveled off after 2 h in both species. Absorption by sugarbeet decreased as oxyethylene content of octoxynol was increased from 5 to 30 units, whereas 16 oxyethylene units induced the greatest uptake by kohlrabi. The effect of oxyethylene content on both the enhancement of glyphosate uptake and wetting characteristics of solutions was similar, but only within species - it differed markedly between species.

The effects of nonionic surfactants on ^{14}C-glyphosate-mono (isopropylammonium) diffusion across isolated tomato fruit cuticles were compared.[14] Glyphosate uptake increased with EO content, reaching an optimum at 17 EO; however, with 40 EO content, uptake decreased below control values. There was a strong influence of the hydrophobe on glyphosate penetration by different surfactants with similar mean EO content (10 EO). Penentration was enhanced most by the primary aliphatic amine followed by the nonylphenol. The aliphatic alcohol showed no improvement on glyphosate transfer across cuticles. Water sorption was greatly enhanced by primary aliphatic amine (10 EO) and nonylphenol (17 EO), but aliphatic alcohol (10 EO) and a shorter-chained nonylphenol (4 EO) did not significantly enhance water sorption.

A trans-membrane electrochemical potential that was capable of driving several proton-amino acid symports was imposed across the membrane of isolated common lambsquarters vesicles.[56] Glyphosate, atrazine, and bentazon transport was tested. Atrazine accumulated inside the vesicles, but flux was not influenced by the trans-membrane ΔpH. Bentazon accumulation was driven by the imposed proton concentration difference while glyphosate flux was very low and unresponsive to the imposed proton motive force. This suggests that the plasma membrane is a barrier to cellular uptake by glyphosate. Surfactants at 0.01% did not impact the imposed pH gradient but stimulated glyphosate transport three to four fold. Both cationic (fatty amine ethoxylates) and nonionic (alcohol ethoxylates) surfactants increased glyphosate influx *in vitro*.

The mobility of 2,4-D in cuticular membranes isolated from bitter orange leaves was increased 25- to 30-fold by 1-heptanol, 1-octanol and 1-nonanol.[62] Increasing the number of carbon atoms in the alcohols decreased their effectiveness as did increasing ethoxylation. Free glycols had no effect on mobility. Alcohols and ethoxylated alcohols having between 7 and 10 carbon atoms are powerful accelerator adjuvants, as long as the degree of ethoxylation is not too high.

HLB of polyethylene-glycol-based surfactants was inversely related to the sorption of those surfactants by isolated cuticles of apple leaves and penetration of 2,4-D.[70] Sorption of 2,4-D by both apple leaf cuticles and dewaxed cuticle membranes was unaffected by surfactant pretreatment.

The effect of nonionic polyoxyethylene surfactants on foliar uptake of ^{14}C-difenzoquat and ^{14}C-2,4-D was studied in wild oat and field bean plants.[31] Aliphatic C_{13}/C_{15} alcohol surfactants generally improved uptake more than did nonylphenol surfactants when used at equivalent concentrations and EO contents. The surfactant threshold for enhancement of difenzoquat uptake in wild oat was much lower (0.05 g/L) than in field bean (> 0.5 g/L). For 2,4-D, surfactants at > 0.5 g/L were needed, but there was little dependence on surfactant EO content for enhancement. Surfactants of low EO content (5 to 6) were less effective for difenzoquat uptake than those of high content (10 to 20), particularly in wild oat. Adjuvants with humectant properties also promoted penetration of difenzoquat, but less so than did alcohol or nonylphenol surfactants.

Uptake and translocation of herbicides applied with a nonionic surfactant to the roots of Virginia buttonweed were greatest with 2,4-D > clopyralid > triclopyr > dichlorprop > fluroxypyr > dicamba.[57]

Marlipal 34 surfactant series with EO content from 6 to 20 were used with cyanazine and some other compounds to study their uptake by field bean and wheat.[69] The surfactants EO content did not influence their ability to promote uptake of cyanazine.

TF8035 (Turbocharge™) consistently enhances the biological activity of tralkoxydim, enabling a 20% reduction in the rate of active ingredient to control wild oat and green foxtail.[8] Improvements in biological activity are unlikely to be due to increased uptake or translocation since time course studies showed no significant differences from the standard adjuvant (Charge™).

C. EFFECTS ON DROPLET SIZE, DRIFT, AND SPREADING

The addition of drift control adjuvants, Sta-Put® (a polyvinyl polymer adjuvant) and Paratac® (an oil soluble polyisobutylene polymer), increased droplet size with two diluents, water and a paraffinic oil (Orchex 796), and decreased the percentage of smaller droplets.[29] Both adjuvants produced large differences in droplet size at lower air pressures, but had little effect at higher nozzle pressures (air pressures were 14, 28, 42, 56, and 84 kPa).

Five drift control agents, Direct, Driftgard, Formula 358, Nalcotrol®, and Target, produced larger droplets and reduced the amount of small airborne droplets.[19] In addition, all five products increased the percentage of droplets around the mean droplet diameter. These drift control agents reduced drift to varying degrees.

Soybean or cotton seed oils did not spread as well as petroleum oils but methylated soybean and sunflower oils had high spread coefficients on both upper and lower leaf surfaces of johnsongrass.[43] Spread of paraffinic oils exceeded that of any water/adjuvant mixture tested. Water droplets with adjuvant had an 80% weight loss after 6 min, but low volatile paraffinic oil droplets had little loss at 2 d after application.

Droplets of several different petroleum-based oils spread much better on leaf surfaces of *Erythoxylum* sp. than either water droplets with adjuvants or droplets of soybean or cotton oil.[44]

D. ADJUVANT EVAPORATION

A simple, rapid assay was devised for measuring relative evaporation rates from individual drops.[28] The evaporation rates of a variety of adjuvants under controlled temperature and humidity conditions were measured. Sta-Put®, Bond®, Intac, and Nalcotrol® (0.24 ml/L) showed the least evaporative loss (40 to 45% loss). Nalcotrol® (0.63 and 0.48 ml/L) and Direct lost 45 to 50%; 38F, Surfix®, 41A, Plex + Windfall, Nu-FilmP®, Agro Flow and tap water lost 50 to 55%, and Prime Oil lost between 55 and 60%. Distilled water and Penetrator® lost slightly more than 60%.

VII. BIOHERBICIDES

Biological organisms were incorporated into granules containing unique inverting oil, sucrose, and water-absorbent starch.[53] The oil component slows the drying process of the living agents, and sucrose prolongs vitality for extended storage periods at room temperature, 3°C, and -10°C. The inverting oil and water absorbent starch allow for reactivation of the biocontrol agents with minimal dew or mist. The resultant product can be formed into various sized granules which can be applied with granule applicators or resuspended in aqueous slurries and sprayed. Greenhouse testing using *Colletotrichum truncatum* aqueous slurry on hemp sesbania and *Sclerotinia sclerotiorum* aqueous slurry and granules on Canada thistle resulted in 80% control of hemp sesbania. Control of Canada thistle was 25% for the aqueous slurry and 75% for the granules. Field studies resulted in complete failure of control.

VIII. SAFENERS

Grain sorghum seeds treated with CGA-133205, oxabetrinil, and flurazole and stored for up to 24 weeks at various humidity levels, were adversely affected by the safeners as reflected by fewer normal seedlings and increases in the number of ungerminated seeds, although the safeners minimized reductions in seedling shoot fresh weight caused by alachlor and metolachlor compared to the no-safener check.[77]

Oxabetrinil, flurazole, naphthalic anhydride, dichlormid, and R-29148 protected against significant injury to corn from the interaction between primisulfuron and terbufos (insecticide) and from metolachlor.[63] CGA-133205 killed corn plants. No primisulfuron-terbufos interaction was observed in sorghum and none of the safeners protected sorghum from primisulfuron injury. Preemergence applications of the antioxidants, piperonyl butoxide and metyrapone, increased primisulfuron injury to corn. The injury was reversed by flurazole and R-29148; however, protection by the safeners against injury from metyrapone was not evident. The increased injury from piperonyl butoxide was reportedly due to inhibition of primisulfuron metabolism; naphthalic anhydride did not reverse the inhibition.

Terbufos inhibited the metabolism of nicosulfuron, but pretreatment of corn seed with naphthalic anhydride decreased the rates of such inhibition in excised corn leaves.[13,64] Field experiments confirmed the protecting effects.

BAS 145138 increased metabolism of primisulfuron and nicosulfuron in corn shoots, but not roots.[55] Herbicide metabolism in BAS 145138-treated tissue was inhibited by bentazon, tetcyclacis, and piperonyl butoxide, but not by tridiphane, and naphthalic anhydride induced a greater increase in sulfonylurea metabolism than did BAS 145138.

Tank-mixes of 2,4-D with nicosulfuron decreased visual injury to corn caused by nicosulfuron/terbufos interaction.[65] Reduction of injury was observed when 2,4-D was applied with or 1 d after nicosulfuron application. Application of 2,4-D at 1 or 2 d before or 2 d after nicosulfuron provided little protection. The metabolism of ^{14}C-nicosulfuron was decreased by the presence of terbufos, but was reversable if 2,4-D was applied immediately before nicosulfuron.

IX. DEFOLIANTS

At 5 d after treatment, leaf drops of cotton (30/21° C day/night temperatures) was 17% with thidiazuron alone, 37% with crop oil concentrate, 40% with ammonium sulfate, and 75% with all adjuvants combined.[66] At 21/13° C day/night temperatures, there was less than 10% leaf drop with all treatments, and at 10 d after treatment there was no difference among any of the treatments at any temperature. Absorption of ^{14}C-thidiazuron by cotton was increased with several adjuvants.

X. GUIDELINES FOR ADJUVANT USE

Agrichemical manufacturers are moving toward recommending specific adjuvants that research has identified as maximizing the efficacy of their products. For example, in 1992 DuPont issued guidelines to manufacturers of adjuvants to insure that its herbicides are applied with adjuvants that will optimize performance.[25] The guidelines address surfactants, petroleum oil concentrates, fertilizer adjuvants, and blends. Formulated surfactant products must contain at least 50% nonionic surfactant and have an HLB from 12 to 17. For oil concentrates, the oil type and a minimum amount of emulsification must be defined. The key component of fertilizer adjuvants is the amount of ammonium. All adjuvants must have established tolerances for the target crops or contain only EPA-exempt ingredients. For DuPont to recommend a new adjuvant, information on its composition and performance is required. More than 130 adjuvants have been approved.[18] If the adjuvant is not approved, DuPont will not assume the risk that the adjuvant-herbicide combination will work appropriately.[25]

Actual field trials are the only means available to both verify and evaluate adjuvant label claims.[58] However, the high costs and time considerations associated with field trials make laboratory evaluations an important screening procedure for adjuvants. The use of standardized laboratory procedures and a universal adjuvant "language" are vital in order to ensure that results obtained are both reproducible and valid. The American Society of Testing and Materials (ASTM) has published a peer-reviewed and approved list of definitions for many, but not all, spray adjuvant types and functions (Table 7). Some applicable definitions for adjuvant product functions used by a major company are presented in Table 8. Selected spray adjuvant laboratory evaluation methods used by a major company are presented in Table 9. Several procedures are available from

Table 7
Standard Terminology Relating to Agricultural Tank Mix Adjuvants[a]

This standard is issued under the fixed designation E 1519; the number immediately following the designation indicates the year of original adoption or, in the case of revision, the year of last revision. A number in parentheses indicates the year of last reapproval. A superscript epsilon ($^{\epsilon}$) indicates an editorial change since the last revision or reapproval.

1. Scope
   1.1 This terminology is used or is likely to be used in test methods, specifications, guides, and practices related to agricultural tank mix ingredients.
   1.2 These definitions are written to ensure that standards related to agricultural tank mix adjuvants are properly understood and interpreted.
2. Referenced Documents
   2.1 ATSM Standards:
   D 483 Test Method for Unsulfonated Residue of Petroleum Plant Spray Oils[b]
   D 2140 Test Method for Carbon-Type Composition of Insulating Oils of Petroleum Origin[c]
3. Terminology
   3.1 Terms and Definitions:

Acidifier - material that can be added to spray mixtures to lower the pH.
Activator - material that increases the biological efficacy of agrichemicals.
Active ingredient - a component of the formulation that produces a specific effect for which the formulation is designed.
Adjuvant - a material added to a tank mix to aid or modify the action of an agrichemical, or the physical characteristics of the mixture.
Amphoteric surfactant - a surface-active agent capable of forming, in aqueous solution, either surface-active anions or surface-active cations depending on the pH.
Anionic surfactant - a surface-active material in which the active portion of the molecule containing the lipophilic segment forms exclusively a negative ion (anion) when placed in aqueous solution.
Antifoaming agent - material used to inhibit or prevent the formation of foam.
Attractant - material that attracts specific pests.

Buffer or buffering agent - a compound or mixture that, when contained in solution, causes the solution to resist change in pH. Each buffer has a characteristic limited range of pH over which it is effective.
Canopy penetrating agent - an adjuvant that increases the penetration of the spray material into the crop canopy. See deposition aid.
Cationic surfactant - a surface-active material in which the active portion of the molecule containing the lipophilic segment forms exclusively a positive ion (cation) when placed in aqueous solution.
Colorant - a material used to alter the color of the tank mix.
Compatibility agent - a surface-active material that allows simultaneous application of liquid fer- tilizer and agrichemical, or two or more agrichemical formulations, as a uniform tank mix, or improves the homogeneity of the mixture and the uniformity of the application.
Crop oil concentrate - an emulsifiable petroleum oil-based product containing 15 to 20% w/w surfactant and a minimum of 80% w/w phytobland oil.
Crop oil (emulsifiable) - an emulsifiable petroleum oil-based product containing up to 5% w/w surfactant and the remainder of a phytobland oil.
Crop oil (non-emulsifiable) - See phytobland oil.
Defoaming agent - material that eliminates or suppresses foam in the spray tank.
Deposition aid - material that improves the ability of pesticide sprays to deposit on targeted surfaces.
Dormant oil - a horticultural spray oil applied during the dormant phase of the targeted plant. (See horticultural spray oil).
Drift control agent - a material used in liquid spray mixtures to reduce spray drift.
Emulsifier - a surfactant that promotes the suspension of one immiscible liquid in another.
Evaporation reduction agent - a material that reduces the evaporation rate of a spray mix during or after application or both.
Extender - material that increases the effective life of an agrichemical after application.
Foam suppressant - See defoamer.
Foaming agent - a material that increases the volume or stability of the foam formed in a spray mixture.
Humectant - a material which increases the equilibrium water content and increases the dyring time of an aqueous spray deposit.
Modified vegetable oil - an oil, extracted from seeds, that has been chemically modified (for example, methylated).
Modified vegetable oil concentrate - an emulsifiable, chemically modified vegetable oil product containing 5 to 20% w/w surfactant and the remainder chemically modified vegetable oil.
Naphtha-based oil - a petroleum oil containing a majority of the naphtha fraction.
Nonionic - a material having no ionizable polar end groups but comprised of hydrophilic and lipophilic segments.
Oil - See petroleum, vegetable, paraffinic, and so forth.
Paraffinic oil - a petroleum oil (derived from paraffin crude) whose paraffinic carbon type content is typically greater than 60%. D 2140
Penetrant - a material that enhances the ability of an agrichemical to enter a substrate or penetrate a surface.
Petroleum oil - oil derived from petroleum; contains a mixture of hydrocarbons that are broadly classified as paraffins, napthenes, aromatics, or other unsaturates, or combination thereof.
Phytobland oil - a highly refined paraffinic material with a minimum unsulfonated residue of 92% v/v. D 483
Spreader - a material which increases the area that a droplet of a given volume of spray mixture will cover on a target.
Spreader/sticker - a material that has the properties of both a spreader and a sticker.
Sticker - a material that assists the spray deposit to adhere or stick to the target and may be measured in terms of resistance to time, wind, water, mechanical action, or chemical action.
Vegetable oil - oil extracted from seeds; typically those of corn, cotton, peanut, rapeseed, sunflower, canola, or soybean.
Vegetable oil concentrate - an emulsifiable vegetable oil product containing 5 to 20% w/w surfactant and a minimum of 80% w/w vegetable oil.
Wetting agent - wetting agents can be considered synonymous with spreading agents in function.

This standard is subject to revision at any time by the responsible technical committee and must be reviewed every five years and if not revised, either reapproved or withdrawn. Your comments are invited either for revision of this standard or for additional standards and should be addressed to ASTM Headquarters. Your comments will receive careful consideration at a meeting of the responsible technical committee, which you may attend. If you feel that your comments have not received a fair hearing, you should make your views known to the ASTM Committee on Standards, 1916 Race St., Philadelphia, PA 19103.

[a]Reproduced by permission. This terminology is under the jurisdiction of ASTM Committee E-35 on Pesticides and are the direct responsibility of Subcommittee E35.22 on Pesticide Formulation and Application Systems.

Current edition approved Feb. 1994. Published April 1994. Originally published at E 1519-93. Last previous edition E 1519-94.

[b]Annual Book of ASTM Standards, Vol. 05.01.

[c]Annual Book of ASTM Standards, Vol. 10.03.

Table 8
Applicable Definitions for Adjuvant Product Function Rankings[a]

WETTING/SPREADING
The ability to reduce interfacial surface tension and increase the area that a given volume of a spray solution will cover on a target. Rankings are based on comparisons of static surface tension, dynamic surface tension, and spray droplet contact angle.

BUFFERING
The ability to reduce and maintain a particular pH range in a spray mix. Rankings determined by comparing buffer capacity and pH modification performance.

WATER CONDITIONING
The capability to modify chemical and physical characteristics of water that result in spray application and/or efficacy problems. This would include pH modifications, sequestration of hard water cations, and evaporation reduction.

DRIFT REDUCTION
The potential to reduce droplet drift by modification of the physical properties of a spray mix.

DEPOSITION
The ability of the product to improve the ability of pesticide sprays to deposit on targeted surfaces. Rankings based on comparisons of adjuvant functions related to evaporation reduction, humectancy, viscosity modifications, and changes in spray mix density.

STICKING
The capability of the adjuvant to improve the adherence of a spray droplet on an application target. Rankings are based on comparisons of deposit resistance to time, wind, water, mechanical, or chemical action.

COMPATIBILITY
The ability of a product to provide the simultaneous application of multiple component spray mixtures of fertilizers and/or agrichemicals. Rankings are based on how the product improves the uniformity, homogeneity, and/or ease of application of mixtures known to cause problems.

DEFOAMING
The ability to eliminate or suppress foam that originates from spray mixes.

TRANSLOCATION
The relative potential for the product to enhance the movement of active ingredients to their site of action. Comparisons are based on literature references and field research with various compounds and application targets. Plant characteristics, environmental conditions, and type of agrichemical formulation may affect these relative rankings.

CUTICULAR PENETRATION
The relative potential for the product to enhance the absorption of active ingredients through the cuticular area of plant surfaces. Comparisons are based on literature references and field research with one or more of the components in the adjuvant formulations. Plant characteristics, environmental conditions, and type of agrichemical formulation may affect these relative rankings.

ANTAGONISMS
The relative potential for the product to eliminate or suppress antagonisms that occur when one or more pesticides are applied simultaneously. In particular, those combinations that result in one active ingredient impeding the absorption and/or translocation of another are of interest. Comparisons are based on literature references and field research.

UV LIGHT INHIBITION
The ability of the adjuvant formulation to prevent or moderate the impact of sunlight on some pesticide active ingredients. Comparisons are based on the potential for the products to inhibit the ability of UV

light to decompose the active ingredients by means of forming a protective film and/or enhancing the rate at which they are absorbed into the application target.

AQUATIC APPLICATIONS
The product has been evaluated for toxicity to aquatic organisms under guidelines described by the U.S. Environmental Protection Agency as described in publication PB86-129277. Based on the results obtained, use of the product at recommended rates would be safe for application in aquatic environments.

[a]Source: Helena Chemical Company, Memphis, TN with permission. Definitions are not standardized definitions used industry wide, but are working definitions used by a major company in the industry.

Table 9
Spray Adjuvant Laboratory Evaluation Methods[a]

PERCENT ACTIVES: ASTM METHOD D176; OVERVIEW: This method uses a regular draft oven operated at 45°C. The sample is heated until it reaches a constant weight. Weight loss is then measured.
SPECIFIC GRAVITY: ASTM METHOD D1122; OVERVIEW: This method employs a hydrometer.
pH: ASTM METHOD E70; OVERVIEW: This method uses a calibrated electronic pH meter and electrode. To have significance, the concentration of the solution which is measured must be reported.
SURFACE TENSION: ASTM METHOD D1331; OVERVIEW: This method is the Du Nuoy Ring Method. A platinum-iridium ring is dropped into the test solution. The force required to pull this ring out of the liquid is then measured on a Du Nuoy Ring Tensiometer.
CONTACT ANGLE: GONIOMETRY; OVERVIEW: This method uses the naval research laboratory goniometer to directly measure the angle of contact of a 4 μl droplet on Parafilm M.
% ALCOHOL AND GLYCOL: GAS CHROMATOGRAPHY; OVERVIEW: This procedure uses the gas liquid chromatograph in conjunction with a flame ionization detector to determine the presence and levels of alcohols (non-ethyoxylated) and glycols.
CORROSIVITY: ASTM METHOD G31-72; OVERVIEW: Metal and plastic strips are partially immersed in pure product in sealed containers. These containers are maintained at a constant temperature of 45°C for 1 mo. After this exposure period, the strips are rinsed and reweighed to measure any weight loss due to corrosivity.
FLAMMABILITY: ASTM METHOD D56; OVERVIEW: The Tagg Closed Cup Method is used.
FOAMING: DIRECT MEASUREMENT; OVERVIEW: 125 ml of a solution of the product to be tested is placed in a 250-ml cylinder. This cylinder is then agitated in a repeatable manner and the foam height measured at time 0 and after 3 min.
DISSOLUTION EASE: OVERVIEW: This is a subjective grading of the product's ability to mix with water. On a scale of 1-5, 1 = mixes very easily, 5 = has trouble going in.

[a]Source: Helena Chemical Company, Memphis, TN with permission. Evaluation methods are not standardized methods used industry wide, but are used by a major company in the industry in the absence of standardized methods.

ASTM[58]. Several parameters are applicable to spray adjuvant evaluation in the laboratory and include: active ingredient content, surface tension reduction (static and dynamic), wetting performance (static-contact angle and spread diameter), wetting performance (dynamic), compatibility, emulsion performance, evaporation reduction, ionic classification, foaming and defoaming.[58] Some modern procedures now in use are: the measurement of dynamic surface tension and contact angle; droplet size and nature of formation (e.g. by laser technology); tracer studies using radiolabeled adjuvants; and others. Several spray adjuvant functional claims lack standardized methodology. Examples include: spray mix, sticking, binding, and rainfastness; penetration; activation of spray mix components; and herbicide safeners. Although there are limitations associated with laboratory evaluations of agricultural spray adjuvants, they provide a means of screening the hundreds of products on the market today, and they are essential in assuring that adjuvant compositions meet the specifications established by agrichemical producers and researchers.[58]

XI. SPRAY ADJUVANT FORMULATION GOALS

The following goals for spray adjuvant formulations were provided by Johnnie Roberts, Director of Product Development and Registration, Helena Chemical Company, Memphis, TN:

(1) Reduce adjuvant use rates
(2) Maximize performance in low volume applications
(3) Lowest possible hazard category
(4) Remove flammable components
(5) Improve shipping, storage, and handling characteristics
(6) Use components with lowest possible toxicity to aquatic organisms
(7) Reduce/eliminate volatile organic constituents and ethylene oxides
(8) All components exempt and active
(9) Minimize use of inerts
(10) Improve compatibility with dry pesticide formulations
(11) Develop "designer" adjuvants
(12) Develop multi-functional adjuvants
(13) Develop new types of adjuvants
(14) Meet EPA-registered pesticide label recommendations
(15) Exceed EPA-registered pesticide label recommendations

XII. REGULATORY MATTERS

EPA initially focused on the potential effects of active ingredients of pesticides.[72] Recognizing that while pesticidally inert some adjuvants may be of toxicological or environmental concerns, the EPA developed in 1987 a policy to begin addressing potential risks from inert ingredients in pesticide products. EPA lists inerts under four categories: (1) toxicological concern; (2) potentially toxic; (3) insufficient data to classify; and (4) minimal concern. There are approximately 100 inerts listed under categories 1 and 2; 1100 under categories 3 and 4. Approximately 100 are GRAS (generally recognized as safe).

EPA does not regulate end-use adjuvants and some states do not require registration or data. Some states require registration but no data; others require data but no registration. However, in California, adjuvants are designated as pesticides and are subject to EPA guidelines. In any case, pesticides may not be used in a manner that is inconsistent with information published on the label --- which often includes instructions for the inclusion of adjuvants.

British and European regulations and requirements state that adjuvants, substances without significant pesticidal properties which enhance the effectiveness of a pesticide, can be used with pesticides only in accordance with the approval of the pesticide or as varied by lists of authorized adjuvants published by Ministers.[11] Data must be supplied for adjuvants to appear on the list.

Chapman and Mason[11] provide a list of authorized adjuvants.

XIII. SUMMARY

Use of surfactants, mineral and vegetable oils, emulsifiers, and fertilizer salts can greatly enhance the activity of foliage-applied herbicides.[20-22,36] Appropriate combinations of herbicides with adjuvants can enhance the rate and efficiency of herbicide delivery to the target site. Variability in response and specificity of requirement, however, underline the need to establish guideline criteria, enabling optimum combinations of herbicides and adjuvants to be made. Considerable progress has been made in several laboratories in (a) characterizing the cuticle and rate-limiting mechanisms influencing the efficiency of transcuticular movements and (b) developing predictive model systems which encompass a range of aspects including spray delivery, deposition, cuticular penetration, and surfactant/herbicide polarity relationships.

Hundreds of pesticide products are currently available for plant protection and growers are realizing that adjuvant products work successfully with the active ingredients of many pesticides, particularly herbicides. In order to respond to economic, environmental, and regulatory pressure, growers have been using the newer high-cost, low-volume, reduced rate pesticides. This challenges the adjuvant industry to provide technology to increase the efficacy of these products, which becomes difficult due to varying physics and chemistry of pesticides, adjuvants, and carriers as well as varying crop species, varieties, pests, growth stages of crops and pests, pest infestations, application technology, and environment. Today's adjuvants must be highly researched, scientifically formulated, and demonstrate that they have added value.

REFERENCES

1. **Anon.**, New adjuvant solves multiple problems, *Farm Chem.*, 156(12), 89, 1993.
2. **Boydton, R. A. and Al-Khatib, K.**, DCX2-5309 organosilicone adjuvant improves control of kochia (*Kochia scoparia*) with bentazon and bromoxynil, *Weed Technol.*, 8, 99, 1994.
3. **Brooks, G. T., Chr.**, Papers from the Third International Symposium on Adjuvants for Agrochemicals, *Pestic. Sci.*, 37, 113, 1993.
4. **Brooks, G. T., Chr.**, Extended summaries: SCI Pesticides Group Symposium: Third International Symposium on Adjuvants for Agrochemicals, *Pestic. Sci.*, 37, 203, 1993.
5. **Brooks, G. T., Chr.**, Papers from the Third International Symposium on Adjuvants for Agrochemicals, *Pestic. Sci.*, 38, 65, 1993.
6. **Brooks, G. T., Chr.**, Extended summaries: SCI Pesticides Group Symposium: Third International Symposium on Adjuvants for Agrochemicals, *Pestic. Sci.*, 38, 247, 1993.
7. **Bruce, J. A., Penner, D., and Kells, J. J.**, Absorption and activity of nicosulfuron and primisulfuron in quackgrass (*Elytrigia repens*) as affected by adjuvants, *Weed Sci.*, 41, 218, 1993.
8. **Bryan, I. B., Murfitt, R. C., Pierce, A., Howell, A., Barrientos, C., and Pastushok, G.**, Adjuvant advances with tralkoxydim, *Abstr. Weed Sci. Soc. Am.*, 34, 17, 1994.
9. **Buick, R. D.**, A mechanistic model to describe organosilicone surfactant promotion of triclopyr uptake, *Pestic. Sci.*, 36, 127, 1992.
10. **Burnside, O. C.**, Weed science - the stepchild, *Weed Technol.*, 7, 515, 1993.
11. **Chapman, P. J. and Mason, R. D.**, British and European regulations and registration requirements for non-pesticidal co-formulants in pesticides and for adjuvants, *Pestic. Sci.*, 37,

167, 1993.

12. **Chow, P. N. P., Grant, C. A., Hinshalwood, A. M., and Simundson, E., Eds.,** *Adjuvants and Agrochemicals,* Vol. I, *Mode of Action and Physiological Activity,* CRC Press, Boca Raton, FL, 1989, 207; Vol. II, *Recent Development, Application and Bibliography of Agro-Adjuvants,* CRC Press, Boca Raton, FL, 1989, 222.
13. **Corbin, F. T., Moreland, D. E., Siminszky, B., and McFarland, J. E.,** Metabolism of primisulfuron in terbufos and/or naphthalic anhydride-treated corn, *Abstr. Weed Sci. Soc. Am.,* 33, 70, 1993.
14. **Coret, J. M. and Chamel, A. R.,** Influence of some nonionic surfactants on water sorption by isolated tomato fruit cuticles in relation to cuticular penetration of glyphosate, *Pestic. Sci.,* 38, 27, 1993.
15. **Cross, J. and Singer, E. J., Eds.,** *Cationic Surfactants: Analytical and Biological Evaluation,* Marcel Dekker, New York, 1994, 392.
16. **Cudney, D. W., Orloff, S. B., and Adams, C. J.,** Improving weed control with 2,4-DB amine in seedling alfalfa (*Medicago sativa*), *Weed Technol.,* 7, 465, 1993.
17. **Dunne, C. L., Gillespie, G. R., and Porpiglia, P. J.,** Primary linear alcohol ethoxylates as adjuvants for primisulfuron, *Weed Sci.,* 42, 82, 1994.
18. **DuPont Agricultural Products,** Approved adjuvant list for use with DuPont row crops and cereal herbicides, Agricultural Bull., E.I. duPont deNemours and Company, Wilmington, DE, 1994, 2.
19. **Fahnestock, A. L.,** Drift control agents work, *Farm Chem.,* 156(11), 24, 1993.
20. **Foy, C. L.,** Economic and right uses of surfactants for increased efficacy of herbicides, *Korean J. Weed Sci.,* 11, 74, 1991.
21. **Foy, C. L., Ed.,** *Adjuvants for Agrichemicals,* CRC Press, Boca Raton, FL, 1992, 735.
22. **Foy, C. L.,** Progress and developments in adjuvant use since 1989 in the USA, *Pestic. Sci.,* 38, 65, 1993.
23. **Gaskin, R. E. and Holloway, P. J.,** Some physicochemical factors influencing foliar uptake of glyphosate-mono (isopropylammonium) by polyoxyethylene surfactants, *Pestic. Sci.,* 34, 195, 1992.
24. **Gillespie, G. R. and Vitolo, D. B.,** Response of quackgrass (*Elytrigia repens*) biotypes to primisulfuron, *Weed Technol.,* 7, 411, 1993.
25. **Green, J. M., Chicoine, T. K., and Dobrotka, M. D.,** Guidelines for adjuvant use with DuPont row crops and cereal herbicides, *Abstr. Weed Sci. Soc. Am.,* 34, 98, 1984.
26. **Green, J. M. and Green, J. H.,** Surfactant structure and concentration strongly affect rimsulfuron activity, *Weed Technol.,* 7, 633, 1993.
27. **Gronwald, J. W., Jourdan, S. W., Wyse, D. L., Somers, D. A., and Magnusson, M. U.,** Effect of ammonium sulfate on absorption of imazethapyr by quackgrass (*Elytrigia repens*) and maize (*Zea mays*) cell suspension cultures, *Weed Sci.,* 41, 325, 1993.
28. **Hall, F. R., Kirchner, L. M., and Downer, R. A.,** Measurement of evaporation from adjuvant solutions using a volumetric method, *Pestic. Sci.,* 40, 17, 1994.
29. **Hanks, J. E. and McWhorter, C. G.,** Increasing droplet size with ULV application, *Abstr. Weed Sci. Soc. Am.,* 33, 96, 1993.
30. **Harker, N. K.,** Effect of various adjuvants on sethoxydim activity, *Weed Technol.,* 6, 865, 1992.
31. **Holloway, P. J. and Edgerton, B. M.,** Effects of formulation with different adjuvants on foliar uptake of difenzoquat and 2,4-D: model experiments with wild oat and field bean, *Weed Res.,* 32, 183, 1992.
32. **Jansen, L. L.,** Enhancement of herbicides by silicone surfactants, *Weed Sci.,* 21, 130, 1973.
33. **Jordan, D. L., Frans, R. E., and McClelland, M. R.,** Influence of application variables on efficacy of postemergence applications of DPX-PE350, *Weed Technol.,* 7, 619, 1993.
34. **Jordon, D. L., Reynolds, D. B., Herrick, J. K., and Wilcut, J. W.,** Influence of application rate and timing and ammonium sulfate on efficacy of glyphosate, *Abstr. Weed Sci. Soc. Am.,* 34, 14, 1994.
35. **Kells, J. J. and Lee, P. K.,** Efficacy of K2000, a soybean based adjuvant, *Abstr. Weed Sci. Soc. Am.,* 34, 14, 1994.
36. **Kirkwood, R. C.,** Use and mode of action of adjuvants for herbicides: a review of some current work, *Pestic. Sci.,* 38, 93, 1993.
37. **Knoche, M. and Bukovac, M. J.,** Interaction of surfactant and leaf surface in glyphosate absorption, *Weed Sci.,* 41, 87, 1993.
38. **Kwon, C.-S. and Penner, D.,** The effects of adjuvants and piperonyl butoxide on the activity of

sulfonylurea herbicides, *Abstr. Weed Sci. Soc. Am.*, 34, 88, 1994.

39. **Lawrie, J. and Clay, V.**, Effects of herbicide mixtures and additives on *Rhododendron ponticum, Weed Res.*, 33, 25, 1993.
40. **Levene, B. C. and Owen, M. D. K.**, Movement of ^{14}C-bentazon with adjuvants into common cocklebur (*Xanthium strumarium*) and velvetleaf (*Abutilon theophrasti*), *Weed Technol.*, 8, 93, 1994.
41. **Manthey, F. A., Szelezniak, E. F., Anyszka, Z. M., and Nalewaja, J. D.**, Foliar absorption and phytotoxicity of quizalofop with lipid compounds, *Weed Sci.*, 40, 558, 1992.
42. **McKay, R. B.**, Ed., *Technological Applications of Dispersions*, Marcel Dekker, New York, 1994, 576.
43. **McWhorter, C. G., Ouzts, C., and Hanks, J. E.**, Spread of water and oil droplets on johnsongrass (*Sorghum halepense*) leaves, *Weed Sci.*, 41, 460, 1993.
44. **McWhorter, C. G. and Ouzts, C.**, Leaf surface morphology of *Erythroxylum* sp. and droplet spread, *Weed Sci.*, 42, 18, 1994.
45. **Meister Publishing Company**, *Farm Chemicals Handbook* 1993, Meister Publishing Co., Willoughby, OH, 1993, 833.
46. **Motooka, P., Ching, L., Powley, J., Onuma, K., and Nagai, G.**, Effect of metsulfuron methyl rates, surfactants and split applications on the control of largeleaf lantana (*Lantana camara L.*), *Abstr. Weed Sci. Soc. Am.*, 33, 30, 1993.
47. **Nalewaja, J. D.**, Oil and surfactant adjuvants with nicosulfuron, *Abstr. Weed Sci. Soc. Am.*, 34, 75, 1994.
48. **Nalewaja, J. D. and Matysiak, R.**, Spray carrier salts affect herbicide toxicity to kochia (*Kochia scoparia*), *Weed Technol.*, 7, 154, 1993.
49. **Nalewaja, J. D. and Matysiak, R.**, Optimizing adjuvants to overcome glyphosate antagonistic salts, *Weed Technol.*, 7, 337, 1993.
50. **Nalewaja, J. D., Matysiak, R., and Freeman, T. P.**, Spray droplet residual of glyphosate in various carriers, *Weed Sci.*, 40, 576, 1992.
51. **Nalewaja, J. D., Matysiak, R., and Manthey, F. A.**, Commercial adjuvants with nicosulfuron and glyphosate, *Abstr. Weed Sci. Soc. Am.*, 33, 105, 1993.
52. **Narayanan, K. S., Paul, S. L., and Chaudhuri, R. K.**, *N*-alkyl pyrrolidones for superior agricultural adjuvants, *Pestic. Sci.*, 37, 225, 1993.
53. **Quimby, P. C., Jr., Caesar, A. J., Bridsall, J. L., and Boyette, C. D.**, Adjuvants and spraying systems for application of bioherbicides, *Abstr. Weed Sci. Soc. Am.*, 34, 104, 1994.
54. **Reeves, B. G.**, The rationale of adjuvant use with agrichemicals, in *Adjuvants for Agrichemicals*, Foy, C. L., Ed., CRC Press, Boca Raton, FL, 1992, chap. 46.
55. **Rehab, I. F., Burton, J. D., Maness, E. P., Monks, D. W., and Robinson, D. A.**, Effects of safeners on nicosulfuron and primisulfuron metabolism in corn, *Abstr. Weed Sci. Soc. Am.*, 33, 70, 1993.
56. **Riechers, D. E., Liebl, R. A., Wax, L. M., and Bush, D. R.**, Effect of surfactants on glyphosate transport into plasma membrane vesicles isolated from common lambsquarters (*Chenopodium album L.*), *Abstr. Weed Sci. Soc. Am.*, 33, 109, 1993.
57. **Rippee, J. H. and Coats, G. E.**, Influence of surrfactant on absorption and translocation of selected herbicides in Virginia buttonweed, *Abstr. Weed Sci. Soc. Am.*, 33, 109, 1993.
58. **Roberts, J. R.**, A review of the methodology employed in the laboratory evaluation of spray adjuvants, in *Adjuvants for Agrichemicals*, Foy, C. L., Ed., CRC Press, Boca Raton, FL, 1992, chap. 48.
59. **Roggenbuck, F. C., Kells, J. J., Lee, P. K., and Penner**, D., New adjuvants derived from soybean enhance efficacy and absorption, *Abstr. Weed Sci. Soc. Am.*, 33, 72, 1993.
60. **Royneberg, T., Balke, N. E., and Lund-Hoie, K.**, Cycloxydim absorption by suspension-cultured velvetleaf (*Abutilon theophrasti* Medic.) cells, *Weed Res.*, 34, 1, 1994.
61. **Royneberg, T., Balke, N. E., and Lund-Hoie, K.**, Effect of adjuvants and temperature on glyphosate absorption by cultured cells of velvetleaf (*Abutilon theophrasti* Medic.), *Weed Res.*, 32, 419, 1992.
62. **Schönherr, J.**, Effects of alcohols, glycols and monodisperse and ethoxylated alcohols on mobility of 2,4-D in isolated plant cuticles, *Pestic. Sci.*, 39, 213, 1993.
63. **Simarmata, M. and Penner, D.**, Protection from primisulfuron injury to corn (*Zea mays*) and sorghum (*Sorghum bicolor*) with herbicide safeners, *Weed Technol.*, 7, 174, 1993.
64. **Siminszky, B., Corbin, F. T., Sheldon, B. S., and Walls, F. R., Jr.**, Nicosulfuron resistance mechanisms and metabolism in terbufos and/or naphthalic anhydride treated corn, *Abstr. Weed*

Sci. Soc. Am., 33, 112, 1993.

65. **Simpson, D. M., Diehl, K. E., and Stoller, E. W.**, Mechanism for 2,4-D safening of nicosulfuron/terbufos interaction in corn, *Abstr. Weed Sci. Soc. Am.*, 33, 112, 1993.
66. **Snipes, C. E. and Wills, G. D.**, Influence of temperature and adjuvants on thidiazuron activity in cotton leaves, *Weed Sci.*, 42, 13, 1994.
67. **Stevens, P. J. G.**, Organosilicone surfactants as adjuvants for agrochemicals, *Pestic. Sci.*, 38, 103, 1993.
68. **Stock, D., Edgerton, B. M., Gaskin, R. E., and Holloway, P. J.**, Surfactant-enhanced foliar uptake of some organic compounds: interactions with two model polyoxyethylene aliphatic alcohols, *Pestic. Sci.*, 34, 233, 1992.
69. **Stock, D., Holloway, P. J., Grayson, B. T., and Whitehouse, P.**, Development of a predictive uptake model to rationalise selection of polyoxyethylene surfactant adjuvants for foliage applied agrochemicals, *Pestic. Sci.*, 37, 233, 1993.
70. **Tan, S. and Crabtree, G. D.**, Effects of nonionic surfactants on cuticular sorption and penetration of 2,4-dichlorophenoxy acetic acid, *Pestic. Sci.*, 35, 299, 1992.
71. **Taylor, J. S., Harker, N., and Roberton, J. M.**, Seaweed extract and alginates with sethoxydim, *Weed Technol.*, 7, 916, 993.
72. **Tinsworth, E. F.**, Regulation of pesticides and inert ingredients in pesticide products, in *Adjuvants for Agrichemicals*, Foy, C. L., Ed., CRC Press, Boca Raton, FL, 1992, chap. 20.
73. **Urvoy, C., Pollacsek, M., and Gauvrit, C.**, Seed oils as additives: penetration of triolen, methyloleate and diclofop-methyl in maize leaves, *Weed Res.*, 32, 375, 1992.
74. **Wanamarta, G., Kells, J. J., and Penner, D.**, Overcoming antagonistic effects of Na-bentazon on sethoxydim absorption, *Weed Technol.*, 7, 322, 1993.
75. **Ware, G. W.**, *The Pesticide Book*, 3rd ed., Thomson Publications, Fresno, CA, 1989, 340.
76. **Wax, L. M.**, Spray additives improve performance, *Farm Chem.*, 152(2), 56, 1989.
77. **Zhang, Z., Coats, G. E., and Boyd, A. H.**, Germination and seedling growth of sorghum (*Sorghum bicolor*) hybrids after seed storage with safeners at varying humidities, *Weed Sci.*, 42, 98, 1994.
78. **Zollinger, R. K.**, Influence of adjuvants and herbicide combinations on imazethapyr efficacy, *Abstr. Weed Sci. Soc. Am.*, 33, 5, 1993.

Index

E

F

N

O

P

Q

R

T

U

V

W

X

Z